Intelligent Systems and Control

PRINCIPLES AND APPLICATIONS

LAXMIDHAR BEHERA   NDRANI KAR

OXFORD
UNIVERSITY PRESS

OXFORD
UNIVERSITY PRESS

Oxford University Press is a department of the University of Oxford.
It furthers the University's objective of excellence in research, scholarship,
and education by publishing worldwide. Oxford is a registered trademark of
Oxford University Press in the UK and in certain other countries

Published in India by
Oxford University Press
22 Workspace, 2nd Floor, 1/22 Asaf Ali Road, New Delhi 110002, India

First Edition published in 2009
Fourth impression 2012
Digitally Printed in 2024

ISBN-13: 978-0-19-806315-5
ISBN-10: 0-19-806315-6

Typeset in Times Roman
by Archetype, New Delhi 110063
Printed in India by Manipal Technologies Limited, Manipal

Preface

Control systems are decision-making systems that are designed to provide autonomy to dynamic systems, such as power plants, chemical processes, aerospace, and robotic systems. With technological advancement, there is urgent need to design control systems that are able to maintain acceptable performance levels under significant unanticipated uncertainties. Recent advances in intelligent control methodologies have enabled us to address this issue better. Different intelligent control paradigms have been developed emulating certain characteristics of intelligent biological systems. Thus, computational intelligent techniques, such as neural networks, fuzzy logic, evolutionary computation, and machine learning, have been integrated with automatic control design to develop intelligent controllers. Such controllers can drive uncertain complex systems, with a greater degree of autonomy than the available classical control schemes.

In the early days of research in intelligent control, some scientists and researchers used to be skeptical that control theory, which has a strong foundation in mathematical treatment and stability analysis, is being made a software technology. After two decades since the advent of this field of intelligent systems and control, there are very few skeptics. Today intelligent control is a well-established field of study, which provides unparalleled mathematical treatment as well as explicit stability analysis for unstructured and poorly modelled nonlinear dynamic and complex systems.

This book is intended as a textbook primarily for senior undergraduate students of electrical and computer science engineering. Postgraduate students will find this book to be a useful reference. Beneficiaries of this book will be practising engineers and scientists who are working in the area of Intelligent Systems and Control. Although a lot of research work and innovations have taken place in this area of intelligent system and control over the last two decades, there is no standard textbook available for graduate level students except some reference books and journal articles. This book was conceptualized when Prof. Behera delivered 32-hour video lectures under the NPTEL (National Programme on Technology Enhanced Learning) programme initiated by the Government of India.

Preface

Control systems are decision-making systems that are designed to provide autonomy to dynamic systems, such as power plants, chemical processes, aerospace, and robotic systems. With technological advancement, there is urgent need to design control systems that are able to maintain acceptable performance levels under significant unanticipated uncertainties. Recent advances in intelligent control methodologies have enabled us to address this issue better. Different intelligent control paradigms have been developed emulating certain characteristics of intelligent biological systems. Thus, computational intelligent techniques, such as neural networks, fuzzy logic, evolutionary computation, and machine learning, have been integrated with automatic control design to develop intelligent controllers. Such controllers can drive uncertain complex systems with a greater degree of autonomy than the available classical control schemes.

In the early days of research in intelligent control, some scientists and researchers used to be skeptical that control theory, which has a strong foundation in mathematical treatment and stability analysis, is being made a software technology. After two decades since the advent of this field of intelligent systems and control, there are very few skeptics. Today intelligent control is a well-established field of study, which provides unparalleled mathematical treatment as well as explicit stability analysis for unstructured and poorly modelled nonlinear dynamic and complex systems.

This book is intended as a textbook primarily for senior undergraduate students of electrical and computer science engineering. Postgraduate students will find this book to be a useful reference. Beneficiaries of this book will be practising engineers and scientists who are working in the area of Intelligent Systems and Control. Although a lot of research work and innovations have taken place in this area of intelligent system and control over the last two decades, there is no standard textbook available for graduate level students except some reference books and journal articles. This book was conceptualized when Prof. Behera delivered 32-hour video lectures under the NPTEL (National Programme on Technology Enhanced Learning) programme initiated by the Government of India.

About the Book

Intelligent Systems and Control: Principles and Applications covers the basics of nonlinear control, neural networks, and fuzzy logic so that the readers can easily follow intelligent control methodologies and applications. Design principles for fuzzy and neural control schemes have been enumerated with an easy understanding for the readers. Stability analysis of control systems has been provided with rigour. The intelligent control systems have been simulated for benchmark nonlinear systems across disciplines such as electrical, electromechanical, and process control systems. Details of real-time experiments for cart-pole inverted pendulum system and 7 degrees of freedom (DOF) robot manipulator using intelligent control schemes have been included in the book to illustrate that these advanced control schemes can be effectively implemented in real time. Many examples with MATLAB codes have been provided for the readers to properly comprehend the subject matter provided in this book. Each chapter is concluded with a set of exercises to be used by the readers for self-practice.

Online Resources

C-codes for selected end-chapter exercises and examples have been included in the Online Resource Centre (ORC) of the book. Simulation results and experimental videos are also included in the ORC.

Content and Coverage

The book is organized into 10 chapters. A brief description of each chapter is as follows.

Chapter 1 presents the fundamentals of nonlinear control. Nonlinear dynamics in continuous and discrete time form have been presented. Stability analyses of nonlinear systems have been presented with illustrative examples. Simple nonlinear strategies have been described for the readers to appreciate the challenges involved with such design. The basic concepts covered in this chapter make the presentation self-sufficient as the readers will find later treatment in the book to be easily understandable.

Neural networks have been widely used in intelligent control. An informative description of neural networks has been presented in *Chapter 2*. This chapter contains architectures and algorithms associated with multi-layered, radial basis function, recurrent, and self-organizing map (SOM) networks. Many illustrative examples from system identification perspectives have been taken to make the readers familiarize with these networks in application to neural control.

Chapter 3 introduces the concept of fuzzy number and fuzzy set. Knowledge representation using rule-base has been discussed. Conceptual paradigms for fuzzy PI and PD controllers have been presented. The approach to system identification as linguistic rules using the popular T–S fuzzy representation has been discussed with illustrative examples.

Chapter 4 describes indirect adaptive control (IAC) schemes using neural networks. This chapter introduces the simultaneous system identification using neural network and control strategies for both affine and non-affine nonlinear systems. Some of the key concepts, such as self-tuning and network inversion, have been used to develop such IAC schemes.

Direct adaptive control using neural networks has been presented in *Chapter 5*. The discussion starts with continuous time single input and single output input-affine nonlinear systems. The methodologies derived have been extended to multi-input multi-output affine systems. Control design concepts for corresponding discrete-time systems have been presented. Each of these concepts is illustrated through examples. The chapter ends with the description of back-stepping design principles for nonlinear systems that can be expressed in strict-feedback form.

Chapter 6 presents approximate dynamic programming approaches to optimal control of nonlinear systems. HJB (Hamilton–Jacobi–Bellman) formulations of both continuous time and discrete time systems have been presented. Some existing schemes, such as HDP, DHP, and SNAC, have been discussed with simulated examples for dynamic nonlinear systems. The chapter ends with the discussion that the global cost function of an input-affine system can be obtained as fuzzy-average of local cost functions just as a nonlinear system can be represented as fuzzy-merging of local linear systems in a Takagi–Sugeno fuzzy model paradigm.

The methodologies for designing the popular fuzzy logic controller are discussed in *Chapter 7*. The fuzzy rules derivation and optimization of fuzzy membership functions have been done using genetic algorithm (GA). Using Lyapunov synthesis approach, generation of fuzzy rules that guarantees stability has been presented. Readers will be able to appreciate these concepts through many examples presented in this chapter.

In *Chapter 8*, control strategies for nonlinear systems through T–S fuzzy parameterization have been developed. Concepts like LMI and robust control frameworks have been used to design such controllers. The readers will find these concepts developed both for continuous and discrete-time applications easy to grasp. Simulated examples will further enhance their understanding.

A real-time implementation for an inverted pendulum system on a cart using fuzzy network inversion has been described in detail in *Chapter 9* while *Chapter 10* discusses real-time implementation schemes for the visual kinematic control of a 7DOF manipulator.

Readers will also find many MATLAB codes given in the book to be extremely helpful to understand and appreciate the concepts presented in this book.

Acknowledgements

Writing a textbook on Intelligent Systems and Control was not an easy task as the field is yet to mature. However, given the interest level among students and researchers alike for this area, we have been inspired to endeavour for such a goal. Numerous e-mails received, after the video lectures given by Prof. Behera were placed in You-tube, also acted as catalyst. We especially thank Mr Swagat Kumar and Mr Prem Kumar, PhD students at IIT Kanpur, for their inputs for Chapters 6 and 8. We also gratefully acknowledge the initial funding for this project from CDTE, IIT Kanpur.

Laxmidhar Behera gratefully acknowledges the coordinators of NPTEL scheme, Prof. P.K. Kalra and Prof. Gautam Biswas, who helped him to embark upon this task. He also gratefully acknowledges the Intelligent Systems Research Center (ISRC), University of Ulster, UK, for allowing him to use professional time to devote on such a project.

Laxmidhar Behera thanks all his family members, specifically his parents for their blessings, his wife Gopali Priyadarsini, for being such a brilliant companion, and his three daughters, Yamuna, Lalita, and Visakha, for bringing aesthetics into his life. Last but not least, he acknowledges the association of Bhakti-Vedanta club at IIT Kanpur for long hour *Hari-kirtans*, which brought sublimity to his life.

Indrani Kar thanks her parents, sister, and husband for their continuous support and inspiration throughout her life and all her colleagues and friends for being always good to her.

Laxmidhar Behera
Indrani Kar

Contents

1

Non-linear Control: Primer

Control theory made significant strides after Maxwell introduced the mathematical model for the Governor[1]. Applied mathematical techniques made it possible to control complex dynamical systems than the original flyball governor. These techniques include developments in optimal control[2] in the 1950s and 1960s, followed by progress in stochastic[3], robust[4], adaptive[5], and optimal[6] control methods in the 1970s and 1980s. Conventional control systems are designed using mathematical models of physical systems. It is necessary to derive the system dynamics using physical laws that govern the dynamics. Usually the system dynamics are represented using differential equations for continuous time systems[7] and difference equation for discrete time systems[8]. Although engineering systems usually operate in continuous time, there exist natural discrete time systems such as a queuing system[9]. However, a given continuous time system can be represented as a discrete time system, and a discrete time controller can be designed. In this book, we will mostly represent the dynamics of engineering systems in state-space model for both the continuous and discrete time design of the controller.

A mathematical model, which captures the dynamical behaviour of interest, is chosen and then control design techniques are applied. The controller is then realized via hardware or software and it is used to control the physical system. The mathematical model of the system must be simple enough, so that it can be analysed with available mathematical techniques, and accurate enough to describe the important aspects of the relevant dynamical behaviour. Applications of control methodology have helped make possible space travel and communication satellites, safer and more efficient aircraft, cleaner auto engines, and cleaner and more efficient chemical processes. However, the application domain of a control system such as power plants, aircraft, spacecraft, chemical processes, robotic systems, and unmanned ballistic missiles are naturally non-linear dynamical systems. Since a non-linear function approximation using neural and fuzzy networks has been achieved quite successfully, design of a

1

Non-linear Control: Primer

Control theory made significant strides after Maxwell introduced the mathematical model for the *Governor* [1]. Applied mathematical techniques made it possible to control complex dynamical systems than the original flyball governor. These techniques include developments in optimal control [2] in the 1950s and 1960s, followed by progress in stochastic [3], robust [4], adaptive [5], and optimal [6] control methods in the 1970s and 1980s. Conventional control systems are designed using mathematical models of physical systems. It is necessary to derive the system dynamics using physical laws that govern the dynamics. Usually the system dynamics are represented using differential equations for continuous time systems [7] and difference equation for discrete time systems[8]. Although engineering systems usually operate in continuous time, there exist natural discrete time systems such as a queuing system [9]. However, a given continuous time system can be represented as a discrete time system, and a discrete time controller can be designed. In this book, we will mostly represent the dynamics of engineering systems in state–space model for both the continuous and discrete time design of the controller.

A mathematical model, which captures the dynamical behaviour of interest, is chosen and then control design techniques are applied. The controller is then realized via hardware or software and it is used to control the physical system. The mathematical model of the system must be *simple enough* so that it can be analysed with available mathematical techniques, and *accurate enough* to describe the important aspects of the relevant dynamical behaviour. Applications of control methodology have helped make possible space travel and communication satellites safer and more efficient aircraft, cleaner auto engines, and cleaner and more efficient chemical processes. However, the application domain of a control system such as power plants, aircraft, spacecraft, chemical processes, robotic systems, and unmanned ballistic missiles are naturally non-linear dynamical systems. Since a non-linear function approximation using neural and fuzzy networks has been achieved quite successfully, design of a non-linear controller using intelligent control schemes has received a lot of attentions from researchers [10, 11]. In this chapter, a general introduction to the non-linear control has been provided, so that readers can effectively understand further developments in non-linear control design using intelligent schemes.

1.1 NORMS OF SIGNALS, VECTORS, AND MATRICES

A *norm* denoted by ∥ ∥ is a real-valued function which must satisfy the following conditions:
1. $\| x \| >= 0$.
2. $\| x \| = 0$ if and only if $x = 0$.
3. $\| ax \| = |a| \| x \|$ for any $a \in \mathcal{R}$, where $\mathcal{R}$ is the set of real numbers.
4. $\| x + y \| <= \| x \| + \| y \|$.

1-norm, *2-norm*, and ∞-*norm* of a continuous signal $x(t)$ are defined as follows:

$$\| x \|_1 = \int_{-\infty}^{\infty} |x(t)| dt \tag{1.1}$$

$$\| x \|_2 = \sqrt{\int_{-\infty}^{\infty} x^2(t) dt} \tag{1.2}$$

$$\| x \|_\infty = \sup_t |x(t)| \tag{1.3}$$

A signal will always be bounded by its norm, i.e., for a signal $x(t)$, $x(t) \leq \|x(t)\|$.

Given a vector $x \in \mathcal{R}^n$, the Euclidean space with elements defined by $x = [x_1, x_2, \ldots, x_n]^T$, where T denotes transpose and *1-norm*, *2-norm*, and ∞-*norm* are defined as

$$\| x \|_1 = \sum_{i=1}^{n} |x_i| \tag{1.4}$$

$$\| x \|_2 = \sqrt{\sum_{i=1}^{n} x_i^2} = \left(x^T x\right)^{\frac{1}{2}} \tag{1.5}$$

$$\| x \|_\infty = \max_i \{|x_i|\} \tag{1.6}$$

Example 1.1 Find out the 1-norm, 2-norm, and ∞-norm of the vector
$$x = \begin{bmatrix} 2 \\ -3 \\ 5 \end{bmatrix}.$$

Solution
The norms are calculated as follows.

Analytic solution	MATLAB code
1-norm: $\|x\|_1 = |2| + |-3| + |5| = 10$	`x=[2;-3;5];`
2-norm: $\|x\|_2 = \sqrt{2^2 + (-3)^2 + 5^2} = 6.15$	`x1 = norm(x,1);`
∞-norm: $\|x\|_\infty = \max\{|2|, |-3|, |5|\} = 5$	`x2 = norm(x,2);`
	`xinf = norm(x,inf);`

For a scalar, all the three norms converge to a single numerical value that is the absolute value of the scalar.

Given a matrix $A \in \mathcal{R}^{n \times m}$ with elements

$$A = \begin{bmatrix} a_{11} & a_{12} & a_{13} & \cdots & a_{1m} \\ a_{21} & a_{22} & a_{23} & \cdots & a_{2m} \\ \vdots & & \cdots & & \vdots \\ a_{n1} & a_{n2} & a_{n3} & \cdots & a_{nm} \end{bmatrix}$$

Its induced 1-norm, 2-norm, and ∞-norm are defined as

$$\| A \|_1 = \max_j \sum_{i=1}^{n} |a_{ij}| \tag{1.7}$$

$$\| A \|_2 = \sqrt{\lambda_{max}(A^T A)} \tag{1.8}$$

$$\| A \|_\infty = \max_i \sum_{j=1}^{m} |a_{ij}| \tag{1.9}$$

where $\lambda_{\max}(\cdot)$ denotes the maximum eigenvalue of a matrix.

Example 1.2 What are the induced 1-norm, 2-norm, and ∞-norm of the matrix

$$A = \begin{bmatrix} 2 & 1 & -4 \\ 3 & 7 & 2 \\ 5 & 8 & 3 \end{bmatrix} ?$$

Solution

The induced norms are calculated as follows:

1-norm: $\| A \|_1 = \max\{2 + 3 + 5, \ 1 + 7 + 8, \ 4 + 2 + 1\} = 16$

2-norm: $\| A \|_2 = \sqrt{\lambda_{\max}(A^T A)} = 12.63$

∞-norm: $\| A \|_\infty = \max\{2 + 1 + 4, \ 3 + 7 + 2, \ 5 + 8 + 3\} = 16$

1.2 POSITIVE DEFINITE FUNCTIONS

A continuously differentiable function $f : R^n \to R^+$ is said to be positive definite in a region $S \in R^n$ that contains the origin if

1. $f(0) = 0$
2. $f(x) > 0; \quad x \in S \quad \text{and} \quad x \neq 0$

The function $f(x)$ is said to be positive semidefinite

1. $f(0) = 0$
2. $f(x) \geq 0; \quad x \in S \quad \text{and} \quad x \neq 0$

If the condition (2) becomes $f(x) < 0$, the function is negative definite and if it becomes $f(x) \leq 0$ it is negative semidefinite.

Example 1.3 Is the function $f(x_1, x_2) = x_1^2 + x_2^2$ positive definite?

Solution

$f(0, 0) = 0$ shows that the first condition is satisfied. $f(x_1, x_2) > 0$ for $x_1, x_2 \neq 0$. Second condition is also satisfied. Hence the function is positive definite.

A function $f(x, t)$ is said to be continuously differentiable if both $f(x, t)$ and $\frac{df}{dt}(x, t)$ are continuous. This is also known as the local Lipschitz property.

1.3 POSITIVE DEFINITE MATRICES

A square matrix P is symmetric if $P = P^T$. A scalar function has a quadratic form if it can be written as $x^T P x$ where $P = P^T$ and x is any real vector of dimension $n \times 1$.

Definition of a positive definite matrix is given below.

A real symmetric matrix P is positive definite, i.e., $P > 0$ if

1. $x^T P x > 0$ for every non-zero x.
2. $x^T P x = 0$ only if $x = 0$.

A real symmetric matrix P is positive semidefinite, i.e., $P \geq 0$ if $x^T P x \geq 0$ for every non-zero x. This implies that $x^T P x = 0$ for some $x \neq 0$.

Theorem 1.1 A symmetric square matrix P is positive definite if and only if any of the following conditions holds.

1. Every eigenvalue of P is positive.
2. All the leading principal minors of P are positive.
3. There exists an $n \times n$ non-singular matrix Q such that $P = Q^T Q$.

Similarly, a matrix A is said to be negative definite if $-A$ is positive definite. When none of these conditions satisfy, the definiteness of the matrix cannot be calculated or in other words it is said to be sign indefinite.

Example 1.4 Consider the following third-order matrices. Determine the sign definiteness of them.

$$A_1 = \begin{bmatrix} 2 & 5 & 7 \\ 1 & 3 & 4 \\ 1 & 2 & 5 \end{bmatrix} \quad A_2 = \begin{bmatrix} 4 & 0 & 0 \\ 0 & 7 & -1 \\ 0 & 0 & -3 \end{bmatrix}$$

Solution
The leading principal minors of the matrix A_1 are 2, 1, and 2, hence the matrix is positive definite. The eigenvalues of the matrix A_2 can be straightaway calculated as 4, 7, and -3, i.e., all the eigenvalues are not positive. Again, the eigenvalues of the matrix $-A_2$ are -4, -7, and 3 and hence the matrix A_2 is sign indefinite.

1.4 CONTINUOUS TIME STATE–SPACE MODEL

In general, a system may have many internal variables of concerns besides input and output variables. The state–space representation takes into account all such internal variables such that each state variable possesses a clear engineering meaning. Thus, *state* is a vector and is defined as follows.

1. The *state* of a dynamic system is the smallest set of variables, $x \in R^n$, such that given $x(t_0)$ and $u(t)$, $t > t_0$, $x(t)$, $t > t_0$ can be uniquely determined.

2. *State equations* describe the evolution of the state variables as a function of the input and state variables, being a set of time-dependent ordinary differential equations.

3. *Output equations* are algebraic equations that relate the value of the output signals to the state and the input signals.

1.4.1 LTI State–Space Model

The simplest state–space model is the class of linear time-invariant (LTI) systems for which the general form of state-space model is as follows:

$$\dot{x}(t) = Ax(t) + Bu(t) \quad \text{(state equation)} \tag{1.10a}$$

$$y = Cx(t) + Du(t) \quad \text{(output equation)} \tag{1.10b}$$

with the initial condition $x(t_0) = x(0)$. $x(t) \in R^n$, $y(t) \in R^m$, and $u(t) \in R^p$ are the state, output, and input vectors, respectively. Similarly, $A \in R^{n \times n}$, $B \in R^{n \times p}$, $C \in R^{m \times n}$, and $D \in R^{m \times p}$ are constant matrices.

Usually a system governed by an nth order differential equation and is expressed in terms of n state variables, i.e. $x = [x_1, x_2, \ldots, x_n]^T$. The following example will make this concept clear.

Example 1.5 Consider an nth-order differential equation

$$\frac{d^n y}{dt^n} + a_1 \frac{d^{n-1} y}{dt^{n-1}} + \cdots + a_n y = u$$

Find out the state–space model.

Solution
Let us define the variables as follows:

$$y = x_1$$

$$\frac{dy}{dt} = x_2$$

$$\vdots = \vdots$$

$$\frac{d^{n-1} y}{dt^{n-1}} = x_n$$

$$\frac{d^n y}{dt^n} = -a_1 x_{n-1} - a_2 x_{n-2} - \cdots - a_n x_1 + u$$

The nth-order differential equation may be written in the form of n first-order differential equations as

$$\dot{x}_1 = x_2$$

$$\dot{x}_2 = x_3$$

$$\vdots = \vdots$$

$$\dot{x}_n = -a_1 x_{n-1} - a_2 x_{n-2} - \cdots - a_n x_1 + u$$

or in matrix form as

$$\dot{x} = Ax + Bu$$

where

$$A = \begin{bmatrix} 0 & 1 & 0 & \cdots & 0 \\ 0 & 0 & 1 & \cdots & 0 \\ \vdots & & & \cdots & \vdots \\ 0 & 0 & 0 & \cdots & 1 \\ -a_n & -a_{n-1} & -a_{n-2} & \cdots & -a_1 \end{bmatrix} \qquad B = \begin{bmatrix} 0 \\ 0 \\ \vdots \\ 0 \\ 1 \end{bmatrix}$$

The output can be one of the states or a combination of many states. Since, $y = x_1$ for this case, the output equation is written as

$$y = [1\ 0\ 0\ 0\ \cdots\ 0]x$$

The LTI state–space model is simple because one can have a closed form solution

$$x(t) = e^{At}x(0) + \int_0^t e^{A(t-\tau)} Bu(\tau)d\tau$$

In the case where the system parameters, i.e. elements of constant matrices A, B, C, and D vary with time, we may generalize the state–space representation in the following way:

$$\dot{x}(t) = A(t)x(t) + B(t)u(t) \quad \text{(state equation)} \tag{1.11a}$$

$$y(t) = C(t)x(t) + D(t)u(t) \quad \text{(output equation)} \tag{1.11b}$$

This representation is popularly known as the linear time varying (LTV) state–space model.

1.5 NON-LINEAR STATE–SPACE MODEL

A general state–space model of a non-linear system is given as

$$\dot{x}(t) = f(t, x(t), u(t)), \quad x(t_0) = x_0 \tag{1.12a}$$

$$y(t) = h(t, x(t), u(t)) \tag{1.12b}$$

where f and h are vector valued functions.

However, many practical systems can be represented in input affine non-linear state–space model

$$\dot{x}(t) = f(x(t)) + \Sigma_{i=1}^p g_i(x(t))u_i(t), \quad x(t_0) = x_0 \tag{1.13a}$$

$$y(t) = h(x(t)) \tag{1.13b}$$

1.5.1   Equilibrium Point and Linearization using First-order Taylor Series

For the non-linear system

$$\dot{x} = f(x, u)$$

where each element $f_i(.)$ of the vector function f is continuously differentiable, the equilibrium point (x_0, u_0) is defined as

$$f(x_0, u_0) = 0$$

Let us write the general form of non-linear system $\dot{x} = f(x, u)$ as

$$\frac{dx_1}{dt} = f_1(x_1, x_2, \dots, x_n, u_1, u_2, \dots, u_m)$$

$$\frac{dx_2}{dt} = f_2(x_1, x_2, \dots, x_n, u_1, u_2, \dots, u_m)$$

$$\vdots = \vdots$$

$$\frac{dx_n}{dt} = f_n(x_1, x_2, \dots, x_n, u_1, u_2, \dots, u_m)$$

Let $u_0 = [u_{10} u_{20} \cdots u_{m0}]^T$ be the reference input vector that forces the system $\dot{x} = f(x, u)$ to settle into a constant equilibrium state $x_0 = [x_{10} x_{20} \cdots x_{n0}]^T$ such that $f(x_0, u_0) = 0$ holds. Then the linearization is the process of approximating the non-linear system model by its linear counterpart in a small region about its equilibrium point. We now perturb the equilibrium state by allowing $x = x_0 + \Delta x$ and $u = u_0 + \Delta u$. The *Taylor's expansion* yields

$$\frac{dx}{dt} = f(x_0 + \Delta x, u_0 + \Delta u)$$

$$= f(x_0, u_0) + \frac{\partial f}{\partial x}(x_0, u_0)\Delta x + \frac{\partial f}{\partial u}(x_0, u_0)\Delta u + \cdots$$

$$(1.14)$$

where

$$\frac{\partial f}{\partial x}(x_0, u_0) = \begin{bmatrix} \dfrac{\partial f_1}{\partial x_1} & \cdots & \dfrac{\partial f_1}{\partial x_n} \\ \vdots & & \vdots \\ \dfrac{\partial f_n}{\partial x_1} & \cdots & \dfrac{\partial f_n}{\partial x_n} \end{bmatrix}_{x_0, u_0}$$

$$\frac{\partial f}{\partial u}(x_0, u_0) = \begin{bmatrix} \dfrac{\partial f_1}{\partial u_1} & \cdots & \dfrac{\partial f_1}{\partial u_m} \\ \vdots & & \vdots \\ \dfrac{\partial f_n}{\partial u_1} & \cdots & \dfrac{\partial f_n}{\partial u_m} \end{bmatrix}_{x_0, u_0}$$

are the Jacobian matrices of f with respect to x and u, evaluated at the equilibrium point, (x_0, u_0). Since x_0 is constant,

$$\frac{dx}{dt} = \frac{dx_0}{dt} + \frac{d(\Delta x)}{dt} = \frac{d(\Delta x)}{dt} \qquad (1.15)$$

Furthermore $f(x_0, u_0) = 0$. Let

$$A = \frac{\partial f}{\partial x}(x_0, u_0) \text{ and } B = \frac{\partial f}{\partial u}(x_0, u_0) \tag{1.16}$$

Neglecting the higher order terms in (1.14), the first-order Taylor series *approximation* leads to the following linear model

$$\frac{d(\Delta x)}{dt} = A\Delta x + B\Delta u \tag{1.17}$$

The output equation of the non-linear system model is described as

$$y_1 = h_1(x_1, x_2, \ldots, x_n, u_1, u_2, \ldots, u_m)$$

$$y_2 = h_2(x_1, x_2, \ldots, x_n, u_1, u_2, \ldots, u_m)$$

$$\vdots = \vdots$$

$$y_p = h_p(x_1, x_2, \ldots, x_n, u_1, u_2, \ldots, u_m)$$

or in vector notation

$$y = h(x, u) \tag{1.18}$$

If we let

$$y = y_0 + \Delta y \tag{1.19}$$

then the first-order Taylor series approximation will lead to the following linear output equation.

$$\Delta y = C\Delta x + D\Delta u \tag{1.20}$$

where

$$C = \frac{\partial h}{\partial x}(x_0, u_0) = \begin{bmatrix} \frac{\partial h_1}{\partial x_1} & \cdots & \frac{\partial h_1}{\partial x_n} \\ \vdots & & \vdots \\ \frac{\partial h_p}{\partial x_1} & \cdots & \frac{\partial h_p}{\partial x_n} \end{bmatrix}_{x_0, u_0} \tag{1.21}$$

$$D = \frac{\partial h}{\partial u}(x_0, u_0) = \begin{bmatrix} \frac{\partial h_1}{\partial u_1} & \cdots & \frac{\partial h_1}{\partial u_m} \\ \vdots & & \vdots \\ \frac{\partial h_p}{\partial u_1} & \cdots & \frac{\partial h_p}{\partial u_m} \end{bmatrix}_{x_0, u_0} \tag{1.22}$$

Example 1.6 Consider a non-linear system

$$\dot{x}_1 = \frac{-1}{x_2^2(t)} \tag{1.23a}$$

$$\dot{x}_2 = u(t)x_1(t) \tag{1.23b}$$

Linearize the system (1.23) about the nominal trajectory $[x_{01}(t), x_{02}(t)]$, which is the solution to the equations with initial condition $x_1(0) = x_2(0) = 1$ and input $u(t) = 0$.

Solution

Integrating both sides of Eqn (8.41b) with respect to t, we have

$$x_2(t) = 1$$

Now, Eqn (8.41a) gives

$$x_1(t) = -t + 1$$

Therefore, the nominal trajectory about which Eqns (8.41a) and (8.41b) are to be linearized is described by

$$x_{01}(t) = 1 - t \tag{1.24a}$$
$$x_{02}(t) = 1 \tag{1.24b}$$

Now, evaluating the coefficients of Eqn (1.16), we get

$$\frac{\partial f_1(t)}{\partial x_1(t)} = 0 \quad \frac{\partial f_1(t)}{\partial x_2(t)} = \frac{2}{x_2^3(t)} \quad \frac{\partial f_2(t)}{\partial x_1(t)} = u(t) \quad \frac{\partial f_2(t)}{\partial x_2(t)} = x_1(t)$$

Hence, the linearized equations are given by

$$\begin{bmatrix} \triangle \dot{x}_1(t) \\ \triangle \dot{x}_2(t) \end{bmatrix} = \begin{bmatrix} 0 & 2 \\ 0 & 0 \end{bmatrix} \begin{bmatrix} \triangle x_1(t) \\ \triangle x_2(t) \end{bmatrix} + \begin{bmatrix} 0 \\ 1 - t \end{bmatrix} \triangle u(t) \tag{1.25}$$

which is a set of linear state equations with time-varying coefficients.

We should note that the above equations are linear in terms of $\triangle x$ and $\triangle u$, not in terms of x and u. When $[x_0, u_0]^T = 0$, the linearized model will also be linear in terms of x and u.

1.5.2 Linearization Technique for Operating Points Other Than the Origin

The Taylor series expansion will yield an offset term at operating points other than 0. We would now provide another technique for linearization [7] which will be used to linearize affine non-linear systems at other operating points.

Consider the following affine system

$$\dot{x} = f(x) + g(x)u$$

We need to construct two matrices A and B such that in a neighbourhood of an operating point x_0,

$$f(x) + g(x)u \approx Ax + Bu$$

and

$$f(x_0) + g(x_0)u = Ax_0 + Bu \quad \text{for all } u$$

Since u is arbitrary, we must have

$$g(x_0) = B$$

Our next task is to find A such that in a neighbourhood of x_0,

$$f(x) \approx Ax$$

and

$$f(x_0) = Ax_0$$

Let a_i^T denotes the ith row of matrix A. Thus,

$$f_i(x) \approx a_i^T x \tag{1.26}$$
$$f_i(x_0) = a_i^T x_0, \quad \text{for } i = 1, 2, \cdots n \tag{1.27}$$

where f_i is the ith component of f. Expanding the left side of (1.26) about x_0 and neglecting the higher order terms,

$$f_i(x_0) + \nabla^T f_i(x_0)(x - x_0) \approx a_i^T x \tag{1.28}$$

where $\nabla^T f_i(x)$ is the gradient (a column vector) of f_i at x. Combining Eqns (1.27) and (1.28) we can write

$$\nabla^T f_i(x_0)(x - x_0) \approx a_i^T (x - x_0) \tag{1.29}$$

We will now determine the constant a_i such that the following cost function is minimized

$$E = \frac{1}{2} \| \nabla^T f_i(x_0) - a_i \|^2$$

subject to the constraint $a_i^T x_0 = f_i(x_0)$. This is a convex optimization problem by solving which we get a_i as

$$a_i = \nabla^T f_i(x_0) + \frac{f_i(x_0) - x_0^T \nabla^T f_i(x_0)}{\|x_0\|^2} x_0, \quad x_0 \neq 0 \tag{1.30}$$

Example 1.7 Consider the following dynamical equations of a VanderPol oscillator [12].

$$\dot{x}_1 = x_2$$
$$\dot{x}_2 = -x_1 + \mu(1 - x_1^2)x_2 + u \tag{1.31}$$

Linearize the system in a form $\dot{x} = Ax + Bu$ around the operating point $x_0 = \begin{bmatrix} 1 \\ 0.5 \end{bmatrix}$.

Solution
This system can be linearized at operating points other than the origin using the above described procedure. Since the system is a second-order system $f(x)$ is a 2×1 vector where $f_1(x) = x_2$ and $f_2(x) = -x_1 + \mu(1 - x_1^2)x_2$. The gradient ∇f_i is computed as $\nabla f_i(x) = \begin{bmatrix} \frac{\partial f_i}{\partial x_1} \\ \frac{\partial f_i}{\partial x_2} \end{bmatrix}$.

Thus, $\nabla f_1(x) = \begin{bmatrix} 0 \\ 1 \end{bmatrix}$, $\nabla f_2(x) = \begin{bmatrix} -1 - 2\mu x_1 x_2 \\ \mu(1 - x_1^2) \end{bmatrix}$

Since $g(x) = \begin{bmatrix} 0 \\ 1 \end{bmatrix}$, we have $B = \begin{bmatrix} 0 \\ 1 \end{bmatrix}$ for all operating points. The matrix A at operating point x_0 can now be computed using Eqn (1.30). When $x_0 = \begin{bmatrix} 1 \\ 0.5 \end{bmatrix}$, the matrix A is found out to be $A = \begin{bmatrix} 0 & 1 \\ -1.2 & 0.4 \end{bmatrix}$ for $\mu = 1$. The MATLAB code to find out the matrices A and B with the corresponding output for $x_0 = \begin{bmatrix} 1 \\ 0.5 \end{bmatrix}$ is given below.

MATLAB Code

```
LinVdpol.m

function out = LinVdpol(x);
mu=1;
f = [x(2);-x(1)+mu*(1-x(1)*x(1))*x(2)];
gradf = [0 -1-2*mu*x(1)*x(2);1 mu*(1-x(1)*x(1))];
for i=1:2
A(i,:)=(gradf(:,i) + ((f(i)-
x'*gradf(:,i))/(norm(x)*norm(x)))*x)';
end
A
B=[0;1]
```

The MATLAB command to execute the above code is $LinVdpol([1;0.5])$ which gives A and B matrices as $\begin{bmatrix} 0 & 1 \\ -1.2 & 0.4 \end{bmatrix}$ and $\begin{bmatrix} 0 \\ 1 \end{bmatrix}$.

1.6 LYAPUNOV STABILITY THEORY

Consider the non-linear system

$$\dot{x} = f(x, t), \quad x(t_0) = x_0 \tag{1.32}$$

where $x \in \mathcal{R}^n$ and $t >= 0$. Let $x = x_e$ is an equilibrium point such that

$$f(x_e) = 0 \tag{1.33}$$

Definition 1.1 The equilibrium point $x = x_e$ of (1.32) is said to be *stable in the sense of Lyapunov* if for every $\epsilon > 0$ and any $t_0 \geq 0$ there exists a $\delta(\epsilon, t_0) > 0$ such that

$$|x_0 - x_e| < \delta(\epsilon, t_0) \Rightarrow |x(t) - x_e| < \epsilon \ \forall \ t \geq t_0 \tag{1.34}$$

where $x(t)$ is the solution of (1.32).

Definition 1.2 The equilibrium point $x = x_e$ of (1.32) is an asymptotically stable point if $x = x_e$ is a stable equilibrium point and

$$|x(t) - x_e| \to 0 \text{ as } t \to \infty \tag{1.35}$$

We call the equilibrium point $x = x_e$ unstable if it is not stable.

How do we determine stability or instability of the equilibrium point $x = x_e$? A.M. Lyapunov introduced two methods to analyse stability. In his first method, *indirect method*, he showed that if the non-linear system is linearized about an equilibrium point, certain conclusions about local stability can be made based on eigenvalues of the linearized system. In his second method, *direct method*, the stability results for an equilibrium point $x = x_e$ of (1.32) depend on the existence of an appropriate *Lyapunov function* [12, 13, 14].

The direct method is very powerful and it has several advantages such as given below:

- answers questions of stability of non-linear systems
- can easily handle time varying systems $\dot{x} = f(x, t)$
- can determine asymptotic stability as well as ordinary stability
- can determine the region of asymptotic stability or the domain of attraction of an equilibrium

- can help to design a control law that guarantees global asymptotic stability, i.e., with infinitely large domain of attraction, for a non-linear system.

The main drawback of the method is that there is no systematic way of obtaining Lyapunov functions, this is more of an art than science. For simple second-order systems a good selection for a Lyapunov function is the total energy of the system (kinetic plus potential energy).

1.6.1   Lyapunov Stability of Time Invariant System

Theorem 1.2   Let $x_e = 0$ be an equilibrium point for a system described by
$$\dot{x} = f(x) \qquad (1.36)$$
where $f : S \to R^n$ is a locally Lipschitz and $S \subset R^n$ is a domain that contains the origin. Let $V : S \to R$ be a continuously differentiable and positive definite function in S.

1. If $\dot{V}(x) = \frac{\partial V}{\partial x} f$ is negative semidefinite, then $x_e = 0$ is a stable equilibrium point.
2. If $\dot{V}(x)$ is negative definite, then $x_e = 0$ is an asymptotically stable equilibrium point.

In both cases above V is called a Lyapunov function. Moreover, if the conditions hold for all $x \in R^n$ and $\| x \| \to \infty$ implies that $V(x) \to \infty$, then $x_e = 0$ is globally stable in case 1 and globally asymptotically stable in case 2. The function $V(x)$ is said to be radially unbounded in this case.

The proof of this theorem can be found in [14]. The importance of this theorem is that the existence of a Lyapunov function is sufficient to prove stability in the sense of Lyapunov in the region S. Here are the salient points of Theorem 1.2:

1. The conditions of Theorem 1.2 are only sufficient.
2. Failure of a Lyapunov function candidate to satisfy the conditions for stability does not mean that the equilibrium is not stable or asymptotically stable.
3. The Lyapunov approach allows us to assess the stability of equilibrium points of a system without solving the differential equations that describe the system.
4. There are no generally applicable methods for finding Lyapunov functions. Heuristics and physics of the dynamical system are often used to find a Lyapunov candidate. The variable gradient method [14] can be used to find the Lyapunov function for some simple cases.

Example 1.8   Consider the scalar system
$$\dot{x} = -x^3, \quad x \in R \qquad (1.37)$$
Investigate the stability of the origin $x_e = 0$.

Solution
Consider the candidate Lyapunov function $V(x) = \frac{1}{2}x^2$

The rate derivative of $V(x)$ is given by

$$\dot{V}(x) = \frac{\partial V}{\partial x} f = \frac{\partial V}{\partial x}(-x^3) = -x^4$$

Since $\dot{V}$ is negative definite for all $x \in R$ and $V(x) \to \infty$, when $x \to \infty$, the origin $x_e = 0$ is globally asymptotically stable as per theorem 1.7 .

Example 1.9   Consider the pendulum equation without friction

$$\dot{x}_1 = x_2$$
$$\dot{x}_2 = -a \sin x_1$$

Study the stability of the equilibrium point at the origin, i.e $x_1 = 0,\ x_2 = 0$.

Solution
A natural Lyapunov function candidate for this system is the energy function

$$V(x) = a(1 - \cos x_1) + \frac{1}{2}x_2^2$$

Here $V(0) = 0$ and $V(x)$ is positive definite over the domain $-\pi < x_1 < \pi$. The derivative of $V(x)$ along the trajectories of the system is

$$\dot{V}(x) = \frac{\partial V}{\partial x_1}\dot{x}_1 + \frac{\partial V}{\partial x_2}\dot{x}_2 = a\,x_2 \sin x_1 - a\,x_2 \sin x_1 = 0$$

Since $V(x)$ is positive definite and $\dot{V}(x) = 0$, then by Theorem 1.1, the origin $x_e = \mathbf{0}$ is stable.

1.6.2   LaSalle's Invariance Theorem

Invariance set   A set Ω is said to be an invariant set [15, 14] of (1.36) if a solution $x(t)$ that belongs to Ω at some time instant t_0 also belongs to Ω for all $t \in R$. If the above said condition holds only for future time, i.e., for $t \geq t_0$ the set Ω is said to be a positive invariant set of (1.36).

Theorem 1.3   Let Ω be a positive invariant set of (1.36) and $V(x)$: $\Omega \to R_+$ be a continuously differentiable function such that $\dot{V}(x) \leq 0$ for all $x \in R^n$. Let $D = \{x \in R^n : \dot{V}(x) = 0$ and M be the largest invariant set contained in D. Then, every bounded solution $x(t)$ starting in Ω converges to M when $t \to \infty$.

Corollary 1.1   Let $x_e = 0$ be an equilibrium point of the system (1.36). Let $V(x) : D \to R$ be a continuously differentiable positive definite function on a domain D containing the origin, such that $\dot{V}(x) \leq 0$ in D. Let $S := \{x \in D : \dot{V}(x) = 0\}$. Suppose that S contains no trajectories other than the trivial trajectory $x = \mathbf{0}$ that stay identically in S. Then the equilibrium point $x_e = 0$ of (1.36) is asymptotically stable.

Corollary 1.2 Let $x_e = 0$ be an equilibrium point of the system (1.36). Let $V(x) : R^n \to R$ be a continuously differentiable, radially unbounded, positive definite function such that $\dot{V}(x) \leq 0$ for all $x \in R^n$. Let $S := \{x \in R^n : \dot{V}(x) = 0\}$. Suppose that S contains no trajectories other than the trivial trajectory $x = 0$ that stay identically in S. Then the equilibrium point $x_e = 0$ of (1.36) is globally asymptotically stable.

1.6.3 Chetaev's Instability Theorem

Theorem 1.4 Let $x_e = 0$ be an equilibrium point of the system (1.36). Let $V : S \to R$ be a continuously differentiable such that $V(0) = 0$ and $V(x_0) > 0$ for some x_0 with arbitrarily small $\|x_0\|$. S is a subset of R^n which contains the origin. Define a set U such that $U = \{x \in B_r ; V(x) > 0\}$ where for $r > 0$ $B_r = \{x \in R^n; \|x\| \leq r\}$. Suppose that $\dot{V}(x) > 0$ in U. Then the equilibrium point $x_e = 0$ is unstable.

1.6.4 Lyapunov Stability of Time Varying System

Theorem 1.5 Let $x_e = 0$ be an equilibrium point for a time varying system described by
$$\dot{x} = f(x, t) \tag{1.38}$$
Let $S \subset R^n$ a domain that contains origin and V be a continuously differentiable function that satisfies

1. $W_1(x) \leq V(x, t) \leq W_2(x)$
2. $\dot{V}(x, t) = \frac{\partial V}{\partial t} + \frac{\partial V}{\partial x} f(x, t) \leq -W_3(x)$

for all $t \geq t_0$, and $x \subset S$, where $W_1(x)$, $W_2(x)$, and $W_3(x)$ are continuous positive definite functions on S. Then $x_e = 0$ is uniformly asymptotically stable and V is called a Lyapunov function. Furthermore, if $W_3(x) = 0$, then $x_e = 0$ is uniformly stable.

Corollary 1.3 Suppose that the assumptions of Theorem 1.5 hold for all $x \in R^n$ and $W_1(x) \to \infty$ for $\| x \| \to$, then $x_e = 0$ is globally uniformly asymptotically stable.

Corollary 1.4 Suppose that the assumptions of Theorem 1.5 are replaced by

1. $c_1 \| x \|^q \leq V(x, t) \leq c_2 \| x \|^q$
2. $\dot{V}(x, t) \leq -c_3 \| x \|^q$

for some positive constants c_1, c_2, c_3, and q. Then $x_e = 0$ is exponentially stable. Furthermore, if the assumptions are satisfied for all $x \in R^n$, then $x_e = 0$ is globally exponentially stable.

Proofs of the above corollaries can be found in [14].

Example 1.10   Consider the scalar system described by

$$\dot{x} = -x^3 + \frac{1}{2}x^3 \sin(t), \quad x(t_0) = x_0 \tag{1.39}$$

Investigate the stability of the system.

Solution

Select the Lyapunov function as $V(x) = \frac{1}{2}x^2$. The derivative of $V(x)$ along the trajectories of the system is

$$\dot{V}(x) = \frac{\partial V}{\partial x}\dot{x}$$

$$= x\left(-x^3 + \frac{1}{2}x^3 \sin(t)\right)$$

$$= -x^4\left(1 - \frac{1}{2}\sin(t)\right)$$

$$\leq -x^4$$

By taking $W_3(x) = x^4$, $\dot{V} \leq W_3(x)$. If we choose $W_1(x) = W_2(x) = V(x)$, then Theorem 1.5 and Corollary 1.3 will yield that the origin $x_e = 0$ is globally asymptotically stable.

1.6.5 Lyapunov's Indirect Method

Theorem 1.6   Let x_e be an equilibrium point for the non-linear system $\dot{x} = f(x)$, where $f : D \rightarrow R^n$ is continuously differentiable and D is a neighbourhood of the origin. Let the Jacobian matrix A at $x_e = 0$ be

$$A = \left.\frac{\partial f}{\partial x}\right|_{x=0} \tag{1.40}$$

Let $\lambda_i, i = 1, \ldots, n$ be the eigenvalues of A. Then

1. The origin is asymptotically stable if $\text{Re}(\lambda_i(A)) < 0$ for all i.
2. The origin is unstable if for any i, $\text{Re}(\lambda_i(A)) > 0$.

Example 1.11   Consider the following non-linear system:

$$\dot{x}_1 = -x_2 + ax_1 x_2^2 \tag{1.41a}$$

$$\dot{x}_2 = x_1 - bx_1^2 x_2 \tag{1.41b}$$

with $a \neq b$. Investigate the stability of the equilibrium point of the system.

Solution

At the equilibrium point,

$$-x_2 + ax_1 x_2^2 = 0$$

$$x_1 - bx_1^2 x_2 = 0$$

Multiplying the first equation by x_1, the second by x_2, we get

$$x_1^2 x_2^2(a - b) = 0$$

from which $x_1 = 0$ or $x_2 = 0$. If $x_1 = 0$ then we see from the first equation that $x_2 = 0$ as well, and similarly if we assume that $x_2 = 0$. Therefore, the unique equilibrium of the system is $x_1 = x_2 = 0$.

The linearized system at $x_1 = x_2 = 0$ is

$$\begin{bmatrix} \dot{x}_1 \\ \dot{x}_2 \end{bmatrix} = \begin{bmatrix} 0 & -1 \\ 1 & 0 \end{bmatrix} \begin{bmatrix} x_1 \\ x_2 \end{bmatrix} \tag{1.42}$$

The characteristic equation is

$$\det \begin{vmatrix} -s & -1 \\ 1 & -s \end{vmatrix} = 0 \implies s^2 + 1 = 0 \implies s = \pm \omega i \tag{1.43}$$

Since the characteristic roots are purely imaginary, we cannot draw any conclusion on the stability of the non-linear system using Lyapunov's indirect method. Using Lyapunov's direct method, let the Lyapunov function be

$$V(\boldsymbol{x}) = \frac{1}{2} x_1^2 + \frac{1}{2} x_2^2 \tag{1.44}$$

We see that $V(\boldsymbol{x}) > 0$ for all x_1, x_2 and $V \to \infty$ when $\boldsymbol{x} \to \infty$. The time derivative of V is

$$\begin{aligned} \dot{V}(\boldsymbol{x}) &= x_1(-x_2 + ax_1x_2^2) + x_2(x_1 - bx_1^2x_2) \\ &= -x_1x_2 + ax_1^2x_2^2 + x_1x_2 - bx_1^2x_2^2 \\ &= (a-b)x_1^2x_2^2 \end{aligned}$$

Therefore, we see that

If $a < b$, the system is stable. Furthermore, using the invariance principles, we can define a set $S := \{(x_1, x_2) \in R^2; \dot{V}(\boldsymbol{x}) = 0\}$. $\dot{V}(\boldsymbol{x})$ will be 0 if any of the states is 0. In both cases, i.e., when $x_1 = 0$ or when $x_2 = 0$, (1.41a) implies that

$$-x_2 + ax_1x_2^2 = 0$$
$$x_1 - bx_1^2x_2 = 0$$

whose unique solution is $x_1 = x_2 = 0$. Thus, S contains only the trivial trajectory $x_1 = x_2 = 0$. Hence the equilibrium point is asymptotically stable.

If $a > b$, $\dot{V} > 0$ for all $\boldsymbol{x} \in R^n$. Thus, applying the Chetaev's instability theorem we can conclude that the equilibrium point is unstable.

This result could not have been obtained by the indirect method.

MATLAB code	
LyapEx.m	Executable commands
```function xdot = LyapEx(t,x,a,b); xdot = [-x(2)+a*x(1)*x(2)*x(2); x(1)-b*x(1)*x(1)*x(2)];```	```a=1;b=5; [t,x]=ode45(@LyapEx,[0 100], [0.15 0.15],OPTIONS,a,b); plot(t,x) a=1;b=0.5; [t,x]=ode45(@LyapEx,[0 100], [0.15 0.15],OPTIONS,a,b); plot(t,x)```

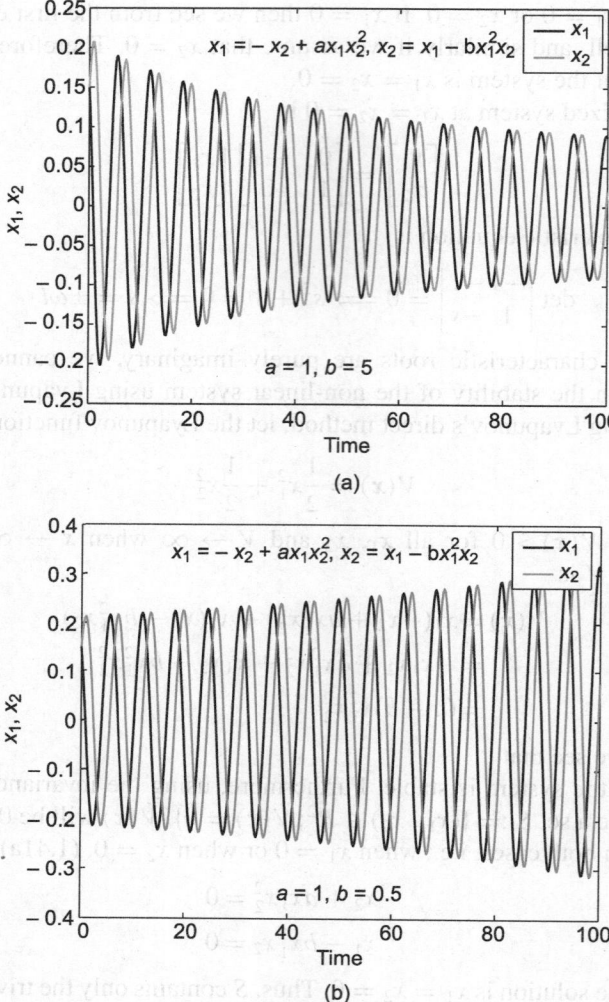

Figure 1.1   Solutions of the system: (a) the system states when $a < b$ and (b) the system states when $a > b$

We will now show through MATLAB simulation that the conclusion we derived for the above system is indeed correct. We will numerically solve the differential Eqns (1.41a) using the *ode45* solver which use the Runge–Kutta fourth-order integration method [16].

It can be seen from Figure 1.1 that when $a < b$, the system states approach zero with time, i.e., the origin $(0, 0)$ is asymptotically stable and when $a > b$, the system states diverge from zero, i.e., the origin is unstable.

Example 1.12   Consider an oscillator with a non-linear spring:

$$\ddot{y} + 3\dot{y} + y^3 = 0 \tag{1.45}$$

Investigate the stability of this system.

Solution

If we were to linearize this system we would get $\ddot{y} + 3\dot{y} = 0$, which has the characteristic equation $s(s+3) = 0$. The -3 characteristic root corresponds to the damping term, but note the existence of a 0 root from the lack of a linear term in the spring restoring force. The linearized version of the system cannot recognize the existence of a non-linear spring term and it fails to produce a non-zero characteristic root related to the restoring force. To see if this nonlinear spring produces a stable or unstable system, we have to resort to the Lyapunov functions. The state space form of the system is

$$\dot{x}_1 = x_2 \tag{1.46a}$$

$$\dot{x}_2 = -3x_2 - x_1^3 \tag{1.46b}$$

with equilibrium $x_1 = x_2 = 0$. Let us try for a Lyapunov function

$$V(x) = \frac{1}{4}x_1^4 + \frac{1}{2}x_2^2 \tag{1.47}$$

We can see that $V(x) > 0$ for all x_1, x_2. The time derivative of V is

$$\dot{V}(x) = \frac{\partial V}{\partial x_1}\dot{x}_1 + \frac{\partial V}{\partial x_2}\dot{x}_2$$

$$= x_1^3 x_2 + x_2(-3x_2 - x_1^3)$$

$$= -3x_2^2$$

$$\leq 0$$

It follows then that the origin $x = 0$ is stable. Further, we see that $\dot{V}(x) = 0$ when $x_2 = 0$. This implies that $x_1 =$ constant. In another way

$$x_2 = 0$$

$$-3x_2 - x_1^3 = 0$$

whose unique solution is $x_1 = x_2 = 0$. Thus, the origin $x = 0$ is asymptotically stable.

1.6.6 Lyapunov Stability for Linear Systems

It is possible to find a Lyapunov function for a linear system in the form

$$\dot{x} = Ax \tag{1.48}$$

Choose as a Lyapunov function the quadratic form

$$V(x) = x^T P x \tag{1.49}$$

where P is a symmetric positive definite matrix. Then we have

$$\dot{V}(x) = \dot{x}^T P x + x^T P \dot{x}$$

$$= (Ax)^T P x + x^T P A x$$

$$= x^T A^T P x + x^T P A x \tag{1.50}$$

$$= x^T(A^T P + PA)x$$

$$= -x^T Q x$$

where

$$A^T P + PA = -Q \tag{1.51}$$

If the matrix Q is positive definite, then the system is asymptotically stable. Therefore, we could pick $Q = I$, as the identity matrix and solve

$$A^T P + PA = -I \tag{1.52}$$

for P and see if P is positive definite (we can do this by looking at the n principal minors of P—Sylvester's criterion). Eqn (1.51) is called Lyapunov's matrix equation and its solution is easy through MATLAB by using the command *lyap*.

Example 1.13 Using the Lyapunov method, determine the stability of the system $\dot{x} = Ax$ where $A = \begin{bmatrix} 0 & 1 & 1 \\ 0.5 & 0 & 2 \\ -3 & -2 & -5 \end{bmatrix}$. Also find out the corresponding Lyapunov matrix.

Solution
Since the MATLAB command *lyap* solves an equation of the form $AX + XA^T = -C$, we have to input the matrix A^T instead of A.

MATLAB code
A=[0 1 1;0.5 0
2;-3 -2 -5];
Q = eye(3,3);
P = lyap(A',Q);
eig(P)

The eigenvalues of P have been computed as 0.1226, 0.8153, and 1.6201. Thus, the solution of the Lyapunov equation is a positive definite matrix. Hence the system is asymptotically stable.

We could argue that having an equation to determine a Lyapunov function for linear systems is useless because for a linear system we can always look at the eigenvalues of A to determine stability or instability. This is true; the usefulness of Lyapunov's matrix equation for linear systems is that it can provide an initial estimate for a Lyapunov function for a non-linear system in the cases where this is done computationally. Furthermore, it can be used to show stability of the linear quadratic regulator design.

Example 1.14 For the following non-linear system

$$\dot{x}_1 = -x_1 + x_2 + x_1(x_1^2 + x_2^2) \tag{1.53a}$$

$$\dot{x}_2 = -x_1 - x_2 + x_2(x_1^2 + x_2^2) \tag{1.53b}$$

show that the origin is stable and find the domain of attraction.

Solution

The easiest way to show the stability is by linearization. The linearized form of the system is

$$\begin{bmatrix} \dot{x}_1 \\ \dot{x}_2 \end{bmatrix} = \begin{bmatrix} -1 & 1 \\ -1 & -1 \end{bmatrix} \begin{bmatrix} x_1 \\ x_2 \end{bmatrix} \qquad (1.54)$$

The characteristic equation is

$$s^2 + 2s + 2 = 0 \qquad (1.55)$$

and we can see that the system is stable, the roots of the characteristic equation have negative real parts. Now since this result is based on linearization, it says that if the initial condition is 'close' to the equilibrium point $(0, 0)$, then the solution will tend to the equilibrium as $t \to \infty$. To find how close is 'close' we need to get an estimate of the *domain of attraction*. We can do this by using the Lyapunov theory. Let us try a Lyapunov function candidate

$$V(x) = \frac{1}{2}x_1^2 + \frac{1}{2}x_2^2 \qquad (1.56)$$

Form

$$\begin{aligned} \dot{V}(x) &= x_1\dot{x}_1 + x_2\dot{x}_2 \\ &= x_1(-x_1 + x_2 + x_1^3 + x_1x_2^2) + x_2(-x_1 - x_2 + x_2x_1^2 + x_2^3) \\ &= -x_1^2 + x_1x_2 + x_1^4 + x_1^2x_2^2 - x_1x_2 - x_2^2 + x_1^2x_2^2 + x_2^4 \\ &= x_1^4 + x_2^4 + 2x_1^2x_2^2 - x_1^2 - x_2^2 \\ &= (x_1^2 + x_2^2)(x_1^2 + x_2^2 - 1) \end{aligned}$$

We can see, therefore, that stability is guaranteed if

$$\dot{V}(x) < 0 \text{ or } x_1^2 + x_2^2 < 1 \qquad (1.57)$$

which means that the domain of attraction of the equilibrium is a circular disc of radius 1. As long as the initial conditions are inside the disc, it is guaranteed that the solution will end up at the stable equilibrium. In the case where the initial conditions lie outside the disc then convergence is not guaranteed. It should be mentioned that the above disc is an estimate of the domain attraction based on the particular Lyapunov function we selected. A different Lyapunov function could have produced a different estimate of the domain of attraction.

1.7 DISCRETE TIME SYSTEMS

The dynamics of discrete time systems are represented by a set of difference equations. In this section, we would provide a brief overview of modelling and stability analysis of discrete time systems.

1.7.1   Discrete Time LTI State–Space Model

The discrete time linear time invariant (LTI) systems are defined by the following state–space model:

$$x(k+1) = Ax(k) + Bu(k) \quad \text{(state equation)} \tag{1.58a}$$

$$y(k) = Cx(k) + Du(k) \quad \text{(output equation)} \tag{1.58b}$$

where $x(0)$ is the initial condition. $x(k) \in R^n$, $y(k) \in R^m$ and $u(k) \in R^p$ are the state, output, and input vectors, respectively. Similarly, $A \in R^{n \times n}$, $B \in R^{n \times p}$, $C \in R^{m \times n}$, and $D \in R^{m \times p}$ are constant matrices.

1.7.2   Discrete Time Non-linear State–Space Model

A general state–space model of a non-linear system is given as

$$x(k+1) = f(k, x(k), u(k)), \quad x(k_0) = x_0 \tag{1.59a}$$

$$y(k) = h(k, x(k), u(k)) \tag{1.59b}$$

where f and h are vector valued functions.

The input affine non-linear discrete time state–space models are defined by

$$x(k+1) = f(x(k)) + g(x(k))u(k), \quad x(k_0) = x_0 \tag{1.60a}$$

$$y(k) = h(x(k)) \tag{1.60b}$$

The linearization techniques, as described in Sections 1.5.1 and 1.5.2, are also valid for non-linear discrete time systems.

1.7.3   ARMAX and NARMAX Models

Another popular method of modelling a discrete time system is using an ARMAX model (auto regression moving average eXogenous inputs) for linear systems and a NARMAX (non-linear auto regression moving average eXogenous inputs) model for non-linear systems. ARMAX and NARMAX models are natural representations of the input–output relationship of linear and non-linear discrete time systems. It is well known for linear discrete time systems that a linear difference equation exists which involves a finite number of calculations at each stage, as

$$y(k) = \sum_{i=1}^{n} a_i y(k-i) + \sum_{i=1}^{m} b_i u(k-i) \tag{1.61}$$

Similarly, the non-linear difference equation model (NARMAX) provides a unified representation for a large class of nonlinear systems. The NARMAX representation of a non-linear discrete time system is as follows.

$$y(k) = f(y(k-1), \ldots y(k-n), u(k-1), \ldots, u(k-m)) \tag{1.62}$$

1.7.4 Lyapunov Stability for Discrete Time Systems

> **Theorem 1.7** Let $x_e = 0$ be an equilibrium point for a time invariant discrete time system described by
> $$x(k+1) = f(x(k)) \tag{1.63}$$
> where $f : S \rightarrow R^n$ is a locally Lipschitz and $S \subset R^n$ is a domain that contains the origin. Let $V : S \rightarrow R$ be a continuously differentiable and positive definite function in S.
>
> 1. If $\Delta V(x(k)) = V(x(k+1)) - V(x(k))$ is negative semidefinite, then $x_e = 0$ is a stable equilibrium point.
> 2. If $\Delta V(x(k))$ is negative definite, then $x_e = 0$ is an asymptotically stable equilibrium point.

The Lyapunov stability analysis for a discrete time system is more difficult to its continuous time counterpart because of the complexity that arises in the time difference of the Lyapunov function.

Lyapunov Stability for Linear Discrete Time Systems

It is possible to find a Lyapunov function for a linear system in the form
$$x(k+1) = Ax(k) \tag{1.64}$$
Choose as Lyapunov function the quadratic form
$$V(x(k)) = x^T(k)Px(k) \tag{1.65}$$
where P is a symmetric positive definite matrix. Then we have
$$\begin{aligned}
\Delta V(x(k)) &= V(x(k+1)) - V(x(k)) \\
&= x^T(k+1)Px(k+1) - x^T(k)Px(k) \\
&= x^T A^T P A x - x^T P x \\
&= x^T(A^T P A - P)x \\
&= -x^T Q x
\end{aligned} \tag{1.66}$$
where
$$A^T P A - P = -Q \tag{1.67}$$
If the matrix Q is positive definite, then the system is asymptotically stable. Therefore, we could pick $Q = I$ as the identity matrix and solve
$$A^T P A - P = -I \tag{1.68}$$
for P and see if P is positive definite. Equation (1.67) is the Lyapunov's matrix equation for discrete time systems and its solution is easy through MATLAB by using the command *dlyap*.

Example 1.15 Using the Lyapunov method, determine the stability of the system $x(k + 1) = Ax(k)$ where $A = \begin{bmatrix} 0 & 0.1 & 0.1 \\ 0.5 & 0 & 0.2 \\ 0.4 & 0.3 & 0.5 \end{bmatrix}$. Also find out the corresponding Lyapunov matrix.

Solution

The MATLAB command *dlyap* solves an equation of the form $AXA^T - X + Q =$. Thus, we have to input the matrix A^T instead of A. The MATLAB solution to the above problem is given below:

MATLAB code
| ```
A=[0 0.1 0.1; 0.5 0 0.2;0.4 0.3 0.5];
Q=eye(3,3);
P = dlyap(A',Q);
eig(P)
``` |

The eigenvalues of P are found to be 1.0001, 1.0978, and 2.3852, thus the solution of the Lyapunov equation is a positive definite matrix. Hence the system is asymptotically stable.

1.8 MODELLING OF DIFFERENT NON-LINEAR SYSTEMS

In this section, we will provide a detailed modelling procedure of four non-linear systems which we will use as simulation examples for different intelligent control strategies in the later chapters of this book.

1.8.1 Inertial Wheel Pendulum

The schematic diagram of an inertial wheel pendulum (IWP) is shown in Figure 1.2. The system dynamics of an IWP can be derived using the Euler–Lagrange equations of motion [7].

Nomenclature

- m_1 (m_p) : Mass of the pendulum
- m_2 (m_r) : Mass of the wheel
- m : Combined mass of the pendulum and wheel
- l_1 : Length of the pendulum
- l_{c1} : Distance from pivot to the centre of mass of the pendulum
- l : Distance from pivot to the centre of mass of pendulum and the wheel
- I_1, J_p : Moment of inertia of the pendulum
- I_2, J_r : Moment of inertia of the wheel
- q_1 : Pendulum angle
- q_2 : Angle of the wheel
- τ : Motor torque input applied on the disc
```

- k : Torque constant of the DC motor
- k_e, k_v : Proportionality constants
- k_u : Output feedback constant

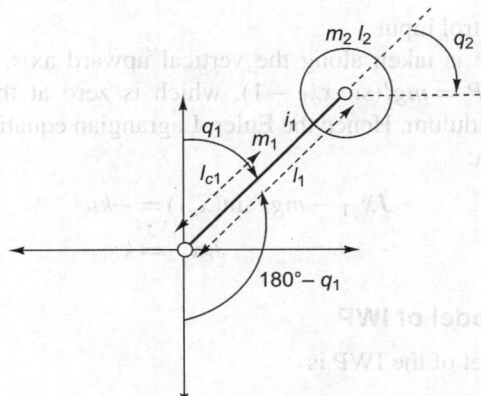

Figure 1.2   Inertial wheel pendulum

The system parameters are redefined as follows:

$$x_{q1} = q_1$$
$$x_{q2} = q_1 + q_2$$
$$m = m_p + m_r$$
$$ml = m_p l_p + m_r l_r$$
$$J = J_p + m_p l_p^2 + m_r l_r^2$$

The kinetic energy K of the system is the sum of the pendulum kinetic energy and the rotor wheel kinetic energy and can be written as

$$K = \frac{1}{2} J \dot{x}_{q1}^2 + \frac{1}{2} J_r \dot{x}_{q2}^2 \qquad (1.69)$$

The potential energy of the system P is

$$P = mgl(1 - \cos x_{q1}) \qquad (1.70)$$

The Lagrangian function L of the system is given by

$$L = K - P = \frac{1}{2} J \dot{x}_{q1}^2 + \frac{1}{2} J_r \dot{x}_{q2}^2 - mgl(1 - \cos x_{q1}) \qquad (1.71)$$

The equations of motion using the Euler–Lagrangian formulation [7] are given as follows:

$$\frac{d}{dt}\left(\frac{\partial L}{\partial \dot{q}}\right) - \frac{\partial L}{\partial q} = \tau \qquad (1.72)$$

where $q = [q_1, q_2]^T$. Neglecting the frictional forces and electrical dynamics of the DC motor placed on the wheel, the torque vector is given as

$$\tau = \begin{bmatrix} 0 \\ ku \end{bmatrix} \tag{1.73}$$

where u is the control input.

If the reference is taken along the vertical upward axis, then the potential energy becomes $P = mgl(\sin x_{q1} - 1)$, which is zero at the vertical upward position of the pendulum. Hence the Euler–Lagrangian equations of motion take the following form:

$$J\ddot{x}_{q1} - mgl\sin(x_{q1}) = -ku \tag{1.74}$$

$$J_r\ddot{x}_{q2} = ku \tag{1.75}$$

State–Space Model of IWP

The dynamic model of the IWP is

$$J\ddot{x}_{q1} = mgl\sin(x_{q1}) - ku \tag{1.76}$$

$$J_r\ddot{x}_{q2} = ku \tag{1.77}$$

The state variables are defined as follows:

$$x_1 = x_{q1}$$

$$x_2 = \dot{x}_{q1}$$

$$x_3 = x_{q2}$$

$$x_4 = \dot{x}_{q2}$$

Using the above state variables, the state–space equations of the system can be written as

$$\left. \begin{aligned} \dot{x}_1 &= x_2 \\ \dot{x}_2 &= \frac{mgl}{J}\sin(x_1) - \frac{k}{J}u \\ \dot{x}_3 &= x_4 \\ \dot{x}_4 &= \frac{k}{J_r}u \end{aligned} \right] \tag{1.78}$$

The control objective is to stabilize the pendulum at the vertically upward position, i.e., $[x_1 \; x_2 \; x_3 \; x_4]^T = [0 \; 0 \; 0 \; 0]^T$.

Linearization Using the Taylor Series Expansion

The non-linear dynamics of the inertial wheel pendulum will be linearized around the vertically upward position using the Taylor series expansion. It is to be noted that the linearized model represents the actual dynamics only in the vicinity of the equilibrium point.

Expanding the system dynamics (**??**) around the equilibrium point

$$x = [x_1 \ x_2 \ x_3 \ x_4]^T = [0 \ 0 \ 0 \ 0]^T$$

we can write the approximated linear model of the system as

$$\dot{x} = \begin{bmatrix} 0 & 1 & 0 & 0 \\ \frac{mgl}{J} & 0 & 0 & 0 \\ 0 & 0 & 0 & 1 \\ 0 & 0 & 0 & 0 \end{bmatrix} x + \begin{bmatrix} 0 \\ -\frac{k}{J} \\ 0 \\ \frac{k}{J_r} \end{bmatrix} u \qquad (1.79)$$

1.8.2 Two Link Manipulator

Consider a planar two link manipulator [17] with two revolute joints as shown in Figure 1.3 where q_1 and q_2 denote joint angles, m_i, l_i, and I_i denote the mass, length and moment of inertia of link i. l_{ci} denotes the distance from the previous joint to the centre of mass of link i.

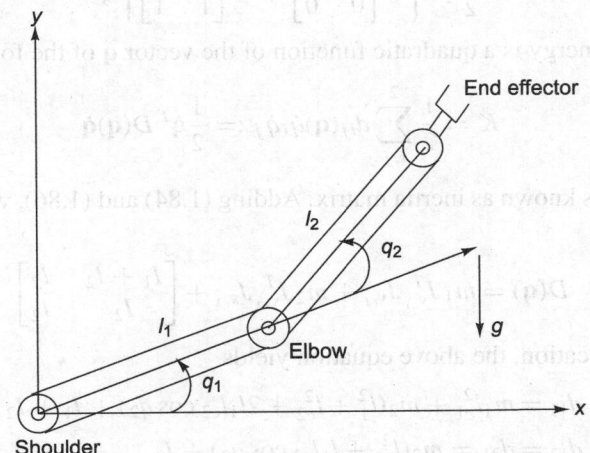

Figure 1.3 Epoch-wise convergence of the mean square error

To derive the Euler–Lagrange equations of motion, we have to first calculate the kinetic and potential energies of the system. The kinetic energy will be a sum of the translational kinetic energy and rotational kinetic energy. The linear velocity of the centre of mass of link 1 can be written as follows:

$$\mathbf{v}_{c1} = J_{\mathbf{v}_{c1}} \dot{\mathbf{q}} \qquad (1.80)$$

where the Jacobian $J_{\mathbf{v}_{c1}}$ can be expressed as [18]

$$J_{\mathbf{v}_{c1}} = \begin{bmatrix} -l_{c1} \sin q_1 & 0 \\ l_{c1} \cos q_1 & 0 \\ 0 & 0 \end{bmatrix} \qquad (1.81)$$

Similarly, the velocity of the centre of mass of link 2 is given as

$$\mathbf{v}_{c2} = J_{\mathbf{v}_{c2}} \dot{\mathbf{q}} \qquad (1.82)$$

where

$$J_{\mathbf{v}_{c2}} = \begin{bmatrix} -l_1 \sin q_1 - l_{c2} \sin(q_1 + q_2) & -l_{c2} \sin(q_1 + q_2) \\ l_1 \cos q_1 + l_{c2} \cos(q_1 + q_2) & l_{c2} \cos(q_1 + q_2) \\ 0 & 0 \end{bmatrix} \quad (1.83)$$

Hence the translational part of the kinetic energy is

$$\frac{1}{2} m_1 \mathbf{v}_{c1} \mathbf{v}_{c1}^T + \frac{1}{2} m_2 \mathbf{v}_{c2} \mathbf{v}_{c2}^T = \frac{1}{2} \dot{\mathbf{q}} \left\{ m_1 J_{\mathbf{v}_{c1}}^T J_{\mathbf{v}_{c1}} + m_2 J_{\mathbf{v}_{c2}}^T J_{\mathbf{v}_{c2}} \right\} \dot{\mathbf{q}} \quad (1.84)$$

To calculate the rotational part, we have to first deal with the angular velocities. The angular velocities of link 1 and 2 are given as

$$\omega_1 = \dot{q}_1 \mathbf{k} \quad \omega_2 = (\dot{q}_1 + \dot{q}_2) \mathbf{k} \quad (1.85)$$

where $\mathbf{k}$ is the unit vector of the base reference frame of the links [18]. The rotational kinetic energy of the overall system can be written as

$$\frac{1}{2} \dot{\mathbf{q}}^T \left\{ I_1 \begin{bmatrix} 1 & 0 \\ 0 & 0 \end{bmatrix} + I_2 \begin{bmatrix} 1 & 1 \\ 1 & 1 \end{bmatrix} \right\} \dot{\mathbf{q}} \quad (1.86)$$

The kinetic energy is a quadratic function of the vector $\dot{\mathbf{q}}$ of the form

$$K = \frac{1}{2} \sum_{i,j}^{2} d_{ij}(\mathbf{q}) \dot{q}_i \dot{q}_j := \frac{1}{2} \dot{\mathbf{q}}^T D(\mathbf{q}) \dot{\mathbf{q}} \quad (1.87)$$

where $D(\mathbf{q})$ is known as inertia matrix. Adding (1.84) and (1.86), we get $D(\mathbf{q})$ as

$$D(\mathbf{q}) = m_1 J_{\mathbf{v}_{c1}}^T J_{\mathbf{v}_{c1}} + m_2 J_{\mathbf{v}_{c2}}^T J_{\mathbf{v}_{c2}} + \begin{bmatrix} I_1 + I_2 & I_2 \\ I_2 & I_2 \end{bmatrix} \quad (1.88)$$

After simplification, the above equation yields

$$d_{11} = m_1 l_{c1}^2 + m_2 (l_1^2 + l_{c2}^2 + 2 l_1 l_{c2} \cos q_2) + I_1 + I_2$$

$$d_{12} = d_{21} = m_2 (l_{c2}^2 + l_1 l_{c2} \cos q_2) + I_2$$

$$d_{22} = m_2 l_{c2}^2 + I_2$$

The potential energy of the system is the sum of those of the two links. For each link, the potential energy is its mass multiplied by the gravitational acceleration and the height of its centre of mass. Thus,

$$V_1 = m_1 g l_{c1} \sin q_1$$

$$V_2 = m_2 g (l_1 \sin q_1 + l_{c2} \sin(q_1 + q_2))$$

and $V = V_1 + V_2 = (m_1 l_{c1} + m_2 l_1) g \sin q_1 + m_2 l_{c2} g \sin(q_1 + q_2)$

Thus, the Lagrangian of the system is $L = K - V$ and the equations of motion are

$$\frac{d}{dt} \left(\frac{\partial L}{\partial \dot{\mathbf{q}}} \right) - \frac{\partial L}{\partial \mathbf{q}} = \tau \quad (1.89)$$

Let us define c_{ijk} and ϕ_k as

$$c_{ijk} := \frac{1}{2}\left\{ \frac{\partial d_{kj}}{\partial q_i} + \frac{\partial d_{ki}}{\partial q_j} - \frac{\partial d_{ij}}{\partial q_k} \right\}$$

$$\phi_k = \frac{\partial V}{\partial q_k}, \quad k = 1, 2$$

Performing the above partial differentiations

$$c_{111} = 0$$

$$c_{121} = c_{211} = -m_2 l_1 l_{c2} \sin q_2$$

$$c_{221} = -m_2 l_1 l_{c2} \sin q_2$$

$$c_{112} = c_{211} = m_2 l_1 l_{c2} \sin q_2$$

$$c_{122} = c_{212} = 0$$

$$c_{222} = 0$$

and

$$\phi_1 = (m_1 l_{c1} + m_2 l_1)g \cos q_1 + m_2 l_{c2} g \cos(q_1 + q_2)$$

$$\phi_2 = m_2 l_{c2} \cos(q_1 + q_2)$$

After doing the required manipulations, the dynamical equations of motion can be expressed as

$$d_{11}\ddot{q}_1 + d_{12}\ddot{q}_2 + (c_{121} + c_{211})\dot{q}_1\dot{q}_2 + c_{221}\dot{q}_2^2 + \phi_1 = \tau_1 \qquad (1.90)$$

$$d_{21}\ddot{q}_1 + d_{22}\ddot{q}_2 + c_{112}\dot{q}_1^2 + \phi_2 = \tau_2 \qquad (1.91)$$

State–Space Model of Two Link Manipulator

Let us consider the state variable as $x_1 = q_1$, $x_2 = \dot{q}_1$, $x_3 = q_2$, and $x_4 = \dot{q}_2$. We can solve Eqns (1.90) and (1.91) to get the expressions of $\ddot{q}_1$ and $\ddot{q}_2$ as

$$\ddot{q}_1 = \frac{d_{12}c_{112}\dot{q}_1^2 - d_{22}c_{221}\dot{q}_2^2 - d_{22}(c_{121} + c_{211})\dot{q}_1\dot{q}_2 + d_{12}\phi_2 - d_{22}\phi_1 + d_{22}\tau_1 - d_{12}\tau_2}{d_{11}d_{22} - d_{12}d_{21}}$$

$$\ddot{q}_2 = \frac{d_{11}c_{112}\dot{q}_1^2 - d_{21}c_{221}\dot{q}_2^2 - d_{21}(c_{121} + c_{211})\dot{q}_1\dot{q}_2 + d_{11}\phi_2 - d_{21}\phi_1 + d_{21}\tau_1 - d_{11}\tau_2}{d_{21}d_{12} - d_{11}d_{22}}$$

Thus, the state space model of the system can be formed as

$$\dot{x}_1 = x_2 \qquad (1.92)$$

$$\dot{x}_2 = \frac{d_{12}c_{112}x_2^2 - d_{22}c_{221}x_4^2 - d_{22}(c_{121} + c_{211})x_2x_4 + d_{12}\phi_2 - d_{22}\phi_1 + d_{22}\tau_1 - d_{12}\tau_2}{d_{11}d_{22} - d_{12}d_{21}}$$

$$(1.93)$$

$$\dot{x}_3 = x_4 \qquad (1.94)$$

$$\dot{x}_4 = \frac{d_{11}c_{112}x_2^2 - d_{21}c_{221}x_4^2 - d_{21}(c_{121} + c_{211})x_2x_4 + d_{11}\phi_2 - d_{21}\phi_1 + d_{21}\tau_1 - d_{11}\tau_2}{d_{21}d_{12} - d_{11}d_{22}}$$

$$(1.95)$$

where ϕ_1 and ϕ_2 are redefined as

$$\phi_1 = (m_1 l_{c1} + m_2 l_1)g \cos x_1 + m_2 l_{c2} g \cos(x_1 + x_3)$$
$$\phi_2 = m_2 l_{c2} \cos(x_1 + x_3)$$

Linearization Using the Taylor Series Expansion

The non-linear dynamics of the two link manipulator system will be linearized around the equilibrium point $x = [x_1 \ x_2 \ x_3 \ x_4]^T = [0 \ 0 \ 0 \ 0]^T$ using the Taylor series expansion.

Expanding the system dynamics (1.92), (1.93), (1.94), and (1.95) around the equilibrium point $x = [x_1 \ x_2 \ x_3 \ x_4]^T = [\frac{\pi}{2} \ 0 \ 0 \ 0]^T$, we can write the approximated linear model of the system as

$$
\begin{bmatrix} \dot{x}_1 \\ \dot{x}_2 \\ \dot{x}_3 \\ \dot{x}_4 \end{bmatrix} =
\begin{bmatrix}
0 & 1 & 0 & 0 \\
\frac{\partial \dot{x}_2}{\partial x_1} & \frac{\partial \dot{x}_2}{\partial x_2} & \frac{\partial \dot{x}_2}{\partial x_3} & \frac{\partial \dot{x}_2}{\partial x_4} \\
0 & 0 & 1 & 0 \\
\frac{\partial \dot{x}_4}{\partial x_1} & \frac{\partial \dot{x}_4}{\partial x_2} & \frac{\partial \dot{x}_4}{\partial x_3} & \frac{\partial \dot{x}_4}{\partial x_4}
\end{bmatrix}_{\pi/2,0,0,0}
\begin{bmatrix} x_1 \\ x_2 \\ x_3 \\ x_4 \end{bmatrix} +
\begin{bmatrix}
0 & 0 \\
\frac{\partial \dot{x}_2}{\partial \tau_1} & \frac{\partial \dot{x}_2}{\partial \tau_2} \\
0 & 0 \\
\frac{\partial \dot{x}_4}{\partial \tau_1} & \frac{\partial \dot{x}_4}{\partial \tau_2}
\end{bmatrix}
\begin{bmatrix} \tau_1 \\ \tau_2 \end{bmatrix} \quad (1.96)
$$

After simplification, the linearized model can be written as

$$
\begin{bmatrix} \dot{x}_1 \\ \dot{x}_2 \\ \dot{x}_3 \\ \dot{x}_4 \end{bmatrix} =
\begin{bmatrix}
0 & 1 & 0 & 0 \\
\frac{d_{22}g(m_1 l_{c1} + m_2 l_1 + m_2 l_{c2}) - d_{12} m_2 l_{c2}}{d_{11} d_{22} - d_{12} d_{21}} & 0 & \frac{d_{22} g m_2 l_{c2} - d_{12} m_2 l_{c2}}{d_{11} d_{22} - d_{12} d_{21}} & 0 \\
0 & 0 & 1 & 0 \\
\frac{d_{21}g(m_1 l_{c1} + m_2 l_1 + m_2 l_{c2}) - d_{11} m_2 l_{c2}}{d_{12} d_{21} - d_{11} d_{22}} & 0 & \frac{d_{21} g m_2 l_{c2} - d_{11} m_2 l_{c2}}{d_{12} d_{21} - d_{11} d_{22}} & 0
\end{bmatrix}_{x_1 = \frac{\pi}{2}}
$$

$$
\begin{bmatrix} x_1 \\ x_2 \\ x_3 \\ x_4 \end{bmatrix} +
\begin{bmatrix}
0 & 0 \\
d_{22} & -d_{12} \\
0 & 0 \\
d_{21} & -d_{11}
\end{bmatrix}
\begin{bmatrix} \tau_1 \\ \tau_2 \end{bmatrix}
$$

1.8.3 An Inverted Pendulum Mounted on a Cart

An inverted pendulum mounted on a motor-driven cart [7] is shown in Figure 1.4. This is a model of the attitude control of a space booster on take-off. The inverted pendulum is unstable in the sense that it may fall over any time in any direction unless a suitable control force is applied. Here we consider only a two-dimensional problem in which the pendulum moves only in the plane of page. The control force is applied to the cart. Assume that the centre of gravity of the pendulum rod is at its geometric centre.

Mathematical Model of the system

Define the angle of the rod from the vertical line as θ. Define also the $(x, \ y)$-coordinates of the centre of gravity of the pendulum rod as $(x_G, \ y_G)$. Then

$$x_G = x + l \sin \theta$$
$$y_G = l \cos \theta$$

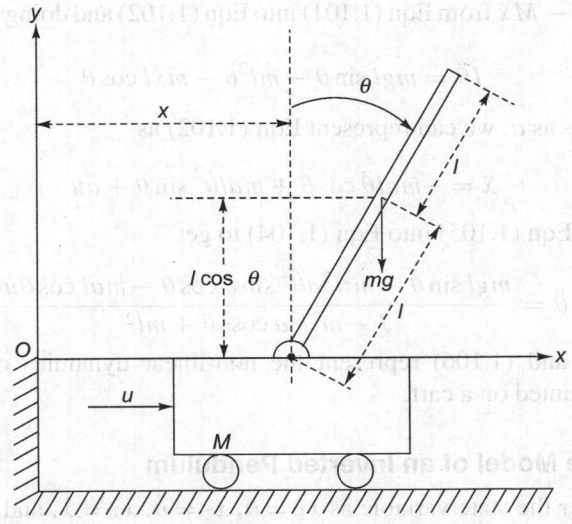

Figure 1.4 An inverted pendulum mounted on a cart

The rotational motion of the pendulum rod about its centre of gravity can be described by

$$I\ddot{\theta} = Vl\sin\theta - Hl\cos\theta \qquad (1.97)$$

where I is the moment of inertia of the rod about its centre of gravity and is given by the following expression:

$$I = \int_{-l}^{l} r^2 dm = \frac{ml^2}{3} \qquad (1.98)$$

The horizontal motion of centre of gravity of the pendulum rod is given by

$$m\frac{d^2}{dt^2}(x + l\sin\theta) = H \qquad (1.99)$$

The vertical motion of centre of gravity of the pendulum rod is

$$m\frac{d^2}{dt^2}(l\cos\theta) = V - mg \qquad (1.100)$$

The horizontal motion of the cart is described by

$$M\frac{d^2x}{dt^2} = u - H \qquad (1.101)$$

Eqns (1.97)–(1.101) describe the motion of an inverted pendulum-on-the cart system. Differentiating (1.99) and combining with (1.101), we get

$$m\ddot{x} + ml\ddot{\theta}\cos\theta - ml\dot{\theta}^2\sin\theta = u - M\ddot{x} \qquad (1.102)$$

Differentiating (1.100) and combining with (1.101) and (1.97), we get

$$I\ddot{\theta} = (mg - ml\dot{\theta}^2\cos\theta - ml\ddot{\theta}\sin\theta)l\sin\theta - (u - M\ddot{x})l\cos\theta \qquad (1.103)$$

Substituting $u - M\ddot{x}$ from Eqn (1.101) into Eqn (1.102) and doing manipulations, we get

$$I\ddot{\theta} = mgl \sin\theta - ml^2\ddot{\theta} - m\ddot{x}l \cos\theta \tag{1.104}$$

Denoting $\frac{1}{m+M}$ as a, we can represent Eqn (1.102) as

$$\ddot{x} = -mal\ddot{\theta} \cos\theta + mal\dot{\theta}^2 \sin\theta + au \tag{1.105}$$

We substitute Eqn (1.105) into Eqn (1.104) to get

$$\ddot{\theta} = \frac{mgl \sin\theta - m^2l^2a\dot{\theta}^2 \sin\theta \cos\theta - mal \cos\theta u}{I - m^2l^2a \cos^2\theta + ml^2} \tag{1.106}$$

Eqns (1.105) and (1.106) represent the non-linear dynamics of the inverted pendulum mounted on a cart.

State–Space Model of an Inverted Pendulum

Let us consider the state variable as $x_1 = \theta$, $x_2 = \dot{\theta}$, $x_3 = x$, and $x_4 = \dot{x}$. Then, using the expression of I, we can rewrite Equation (1.105) in the state–space form as

$$\dot{x}_1 = x_2 \tag{1.107}$$

$$\dot{x}_2 = \frac{g \sin x_1 - mlax_2^2 \sin x_1 \cos x_1}{\frac{4l}{3} - mla \cos^2 x_1} - \frac{mla \cos x_1 u}{\frac{4ml^2}{3} - m^2l^2a \cos^2 x_1} \tag{1.108}$$

Substituting $\ddot{\theta}$ into the expression $\ddot{x}$, we get

$$\ddot{x} = \frac{-mag \sin x_1 \cos x_1 + \frac{4mla}{3}x_2^2 \sin x_1 + \frac{4au}{3}}{\frac{4}{3} - ma \cos^2 x_1} \tag{1.109}$$

The above equation can be written in the state space form as

$$\dot{x}_3 = x_4 \tag{1.110}$$

$$\dot{x}_4 = \frac{-mag \sin x_1 \cos x_1 + \frac{4mla}{3}x_2^2 \sin x_1 + \frac{4au}{3}}{\frac{4}{3} - ma \cos^2 x_1} \tag{1.111}$$

Thus Eqns (1.107), (1.108), (1.110), and (1.111) denote the state–space model of the inverted pendulum system.

Linearization Using the Taylor Series Expansion

The non-linear dynamics of the inverted pendulum system will be linearized around the vertically upward position using the Taylor series expansion.

Expanding the system dynamics Eqns (1.107), (1.108), (1.110), (1.111) around the equilibrium point $x = [x_1\ x_2\ x_3\ x_4]^T = [0\ 0\ 0\ 0]^T$, we can write the approximated linear model of the system as

$$\dot{x}_1 = x_2$$

$$\dot{x}_2 = \frac{\partial}{\partial x_1} \left\{ \frac{g \sin x_1 - mlax_2^2 \sin x_1 \cos x_1}{\frac{4l}{3} - mla \cos^2 x_1} \right\} \Bigg|_{0,0,0,0}$$

$$+ \frac{\partial}{\partial x_2} \left\{ \frac{-mlax_2^2 \sin x_1 \cos x_1}{\frac{4l}{3} - mla \cos^2 x_1} \right\} \Bigg|_{0,0,0,0}$$

$$+ \frac{\partial}{\partial u} \left\{ -\frac{mla \cos x_1 u}{\frac{4ml^2}{3} - m^2 l^2 a \cos^2 x_1} \right\} \Bigg|_{0,0,0,0}$$

$$\dot{x}_3 = x_4$$

$$\dot{x}_4 = \frac{\partial}{\partial x_1} \left\{ \frac{-mag \sin x_1 \cos x_1 + \frac{4mla}{3} x_2^2 \sin x_1}{\frac{4}{3} - ma \cos^2 x_1} \right\} \Bigg|_{0,0,0,0}$$

$$+ \frac{\partial}{\partial x_1} \left\{ \frac{\frac{4mla}{3} x_2^2 \sin x_1}{\frac{4}{3} - ma \cos^2 x_1} \right\} \Bigg|_{0,0,0,0}$$

$$+ \frac{\partial}{\partial u} \left\{ \frac{\frac{4au}{3}}{\frac{4}{3} - ma \cos^2 x_1} \right\} \Bigg|_{0,0,0,0}$$

Simplifying the above expression, we get the linearized dynamics around the equilibrium point $x = [0\ 0\ 0\ 0]^T$ as

$$\begin{bmatrix} \dot{x}_1 \\ \dot{x}_2 \\ \dot{x}_3 \\ \dot{x}_4 \end{bmatrix} = \begin{bmatrix} 0 & 1 & 0 & 0 \\ \frac{3g}{4l-3mla} & 0 & 0 & 0 \\ 0 & 0 & 1 & 0 \\ \frac{-3mag}{4-3ma} & 0 & 0 & 0 \end{bmatrix} \begin{bmatrix} x_1 \\ x_2 \\ x_3 \\ x_4 \end{bmatrix} + \begin{bmatrix} 0 \\ \frac{3a}{3mla-4l} \\ 0 \\ \frac{4a}{4-3ma} \end{bmatrix} u \qquad (1.112)$$

1.8.4 Induction Motor

In an induction motor, the stator is directly supplied with an alternating current which induces an alternating current in the rotor, thus producing a torque to move the rotor. Figure 1.5 shows a schematic diagram of a three-phase induction motor. The three stator phases are separated by an angle of 120° in the space. The voltages in the three phases are V_a, V_b, and V_c and the corresponding alternating currents are i_a, i_b, and i_c, respectively. The rotor windings are shorted. The stator currents are considered to be balanced, i.e.,

$$i_a = I_m \sin(\omega t)$$
$$i_b = I_m \sin(\omega t - 120°)$$
$$i_c = I_m \sin(\omega t + 120°)$$

These currents will generate a rotating magnetic flux which can be represented by a coplanar vector. Since any coplanar vector is uniquely identified by two orthogonal components, we can draw an equivalent two-phase diagram as shown in Figure 1.5(b) where the two virtual windings α and β are separated by an angle of 90°.

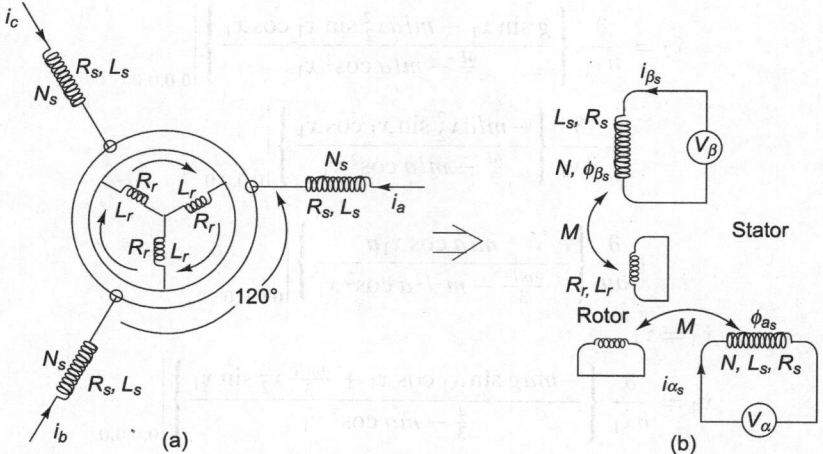

Figure 1.5 Schematic diagram of an induction motor: (a) the actual connection and (b) the two-phase equivalent

If the number of turns in the actual stator windings is N_s and the number of turns in the two-phase equivalent is N, then one can project the stator currents in the direction of α and β as

$$Ni_\alpha = N_s i_a + N_s i_b \cos(120°) + N_s i_c \cos(240°)$$

$$Ni_\beta = N_s i_b \cos(30°) + N_s i_c \cos(150°)$$

The flux ϕ_α will interact with i_β to produce a torque. Similarly, the interaction between ϕ_β and i_α will produce a torque in the opposite direction. Thus, the torque equation can be written as

$$J\dot{\omega} + T_L = K(\phi_\alpha i_\beta - \phi_\beta i_\alpha)$$

$$\text{or, } \dot{\omega} = \frac{K}{J}(\phi_\alpha i_\beta - \phi_\beta i_\alpha) - \frac{T_L}{J} \qquad (1.113)$$

where J is the moment of inertia of the rotor, ω is the electrical angular speed, n_p is the number poles, and T_L is the load torque. K is some proportionality constant. The viscous friction has been neglected here. Since the rotor windings are short circuited, the voltage equations for the rotor windings can be written as

$$0 = R_r i_\alpha + L_r \dot{i}_\alpha + M \dot{i}_\alpha + n_p \omega(L_r i_\beta + M i_\beta) \qquad (1.114)$$

$$0 = R_r i_\beta + L_r \dot{i}_\beta + M \dot{i}_\beta - n_p \omega(L_r i_\alpha + M i_\alpha) \qquad (1.115)$$

where R_r and L_r are rotor resistance and inductance, and M is the mutual inductance between the rotor and stator windings. The flux equations can be written as

$$L_r i_\alpha + M i_\alpha = \phi_\alpha \qquad (1.116)$$

$$L_r i_\beta + M i_\beta = \phi_\beta \qquad (1.117)$$

Combining Eqns (1.114), (1.116), (1.115), and (1.117), we can write

$$\dot{\phi}_\alpha = -\frac{R_r}{L_r}\phi_\alpha + \frac{R_r M}{L_r}i_\alpha - n_p\omega\phi_\beta \tag{1.118}$$

$$\dot{\phi}_\beta = -\frac{R_r}{L_r}\phi_\beta + \frac{R_r M}{L_r}i_\beta + n_p\omega\phi_\alpha \tag{1.119}$$

Thus, Eqns (1.113), (1.118), and (1.119) represent the dynamics of the system where i_α and i_β are the two inputs to the system and ω is the output of the system. K in Eqn (1.113) equals $\frac{n_p M}{L_r}$. The linearized model becomes

$$\begin{bmatrix} \dot{\omega} \\ \dot{\phi}_\alpha \\ \dot{\phi}_\beta \end{bmatrix} = \begin{bmatrix} 0 & \frac{n_p M}{L_r}i_b & -\frac{n_p M}{L_r}i_\alpha \\ -n_p\phi_\beta & -\frac{R_r}{L_r} & -n_p\omega \\ n_p\omega & -\frac{R_r}{L_r} & n_p\phi_\alpha \end{bmatrix}\begin{bmatrix} \omega \\ \phi_\alpha \\ \phi_\beta \end{bmatrix} + \begin{bmatrix} -\frac{n_p M}{L_r}\phi_\beta & \frac{n_p M}{L_r}\phi_\alpha \\ \frac{R_r}{L_r}M & 0 \\ 0 & \frac{R_r}{L_r}M \end{bmatrix} \tag{1.120}$$

1.9 NON-LINEAR CONTROL STRATEGIES

Given a non-linear system, the control problem may be either a regulation or tracking problem. In a regulation problem, the controller regulates the system states around the equilibrium point or a set point. Examples of regulation problems are position control of a robot arm, voltage control of an infinite bus-bar, altitude control of a satellite, and room temperature control. In the case of a tracking problem, the controller makes the system to follow a pre-specified time varying trajectory. When the end effector of a robot manipulator is supposed to follow a pre-specified trajectory such as a circle or a straight line, then this is a case of tracking problem. Mathematically the regulation problem is defined as follows:

Given a non-linear dynamic system described by

$$\dot{x} = f(x, u, t) \tag{1.121}$$

find a control law u such that, starting from anywhere in a region S, the state x converges to x^d, the desired state vector which is a constant vector.

In contrast, the tracking problem can be defined as follows:

Given a non-linear dynamic system described by

$$\dot{x} = f(x, u, t) \tag{1.122}$$

find a control law u such that, starting from anywhere in a region S, the state x converges to x^d, the desired state vector, which is a time-varying vector.

1.9.1 Feedback Linearization

If the closed loop dynamics of a non-linear system becomes linear through a suitable control law, then the approach is known as feedback linearization. In fact, there exists a class of non-linear systems for which feedback linearization is possible.

Consider a class of single input affine non-linear system defined as

$$\frac{d}{dt}\begin{bmatrix} x_1 \\ x_2 \\ \cdots \\ x_{n-1} \\ x_n \end{bmatrix} = \begin{bmatrix} x_2 \\ x_3 \\ \cdots \\ x_n \\ f(\boldsymbol{x}) + g(\boldsymbol{x})u \end{bmatrix} \tag{1.123}$$

$$y = x_1 \tag{1.124}$$

Assume that $g(x) \neq 0$. If we choose the control input $u = \frac{1}{g(x)}[-f(x) + k_v r + \lambda_1 e^{(n-1)} + \cdots + \lambda_{n-1} e^{(1)} + \cdots x_{nd}]$ where $e = y^d - y$ is the output tracking error and $r = e^{(n-1)} + \lambda_1 e^{(n-2)} + \cdots + \lambda_{n-1} e$ (power denotes respective derivatives), the closed loop error dynamics becomes $\dot{r} = -k_v r$ which is linear as well as stable. k_v and λs are positive design parameters. Such design techniques are known as feedback linearization techniques [15].

Example 1.16 Consider a single link manipulator dynamics

$$ml^2 \ddot{q} + b\dot{q} + mgl \sin q = \tau \tag{1.125}$$

where $m = 1$ kg , $l = 1$ m and $g = 10$ m/s^2, $b = 0.1$. Propose a control law that would make the closed loop error dynamics linear.

Solution
It is desired that the manipulator tracks a desired trajectory q^d. Define the tracking error as $e = q^d - q$.

Let the control law be $\tau = a(\ddot{q}^d + K_D \dot{e} + K_P e) + b\dot{q} + c \sin q$ where $a = ml^2$, and $c = mgl$. Replacing the τ in actual dynamics, we obtain

$$a\ddot{q} + b\dot{q} + mgl \sin q = a(\ddot{q}^d + K_D \dot{e} + K_P e) + b\dot{q} + c \sin q \tag{1.126}$$

which reduces to

$$\ddot{e} + K_D \dot{e} + K_P e = 0 \tag{1.127}$$

Thus the closed loop error dynamic is linear and is stable for positive values of K_D and K_P.

1.9.2 Back-Stepping Design

Consider a second order non-linear system of the form

$$\dot{x}_1 = f_1(x_1) + g_1(x_1)x_2 \tag{1.128}$$

$$\dot{x}_2 = f_2(x_1, x_2) + u \tag{1.129}$$

which is usually known as strict feedback form. In the back-stepping approach [15, 19], the variable x_2 in (1.128) is considered as a virtual control and is computed to make this sub-system stable. Let this control law be x_2^d. Then the control action u is computed so that the second sub-system (1.129) is stable as well as the state x_2 follows x_2^d.

Example 1.17 Consider the single link manipulator system

$$\dot{x}_1 = x_2 \tag{1.130}$$

$$\dot{x}_2 = -10\sin(x_1) + u \tag{1.131}$$

Design a back-stepping controller for the above system.

Solution

Let us consider the state x_2 in (1.130) as a virtual control to the sub-system (1.130). Define the error variable for sub-system (1.130) as $e_1 = x_1 - x_{1d}$, where x_{1d} is the desired output. If we choose the virtual control input $x_2 = -k_1 e_1 + \dot{x}_{1d}$, Eqn (1.130) becomes

$$\dot{e}_1 = -k_1 e_1 \tag{1.132}$$

which is stable for a positive k_1. Thus, the desired output for sub-system (1.131) becomes $x_{2d} = -k_1 e_1 + \dot{x}_{1d}$. If the control input u to the second sub-system is chosen as $u = 10\sin(x_1) - k_2 e_2 + \dot{x}_{2d}$, Eqn (1.131) becomes

$$\dot{e}_2 = -k_2 e_2 \tag{1.133}$$

which is a stable dynamics for $k_2 > 0$. Thus, e_2 will converge to zero. In other words, x_2 will follow x_{2d}, hence x_1 will follow x_{1d}.

To generalize the concept of back stepping, let us consider the following non-linear system

$$\dot{x}_1 = f_1(x_1) + g_1(x_1)x_2$$

$$\dot{x}_2 = f_2(x_1, x_2) + g_2(x_1, x_2)x_3$$

$$\dot{x}_3 = f_3(x_1, x_2, x_3) + g_3(x_1, x_2, x_3)x_4$$

$$\cdots = \cdots$$

$$\dot{x}_m = f_m(x_1, x_2, \ldots, x_m) + g_m(x_1, x_2, \ldots, x_m)u \tag{1.134}$$

where $x_i \in R^n$, $i = 1, 2, \ldots, m$ denote the states of the system, $u \in R^n$ is the vector of control inputs. $f_i, g_i \in R^{n \times n}$, $i = 1, 2, \ldots, m$ are non-linear functions that contain both parametric and non-parametric uncertainties, and g_is are known and invertible.

The back-stepping design can be applied to the class of non-linear systems (1.134) as long as the *internal dynamics are stabilizable*. In this method, first select a desirable value of x_2, possibly a function of x_1, denoted by x_{2d}, such that in the ideal system $\dot{x}_1 = f_1(x_1, x_{2d})$, one has stable tracking by $x_1(t)$ of x_{1d}. Then in second step, select x_3 to be x_{3d} so that x_2 tracks x_{2d} and this process is repeated. Finally, select $u(t)$ such that x_m tracks x_{md}.

Integrator Back-Stepping [13, 15]

Integrator back-stepping is a special case where the control input is connected to the system via an integrator or a chain of integrators. Consider the following system:

$$\dot{x} = f(x) + g(x)\xi, \quad f(0) = 0 \tag{1.135}$$

$$\dot{\xi} = u \tag{1.136}$$

where ξ is known as a virtual control. Suppose that the control law $u = \alpha(x)$ stabilizes the system $\dot{x} = f(x) + g(x)u$. Define an error variable z which is a difference between the virtual control and the desired value of the control, i.e., $z = \xi - \alpha(x)$. Let the Lyapunov function of the first sub-system is $V(x)$. The augmented Lyapunov function can be written as

$$V_a(x, z) = V(x) + \frac{1}{2}z^2 \tag{1.137}$$

Taking the time derivative of $V_a(x, \xi)$,

$$
\begin{aligned}
\dot{V}_a(x, z) &= \dot{V}(x) + z\dot{z} \\
&= \frac{\partial V}{\partial x}(f + g\alpha + gz) + z\left(u - \frac{\partial \alpha}{\partial x}(f + g\alpha + gz)\right) \\
&= \frac{\partial V}{\partial x}(f + g\alpha) + z\left(u - \frac{\partial \alpha}{\partial x}(f + g\alpha + gz) + \frac{\partial V}{\partial x}g\right) \\
&\leq -W(x) + z\left(u - \frac{\partial \alpha}{\partial x}(f + g\alpha + gz) + \frac{\partial V}{\partial x}g\right)
\end{aligned}
$$

where u can be chosen such that the derivative of the augmented candidate Lyapunov function is negative definite.

Example 1.18   Let us consider the following system:

$$\dot{x}_1 = -x_1^3 + x_2 \tag{1.138}$$

$$\dot{x}_2 = x_2^2 + u \tag{1.139}$$

Design a stabilizing controller using the back-stepping design.

Solution
The steps involved are as follows:

1. Here x_2 can be treated as the control input to sub-system (1.138). Now define a virtual control α for (1.138) and an error variable z such that $z = x_2 - \alpha$. Eqn (1.138) can thus be written as
$$\dot{x}_1 = -x_1^3 + z + \alpha \tag{1.140}$$

2. Define a candidate Lyapunov function for this sub-system as $V(x_1) = \frac{1}{2}x_1^2$. This implies
$$\dot{V}(x_1) = x_1(-x_1^3 + z + \alpha) \tag{1.141}$$

3. Select a virtual control $\alpha = -k_1 x_1$, where $k_1 \geq 0$. This implies
$$\dot{V}(x_1) = -x_1^4 - k_1 x_1^2 + x_1 z \tag{1.142}$$
and
$$\dot{\alpha} = -k_1(-x_1^3 + x_2) \tag{1.143}$$

4. The virtual state equation is given as

$$\dot{z} = \dot{x}_2 - \dot{\alpha} = x_2^2 + u + k_1(-x_1^3 + x_2) \tag{1.144}$$

5. The augmented candidate Lyapunov function is chosen as $V_a(x, z) = V + \frac{1}{2}z^2$, which implies that

$$\dot{V}_a = \dot{V} + z\dot{z}$$
$$= -x_1^4 - k_1 x_1^2 + x_1 z + z(x_2^2 + u + k_1(-x_1^3 + x_2))$$
$$= -x_1^4 - k_1 x_1^2 + z(x_1 + x_2^2 + u + k_1(-x_1^3 + x_2))$$

6. We can now select an appropriate control action u, which will make the second-order system stabilizable. The simplest choice of u is

$$u = -k_2 z - (x_1 + x_2^2 + k_1(-x_1^3 + x_2))$$
$$= -k_2(x_2 + k_1 x_1) - (x_1 + x_2^2 + k_1(-x_1^3 + x_2))$$

1.9.3 State Feedback Linearizable Systems

Consider the non-linear system of the form

$$\dot{x} = Ax + B(f(x) + g(x)u) \tag{1.145}$$

where $x \in R^n, u \in R^m$, and (A, B) form a controllable pair. The functions $f : R^n \rightarrow R^m$ and $g : R^n \rightarrow R^{m \times m}$ are assumed to be a Lipschitz and $g(x)$ is invertible.

It is desired that the control law should drive x to a state vector x^d, while the closed loop error dynamics remains linear.

Let us define $e = x - x^d$. For a set point regulation problem, x^d is a constant vector. Let us select the control law as

$$u = \frac{1}{g(x)}(-f(x) + Ke) \tag{1.146}$$

where K is so chosen that $A_k = A + BK$ is a Hurwitz. This reduces the term $f(x) + g(x)u$ to Ke. Thus, the closed loop error dynamics is obtained as

$$\dot{e} = \dot{x} - \dot{x}^d \tag{1.147}$$
$$= \dot{x} \quad \text{as} \quad \dot{x}^d = 0 \tag{1.148}$$
$$= Ax + B(f(x) + g(x)u) \tag{1.149}$$
$$= Ae + B(f(x) + g(x)u) + Ax^d \tag{1.150}$$
$$= Ae + BKe + Ax^d \quad \text{as} \quad f(x) + g(x)u = Ke \tag{1.151}$$
$$= A_k e + Ax^d \quad \text{as} \quad A_k = A + BK \tag{1.152}$$

The closed loop error dynamics is linear, and the error e will converge to zero if either $Ax^d = 0$ or $x^d = 0$.

Example 1.19 Consider a single link pendulum example that can be expressed as

$$\dot{x}_1 = x_2$$
$$\dot{x}_2 = -a \sin(x_1) + \tau$$

Implement the control law (1.146) where $\begin{bmatrix} x_1^d \\ x_2^d \end{bmatrix} = \begin{bmatrix} \frac{\pi}{4} \\ 0 \end{bmatrix}$.

Solution

One can represent the dynamics as $\dot{x} = Ax + B(f(x) + g(x)u)$ where $A = \begin{bmatrix} 0 & 1 \\ 0 & 0 \end{bmatrix}$, $B = \begin{bmatrix} 0 \\ 1 \end{bmatrix}$, $f(x) = -a\sin(x_1)$ and $g(x) = 1$.

We can note that $Ax^d = 0$. By placing poles at -1 and -2, the controller gain is found out to be $K = [-2 \quad -3]$. Thus, the controller form looks as

$$u = a\sin(x_1) - 2(x_1 - \frac{\pi}{4}) - 3x_2 \qquad (1.153)$$

With the above control law, the single link manipulator system has been simulated in MATLAB using the Euler integration method. The results are shown in Figure 1.6 which shows that the set point is tracked within 6 s.

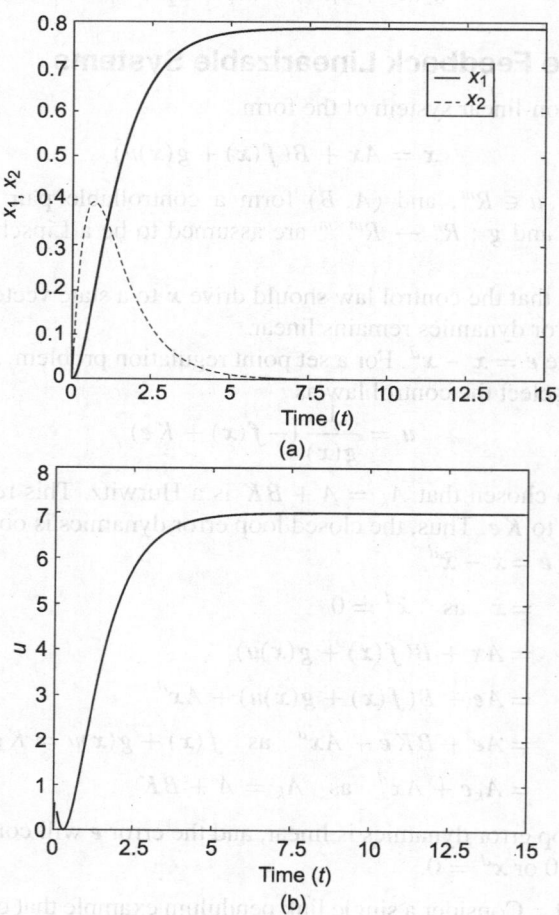

Figure 1.6 Single link manipulator—state feedback linearization method: (a) the system states and (b) the control input

MATLAB code

```
stfeedex.m

clear
a=10;
DT = 0.01;
x(1,1) = 0;
x(1,2) = 0;
xd(1,1) = pi/4;
xd(1,2) = 0;
for i=1:1500
t(i) = DT * i;
e(i,1) = x(i,1) - xd(1,1);
e(i,2) = x(i,2) - xd(1,2);
u(i) = a * sin(x(i,1)) - 2 * e(i,1) - 3 * e(i,2) ;
x(i+1,1) = x(i,1) + DT * x(i,2);
x(i+1,2) = x(i,2) + DT * ( - a * sin(x(i,1)) + u(i) );
end
z = x(1:1500,:);
plot(t,z(:,1),'r',t,z(:,2),'b')
```

SUMMARY

Non-linear systems and systems which are complex as well as mathematically ill-defined have been the subject of interest in intelligent control paradigm. There exists no generic control law that can guarantee stability of any arbitrary non-linear system. Thus, the control solutions have been system specific. Use of the neural and fuzzy networks in the design of the controllers have increased the scope of finding a more robust as well as real-time implementable solutions. However, these intelligent schemes are developed using similar control theoretic tools that have been in use for non-linear systems. Thus, in this chapter, we have introduced relevant concepts in non-linear control. They include representation of non-linear systems in the state–space model, linearization, the Lyapunov stability analysis, and preliminary concepts of the nonlinear control design. Readers will find this chapter to be handy as the discussion starting from Chapter 4 onwards requires the basic concepts presented in this chapter for a thorough understanding.

EXERCISES

1. Consider

$$V(x) = ax_1^2 + 2x_1x_3 + ax_2^2 + 4x_2x_3 + ax_3^2$$

Express it in the form $x^T P x$ and determine the conditions for positive definiteness, negative definiteness, and indefiniteness.

2. Consider the pendulum dynamic equation without friction

$$\ddot{\theta} + a\sin\theta = 0$$

Write down the state equations and using $V(x) = a(1 - \cos x_1) + \frac{1}{2}x_2^2$, examine the stability of the equilibrium point at the origin.

3. Consider again the pendulum dynamic equation, but this time with friction i.e.,

$$\ddot{\theta} + b\dot{\theta} + a \sin \theta = 0$$

Write down the state equations and using the same Lyapunov function as in the previous problem, examine the stability of the equilibrium point at the origin. Now replace $\frac{1}{2}x_2^2$ by the more general quadratic form $\frac{1}{2}x^T P x$ where $P = \begin{bmatrix} p_{11} & p_{12} \\ p_{12} & p_{22} \end{bmatrix}$ and find the elements of P such that the origin is asymptotically stable.

4. Consider the state equations of the earlier example. What are the two equilibrium points of the system? Examine the stability of both equilibrium points by the linearization technique.

5. Consider the first-order system

$$\dot{y} = ay + u$$

together with the adaptive control law

$$u = -ky, \quad k = \gamma y^2, \quad \gamma > 0$$

Taking $x_1 = y$ and $x_2 = k$, examine the stability of the equilibrium set (set of all equilibrium points).

6. For each of the following systems, use a quadratic Lyapunov function candidate to show that the origin is asymptotically stable:
   (a) $\dot{x}_1 = -x_2 - x_1(1 - x_1^2 - x_2^2), \quad \dot{x}_2 = x_1 - x_2(1 - x_1^2 - x_2^2)$
   (b) $\dot{x}_1 = -x_1 - x_2, \quad \dot{x}_2 = 2x_1 - x_2^3$
   Investigate whether the origin is globally asymptotically stable.

7. Consider the system $\dot{x}_1 = x_2, \dot{x}_2 = -h_1(x_1) - x_2 - h_2(x_3), \dot{x}_3 = x_2 - x_3$, where $h_1 = x_1^3$ and $h_2 = \text{sgn}(x_3)$.
   (a) Show that the system has an equilibrium point at the origin.
   (b) Show that $V(x) = \int_0^{x_1} h_1(y)dy + \frac{1}{2}x_2^2 + \int_0^{x_3} h_2(y)dy$ is positive definite for all $x \in R^3$.
   (c) Show that the origin is asymptotically stable.
   (d) Under what conditions on h_1 and h_2, can you show that the origin is globally asymptotically stable?

8. Consider the Lienard equation

$$\ddot{y} + h(y)\dot{y} + g(y) = 0$$

where $g = \sin(y)$ and $h = 1 + y^2$ are continuously differentiable.
   (a) Using $x_1 = y$ and $x_2 = \dot{y}$, write the state equation and find conditions on g and h to ensure that the origin is an isolated equilibrium point.

(b) Using $V(x) = \int_0^{x_1} g(y)dy + \frac{1}{2}x_2^2$ as a Lyapunov function candidate, find conditions on g and h to ensure that the origin is asymptotically stable.

(c) Repeat part (b) using $V(x) = \int_0^{x_1} g(y)dy + \frac{1}{2}(x_2 + \int_0^{x_1} h(y)dy)^2$.

9. A mass–spring system is modelled by

$$M\ddot{y} = Mg - ky - c_1\dot{y} - c_2\dot{y}|\dot{y}|$$

Show that the system has a globally asymptotically stable equilibrium point.

10. (Simple continuous fermenter) It is assumed that a bio-reactor is perfectly stirred. An unstructured bio-mass growth rate model with substrate inhibition kinematics is chosen. The variables and parameters are defined in Table 1.1.

Table 1.1 Variables and parameters of the bio-reactor

X	Biomass concentration (g/l)	
S	Substrate concentration (g/l)	
F	Feed flow rate (l/h)	
V	Volume (l)	4.0
S_F	Substrate feed concentration (g/l)	10.0
Y	Yield coefficient	0.5
μ_{max}	Maximal growth rate(l/h)	1.0
K_1	Saturation parameter (g/l)	0.03
K_2	Inhibition parameter (l/g)	0.5

The biomass component mass balance equation is

$$\frac{dX}{dt} = \mu(S)X - \frac{XF}{V}$$

The substrate mass balance equation is

$$\frac{dS}{dt} = -\frac{\mu(S)X}{Y} + \frac{(S_F - S)F}{V}$$

where $\mu(S) = \mu_{max}\frac{S}{K_2 S^2 + S + K_1}$.

Let X_0, S_0, F_0 be steady state operating points.

(a) Defining the centred state vector can be assumed as $[x_1 \ x_2]^T = [X - X_0 \ S - S_0]^T$, and the manipulable input variable as $u == F - F_0$, show that the state equations become

$$\frac{dx_1}{dt} = \mu(x_2 + S_0)(x_1 + X_0) - \frac{(x_1 + X_0)F_0}{V} - \frac{(x_1 + X_0)}{V}u$$

$$\frac{dx_2}{dt} = -\frac{\mu(x_2 + S_0)(x_1 + X_0)}{Y} + \frac{(S_F - (x_2 + S_0))F_0}{V}$$

$$+ \frac{(S_F - (x_2 + S_0))}{V}u$$

(b) The maximal biomass productivity XF is selected as the desired optimal operating point, i.e., the substrate cost is assumed to be negligible. Show that the equilibrium point of the system is

$$S_0 = \frac{1}{2} \frac{-2K_1 + 2\sqrt{K_1^2 + S_F^2 K_1 K_2 + S_F K_1}}{S_F K_2 + 1}$$

$$X_0 = (S_F - S_0)Y$$

$$F_0 = \mu(S_0)V$$

(c) Substituting the parameter values $S_0 = 0.2187$ and $X_0 = 4.8907$ g/l, $F_0 = 3.2029$ l/h, find the linearized model A and B, where

$$\begin{bmatrix} \dot{x}_1 \\ \dot{x}_2 \end{bmatrix} = A \begin{bmatrix} x_1 \\ x_2 \end{bmatrix} + Bu$$

11. Consider the dynamics of a surge tank as represented by the following differential equation:

$$\frac{dh(t)}{dt} = \frac{-\sqrt{2gh(t)}}{A(h(t))} + \frac{1}{A(h(t))}u(t)$$

where $u(t)$ is the input flow, $h(t)$ is the liquid level of the tank (output of the system), $A(h(t)) = \sqrt{ah(t) + b}$ is the cross-sectional area of the tank. The input $u(t)$ can be both positive or negative, i.e., it can pull liquid out of the tank as well as put it in the tank. The parameters a and b are given as $a = 3$ and $b = 1$.

Design a controller $u(t)$ using the feedback linearization technique such that the system output follows a desired trajectory $h_d(t)$ where $h_d(t)$ is a square wave whose frequency is 1 Hz and amplitude flips between 3 and -3.

12. Consider the following dynamics of a magnetically levitated system:

$$\dot{x} = v$$

$$\dot{v} = g - \frac{k_m k_a^2 V^2}{mx^2}$$

where x is the position and v is the velocity of the system. $g = 9.81$ m/s^2 is the acceleration due to gravity, $k_m = 0.0008$ N (m/A)^2 is the magnetic constant, $k_a = 0.397$ f is the conductance, $m = 0.02$ kg is the mass, and V is the input voltage.

Taking $u = V^2$, design a control law for u using
(a) Feedback linearization
(b) Back-stepping design
such that x converges to 0.

13. The dynamics of an automobile is given by

$$\dot{v}(t) = \frac{1}{m}(-A_\rho v^2(t) - d + f(t))$$

$$\dot{f}(t) = \frac{1}{\tau}(-f(t) + u(t))$$

where v is the vehicle speed and u is the control input. $u > 0$ represents a throttle input and $u < 0$ represents a brake input. m is the mass of the vehicle, A_ρ is the aerodynamic drag, d is a constant frictional force, f is the driving or braking force, and τ is the engine/brake time constant. Assume that $u \in [-1000, \ 1000]$. Take the system parameters as $m = 1300$ kg, $A_\rho = 0.3 \ \mathrm{N \ s^2/m^2}$, $d = 100 \ \mathrm{N}$ and $\tau = 0.2 \ \mathrm{s}$.

Design a control law $u(t)$ using back stepping so that $v(t)$ follows a desired velocity $r = 50$ km/h.

CHAPTER

2

Neural Networks

Artificial neural networks (ANN) have played a major role in the development of intelligent control schemes. These networks are used exclusively for system identification as well as for controller parametrization. In this chapter, multi-layered networks, radial basis function, recurrent networks, and self-organizing map networks will be discussed from a function approximation perspective. Multi-layered networks and radial basis function have feed-forward connection structures. When these networks are represented as controllers or as system models, it becomes relatively easy to perform a closed-loop analysis of a control system in terms of stability and convergence. Thus, extensive use of feed-forward networks in intelligent control system development in the existing literature is not surprising.

2.1 FEED-FORWARD NETWORKS

The simplest form of an artificial neural network is called *a perceptron*. The single layer perceptron model was first developed by Rosenblatt in 1958 [20]. The basic model consists of a single neuron with adjustable synaptic weights and bias. However, in 1969, Marvin Minsky and Seymour Papert wrote a book called 'Perceptrons' [21] in which they proved mathematically that single-layer perceptrons could only classify linearly separable patterns. This book made a turning point in the field of artificial neural networks by showing the limitations of single-layer perceptrons. Though the idea of multi-layer perceptrons was there in the minds of people, but they did not know how to train multi-layer perceptrons. As a result, researchers almost lost their interests in neural networks until about the mid-eighties, when the back-propagation algorithm for training multi-layer perceptrons was discovered. The development of back-propagation algorithm, along with its use in machine learning, was reported by Rumelhart, Hinton, and Williams [22] in 1986. Werbos [23] and Parker [24] have also independently proposed back-propagation algorithm in 1974 and 1985. In 1988, Broomhead and Lowe [25] described the design of a layered feed-forward networks using radial basis functions which came as an alternative to multi-layer perceptrons. Basically a feed-forward network consists of multiple layers of neurons where the neurons in one layer are forward-connected to the neurons of the next layer. Although, there are many variants of a feed-forward networks, multi-layered networks, and radial basis function networks have been used in many applications. These two networks can act as universal approximators, i.e., any non-linear function can be approximated with an arbitrary accuracy using a properly tuned multi-layered network or a radial basis function network.

2.2 MULTI-LAYERED NEURAL NETWORKS

A multi-layered neural network (MNN) as shown in Figure 2.1 usually consists of an input layer of a set of sensory units or source nodes, $L - 1$ hidden layers of neurons, and an output layer of neurons. The input signal propagates through the network on a layer-to-layer basis in the forward direction. Each neuron actuates a response using sigmoidal activation function. As mentioned earlier, MNNs have been successfully applied in solving some difficult problems by training them in a supervised manner with the error back-propagation algorithm. The back-propagation algorithm uses the principle of gradient descent to train the network parameters. The synopsis of the notations used to represent an MLN

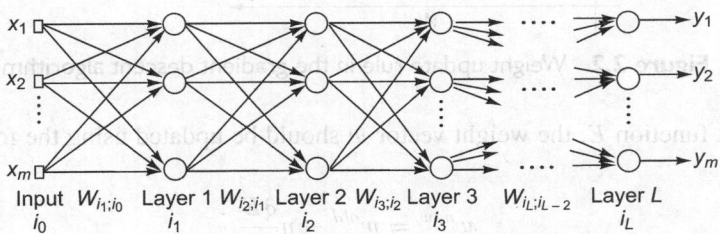

Figure 2.1 An L-layered network

with L layers is as follows:

x: $m \times 1$ input vector

y: $n \times 1$ output vector

i_k; $k = 1, \ldots, L$: index for representing a neuron in the kth layer

i_0: index for representing a neuron in the input layer

h_{i_k}: Weighted sum of the input stimuli to i_k^{th} neuron in the kth layer

v_{i_k}: response of the i_k^{th} neuron in the kth layer

$W_{i_k i_{k-1}}$: Weight connecting i_k^{th} neuron of the kth layer and i_{k-1}^{th} neuron of $k - 1$th layer

$W_{i_2 i_1}$: Weight connecting the i_2^{th} neuron of the 2nd layer and the i_1^{th} unit of the 1st layer

$W_{i_1 i_0}$: Weight connecting the i_1^{th} unit of the 1st layer and the i_0^{th} unit of the input layer

2.2.1 Principle of Gradient Descent

Suppose that a function $y = f(x)$ has to be approximated by a neural network. If the number of training patterns available is N, then the cost function to be minimized to learn this mapping is defined as follows:

$$E = \sum_{p=1}^{p=N} E^p \tag{2.1}$$

where $E^p = \frac{1}{2}(y_p^d - y_p)^2$. y_p and y_p^d denote the pth actual and desired pattern. Since y_p is a function of the network weight vector w, E can also be expressed as a function of w. Consider a typical relationship between the cost function E and the weight vector w, as shown in Figure 2.2, where the cost function has only one global minimum. The principle of gradient descent tells us that to minimize

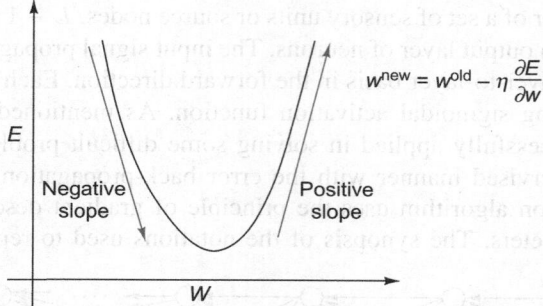

Figure 2.2 Weight update rule in the gradient descent algorithm

the cost function E, the weight vector w should be updated using the following rule:

$$w^{new} = w^{old} - \eta \frac{\partial E}{\partial w}$$

where η is the learning rate. Figure 2.2 shows that E attains its minimum value at $w = w_{min}$. One should notice that when w is less than w_{min}, the slope $\frac{\partial E}{\partial w}$ is negative thus the change in w is positive which will move w towards w_{min}. Similarly when w is greater than w_{min}, the change in w is negative which makes w to move in the left direction, i.e., towards the direction of w_{min}. In both the cases, w will tend to w_{min} in a number of steps depending on the learning rate η.

Batch Update: When the weight vector w is updated such that $\Delta w = w^{new} - w^{old}$ is a function of the overall cost function E, i.e., $\Delta w = -\eta \frac{\partial E}{\partial w}$, the update rule is termed as a *batch update* which is an offline technique of weight update.

Instantaneous Update: When the weight vector w is updated such that $\Delta w = w^{new} - w^{old}$ is a function of the instantaneous cost function E^p, i.e., $\Delta w = -\eta \frac{\partial E^p}{\partial w}$, the update rule is termed as an *instantaneous update*.

2.2.2 Derivation of Back Propagation Algorithm

The back propagation (BP) algorithm [26, 27] offers an effective approach to the computation of the gradients. This can be applied to any optimization formulation, i.e., any type of energy function, either maximization or minimization problem.

Let us consider a two-layered network as shown in Figure 2.3 where i_2, i_1, and i_0 refer to the output layer, hidden layer, and input layer, respectively.

The objective of the back propagation algorithm is to adjust the weights $W_{i_2 i_1}$ and $W_{i_1 i_0}$ so as to minimize a cost function E which will train the network mapping of the required function.

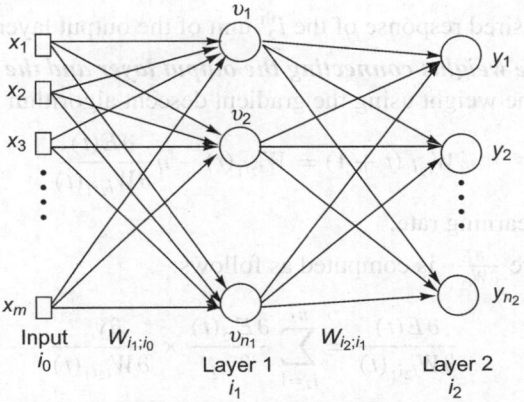

Figure 2.3 Two-layered network

Forward Phase

As shown in Figure 2.3, the input to the i_1^{th} neuron of the hidden layer is given by

$$h_{i_1} = \sum_{i_0=1}^{m} W_{i_1 i_0} x_{i_0} \tag{2.2}$$

where p is the dimension of the input vector x.

Output of the i_1^{th} neuron of the hidden layer is given as

$$v_{i_1} = \Psi(h_{i_1}) = \frac{1}{1 + e^{-h_{i_1}}} \tag{2.3}$$

where $\Psi(.)$ is the sigmoidal activation function. The final response (actual output) of the i_2^{th} neuron in the output layer is given as

$$y_{i_2} = v_{i_2} = \Psi(h_{i_2}) = \frac{1}{1 + e^{-h_{i_2}}} \tag{2.4}$$

where

$$h_{i_2} = \sum_{i_1=1}^{n_1} W_{i_2 i_1} v_{i_1} \tag{2.5}$$

Back Propagation of Error

The weights of a multi-layered network can be updated using either batch mode or instaneneous mode. However, for most applications including control applications that need real-time implementations, it is beneficial to perform instantaneous update. Thus, the instanteneous back propagation algorithm will be derived in this section. The instantaneous cost function in error can be expressed as

$$E(t) = \sum_{i_2=1}^{n_2} E_{i_2}(t) = \frac{1}{2} \sum_{i_2=1}^{n_2} (y_{i_2}^d(t) - y_{i_2}(t))^2 \qquad (2.6)$$

where

$$E_{i_2}(t) = \frac{1}{2}(y_{i_2}^d(t) - y_{i_2}(t))^2 \qquad (2.7)$$

and $y_{i_2}^d$ is the desired response of the i_2^{th} unit of the output layer.

Update of the weights connecting the output layer and the hidden layer

The update of the weight using the gradient descent algorithm looks as

$$W_{i_2 i_1}(t+1) = W_{i_2 i_1}(t) - \eta \frac{\partial E(t)}{\partial W_{i_2 i_1}(t)} \qquad (2.8)$$

where η is the learning rate.

The derivative $\frac{\partial E}{\partial W_{i_2 i_1}}$ is computed as follows:

$$\frac{\partial E(t)}{\partial W_{i_2 i_1}(t)} = \sum_{i_2=1}^{n_2} \frac{\partial E_{i_2}(t)}{\partial y_{i_2}} \times \frac{\partial y_{i_2}}{\partial W_{i_2 i_1}(t)} \qquad (2.9)$$

where the first term can be computed using Eqn (2.7)

$$\frac{\partial E_{i_2}(t)}{\partial y_{i_2}} = -\frac{1}{2} \times 2(y_{i_2}^d - y_{i_2}) \qquad (2.10)$$

and the second term is computed as

$$\frac{\partial y_{i_2}}{\partial W_{i_2 i_1}} = \frac{\partial y_{i_2}}{\partial h_{i_2}} \times \frac{\partial h_{i_2}}{\partial W_{i_2 i_1}} \qquad (2.11)$$

Following relations are obtained using Eqns (2.4) and (2.5), respectively

$$\frac{\partial y_{i_2}}{\partial h_{i_2}} = \frac{\partial}{\partial h_{i_2}} \left[\frac{1}{1 + e^{-h_{i_2}}} \right] = y_{i_2}(1 - y_{i_2}) \qquad (2.12)$$

and $$\frac{\partial h_{i_2}}{\partial W_{i_2 i_1}} = v_{i_1} \qquad (2.13)$$

The computable form of the gradient (2.9) is obtained usings Eqns (2.10), (2.12), and (2.13) as

$$\frac{\partial E(t)}{\partial W_{i_2 i_1}(t)} = -(y_{i_2}^d - y_{i_2})y_{i_2}(1 - y_{i_2})v_{i_1} \qquad (2.14)$$

Thus, the update law turns out to be

$$W_{i_2 i_1}(t+1) = W_{i_2 i_1}(t) + \eta(y_{i_2}^d - y_{i_2})y_{i_2}(1 - y_{i_2})v_{i_1}$$
$$= W_{i_2 i_1}(t) + \eta \delta_{i_2} v_{i_1} \qquad (2.15)$$

where $\delta_{i_2} = y_{i_2}(1 - y_{i_2})(y_{i_2}^d - y_{i_2})$ is the error back propagated from the output layer.

Update of the weights connecting the hidden layer and the input layer

Unlike a neuron in the output layer, a neuron in the hidden layer has no specified desired response. Thus, the output error has to be back-propagated so that weights connecting to the hidden layer from the input layer can be updated.

Update of the weight using the gradient descent algorithm looks as

$$W_{i_1 i_0}(t+1) = W_{i_1 i_0}(t) - \eta \frac{\partial E(t)}{\partial W_{i_1 i_0}(t)} \tag{2.16}$$

The derivative term $\frac{\partial E}{\partial W_{i_1 i_0}}$ is computed as

$$\frac{\partial E(t)}{\partial W_{i_1 i_0}(t)} = \sum_{i_2=1}^{n_2} \frac{\partial E_{i_2}(t)}{\partial y_{i_2}} \times \frac{\partial y_{i_2}}{\partial W_{i_1 i_0}(t)} \tag{2.17}$$

where the first term is computed using Eqn (2.10) and the second term is computed as follows:

$$\frac{\partial y_{i_2}}{\partial W_{i_1 i_0}} = \frac{\partial y_{i_2}}{\partial v_{i_1}} \times \frac{\partial v_{i_1}}{\partial W_{i_1 i_0}} \tag{2.18}$$

In this equation the first term is computed using Eqns (2.4) and (2.5)

$$\frac{\partial y_{i_2}}{\partial v_{i_1}} = \frac{\partial y_{i_2}}{\partial h_{i_2}} \times \frac{\partial h_{i_2}}{\partial v_{i_1}} = y_{i_2}(1 - y_{i_2})W_{i_2 i_1} \tag{2.19}$$

while the second term is computed using Eqns (2.2) and (2.3)

$$\frac{\partial v_{i_1}}{\partial W_{i_1 i_0}} = \frac{\partial v_{i_1}}{\partial h_{i_1}} \times \frac{\partial h_{i_1}}{\partial W_{i_1 i_0}} = v_{i_1}(1 - v_{i_1})x_{i_0} \tag{2.20}$$

Thus, the final computable expression for the gradient term (2.17) is obtained using Eqns (2.10), (2.19), and (2.20):

$$\frac{\partial E}{\partial W_{i_1 i_0}} = -\sum_{i_2=1}^{n_2} (y_{i_2}^d - y_{i_2})y_{i_2}(1 - y_{i_2})W_{i_2 i_1} v_{i_1}(1 - v_{i_1})x_{i_0}$$

$$= -v_{i_1}(1 - v_{i_1})x_{i_0} \sum_{i_2=1}^{n_2} \delta_{i_2} W_{i_2 i_1}$$

Thus the update law (2.16) has the final computable form

$$W_{i_1 i_0}(t+1) = W_{i_1 i_0}(t) + \eta \delta_{i_1} x_{i_0} \tag{2.21}$$

where $\delta_{i_1} = v_{i_1}(1 - v_{i_1}) \sum_{i_2=1}^{n_2} \delta_{i_2} W_{i_2 i_1}$

The above recursive formula is the key to back propagation learning. It allows error signal of a lower layer to be computed as a linear combination of the error signal of the upper layer. In this manner, the error signals are back propagated through all the layers from the top down. This implies that the influence from an upper layer to a lower layer and vice versa can only be effected via the error signals of the intermediate layer.

The BP algorithm is so popular because of the fact that each weight can be simultaneously updated as the factors involved are all local.

2.2.3 Generalized Delta Rule

The delta rule discussed in the previous section can now be generalized to a generic multi-layer network with L layers as shown in Figure 2.1.

Let $W_{i_k,i_{k-1}}$ denote the synaptic weight connecting the ith neuron of layer k to that of layer $k - 1$. Considering the sigmoid function as the activation function for each layer, the weight update law can be written as

$$W_{i_k,i_{l-1}}(t + 1) = W_{i_l,i_{k-1}}(t) + \eta \delta_{i_l} v_{i_{k-1}} \qquad (2.22)$$

where $v_{i_{k-1}}$ is the output of the ith neuron of layer $k - 1$. The term δ_{i_k} for each layer is expressed as

$$\delta_{i_L} = (y_{i_L}^d - y_{i_L})y_{i_L}(1 - y_{i_L}) \quad \text{for the output layer L} \qquad (2.23)$$

$$\delta_{i_l} = v_{i_l}(1 - v_{i_l}) \sum_{i=1}^{n_k} \delta_{i_{l+1}} W_{i_{l+1},i_l} \quad \text{for other hidden layers} \qquad (2.24)$$

All the weights can be updated in a similar manner. Thus for example, if we consider a four layered network, the weight update rule for the 4th layer or the output layer will be

$$W_{i_4,i_3}(t + 1) = W_{i_4,i_3}(t) + \eta \delta_{i_4} v_{i_3}$$

$$\text{where } \delta_{i_4} = (y_{i_4}^d - y_{i_4})y_{i_4}(1 - y_{i_4})$$

y_{i_4} is the ith unit of the output layer and $y_{i_4}^d$ is the corresponding desired output. Similarly, the weight update law for the 3rd layer is derived as follows:

$$W_{i_3,i_2}(t + 1) = W_{i_3,i_2}(t) + \eta \delta_{i_3} v_{i_2}$$

$$\text{where } \delta_{i_3} = v_{i_3}(1 - v_{i_3}) \sum_{i=1}^{n_4} \delta_{i_4} W_{i_4,i_3}$$

The weight update law for the 2nd layer

$$W_{i_2,i_1}(t + 1) = W_{i_2,i_1}(t) + \eta \delta_{i_2} v_{i_1}$$

$$\text{where, } \delta_{i_2} = v_{i_2}(1 - v_{i_2}) \sum_{i=1}^{n_3} \delta_{i_3} W_{i_3,i_2}$$

It has been assumed that the number of neurons in each layer is n. The weight update law for the 1st layer

$$W_{i_1,i_0}(t + 1) = W_{i_1,i_0}(t) + \eta \delta_{i_1} x_{i_0}$$

$$\text{where } \delta_{i_1} = v_{i_1}(1 - v_{i_1}) \sum_{i=1}^{n_2} \delta_{i_2} W_{i_2,i_1}$$

where x_{i_0} represents a typical signal in the input layer.

Example 2.1   Let us consider the following discrete time dynamical system:

$$y(k+1) = \frac{y(k)}{1 + y^2(k)} + u^3(k) \qquad (2.25)$$

Identify the system using a multi-layered network.

Solution

Since a feed-forward network is a static network, when a dynamical system is identified using this kind of a network, the previous state is also input to the network along with the original system input. The system being a stable one, the input is generated randomly between ±1. A total of 1000 data pairs are generated with this input range using Eqn (2.25). An MLN is considered with one input layer, one hidden layer, and one output layer. The number of neurons in the hidden layer is taken as 15 and the sigmoid function is taken as the activation function. An instantaneous update using the back propagation algorithm is employed to train the network. The training results are shown in Figure 2.4 where the left figure shows the input–output relationship of the training data as well as the network output after training. Figure 2.4(b) shows the desired and actual output of the network at every sampling instant.

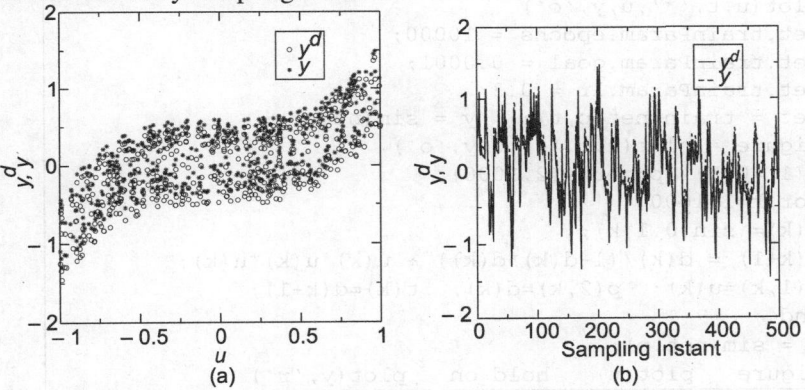

Figure 2.4   System identification results after training

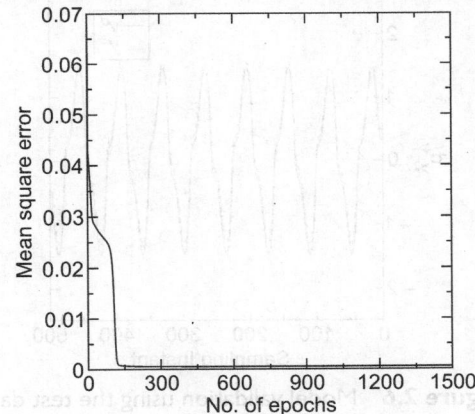

Figure 2.5   Epoch-wise convergence of the mean square error

The convergence of mean square error over 1500 epochs is shown in Figure 2.5. Once the training is over, 1000 new data points are generated to validate the identified model while taking the input to the system as $u(k) = \sin(0.1\,k)$. The result of model validation is shown in Figure 2.6. The RMS error for the training data set is found to 0.022, while for the testing data set it is 0.034. The MATLAB code for the above example is given as follows.

```
mlpsiso.m
clear
d(1)=0.5;  p=zeros(2,1000);
for k=1:1000
u(k)= 2*rand - 1;
d(k+1) = d(k)/(1+d(k)*d(k)) + u(k)*u(k)*u(k);
p(1,k)=u(k);  p(2,k)=d(k);  t(k)=d(k+1);
end
pr = [-1 1;-1.5 1.5];
net = newff(pr,[50 1],'radbas' 'purelin','traingdx');
y = sim(net,p);
plot(u,t,'*',u,y,'o')
net.trainParam.epochs = 10000;
net.trainParam.goal = 0.0001;
net.trainParam.lr = 0.3;
net = train(net,p,t);    y = sim(net,p);
figure   plot(u,t,'*',u,y,'o')
d(1)=0.5;  p=zeros(2,1000);
for k=1:1000
u(k)= sin(0.1*k);
d(k+1) = d(k)/(1+d(k)*d(k)) + u(k)*u(k)*u(k);
p(1,k)=u(k);  p(2,k)=d(k);  t(k)=d(k+1);
end
y = sim(net,p);
figure   plot(t)   hold on   plot(y,'r')
```

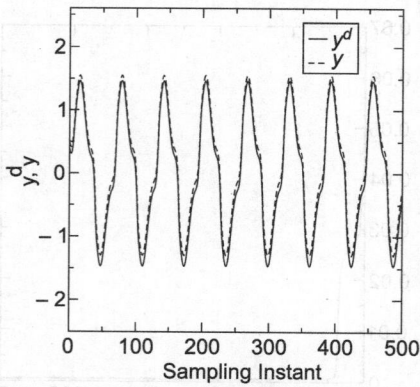

Figure 2.6 Model validation using the test data

2.2.4 Convergence of the BP Learning Algorithm

The weight update law in back propagation is derived using the gradient descent algorithm where a cost function E has to be minimized with respect to the weight vector W. The gradient–descent algorithm ensures that the first-order approximation of $E(\mathbf{W})$ reaches its global minimum, but it does not ensure global convergence of the original cost function $E(\mathbf{W})$. The is given as

$$\mathbf{W}(t+1) = \mathbf{W}(t) - \eta \nabla E(t)$$

where $\eta > 0$ is the learning rate and ∇E is the gradient of the cost function E. Learning rate η determines the speed of convergence. Convergence depends on a proper choice of initial conditions. Let us consider minimization of the following function:

$$E(W) = 0.5W^2 - 8 \sin W + 7$$

It is a multi-modal function with two local minima and a global minimum.

The gradient–descent update law for parameter W is obtained as

$$\Delta W = -\eta \frac{\partial E}{\partial W} = -\eta (W - 8 \cos W)$$

Figure 2.7 shows the convergence of the cost function with respect to the weight W.

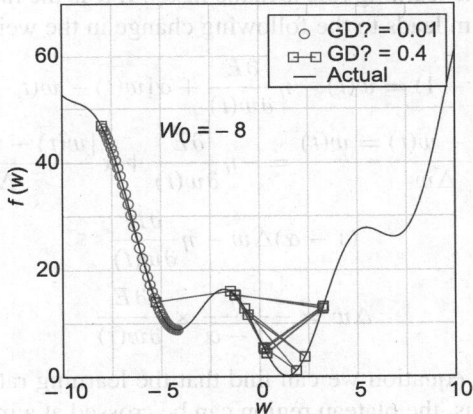

Figure 2.7 Convergence of the gradient–descent algorithm

It can be seen from the figure that when $\eta = 0.01$ and $W_0 = -8$, the final weight $W = -4$ corresponds to a local minimum. When $\eta = 0.4$ and $W_0 = -8$, the local minimum can be avoided and the final weight ultimately settles down at the global minimum. One can conclude that for a small step size, the BP gets stuck at a local minimum. For a larger step size, it may come out of local minima and get into global minimum, but the algorithm may not converge for a larger step size. Moreover, when the step size is large, it zigzags its way about the true direction to the global minimum, thus leading to a slow convergence. There is no comprehensive method to select the initial weight W_0 and learning rate η for a given problem so that the optimal weight vector can be reached.

2.2.5 Variations in the Back Propagation Algorithm

As discussed earlier, the back propagation algorithm (BPA) is extremely slow to converge. Several variations have been suggested to improve the speed and also to account for other factors such as generalization ability and avoidance of local minima. Since the BPA is a gradient–descent algorithm, all techniques available for improving upon the gradient–descent procedure apply to the BPA. Those include making the learning rate η and the coefficient for the momentum term either gradually decay or adaptive. Also, any second derivative based non-linear optimization method, such as the Newton–Raphson method or the conjugate gradient method, may be adopted for the training of feed-forward networks. However, these variants also share problems present in the standard BPA and may converge faster in some cases and slower in others.

Adding a Momentum Term

The problem arises from error surfaces with step sides and shallow valley floor. One simple but effective way to deal with this problem is addition of a *momentum factor*. This allows us to safely increase the learning rate and to avoid long flat surfaces.

Adding a momentum factor simply means that one is giving a momentum to the learning, i.e., the learning rate is made faster. If α is the momentum rate, then adding a momentum leads to the following change in the weight update law:

$$w(t + 1) = w(t) - \eta \frac{\partial E}{\partial w(t)} + \alpha[w(t) - w(t - 1)]$$

$$\frac{w(t + 1) - w(t) = w(t)}{\Delta w} = -\eta \frac{\partial E}{\partial w(t)} + \alpha \frac{[w(t) - w(t - 1)]}{\Delta w}$$

$$(1 - \alpha)\Delta w - \eta \frac{\partial E}{\partial w(t)}$$

$$\Delta w = -\frac{\eta}{1 - \alpha} \times \frac{\partial E}{\partial w(t)} \tag{2.26}$$

From the above equation we can find that the learning rate is increased by a factor of $\frac{1}{(1-\alpha)}$. Thus, the plateau region can be crossed at a much faster rate.

2.3 RADIAL BASIS FUNCTION NETWORKS

The popularity of a radial basis function network (RBFN) is mainly due to its simple architecture with a single hidden layer, which is especially good in applications requiring locally tunable properties [28]. As the name implies, this network makes use of radial functions to represent an input in terms of radial centres. However, this network is very attractive for intelligent control applications since the response of this network is linear in terms of weights. Thus, weight update rules for such networks within intelligent control paradigm become easy to derive through the Lyapunov stability analysis as the readers will notice in Chapters 4 and 5. The network architecture of a RBFN is shown in Figure 2.8.

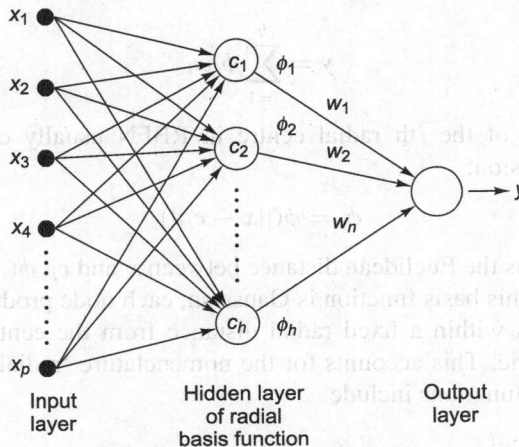

Figure 2.8 Radial basis function network

The RBFN consists of three layers:

- An input layer
- A hidden layer
- An output layer

The hidden units provide a set of functions that constitute an arbitrary basis for the input patterns.

1. Hidden units are known as radial centres. Each radial centre is represented by a vector c_i, $i = 1, \ldots, h$, where h is the number of radial centres in the hidden layer.

2. The transformation from the input space to the hidden unit space is non-linear whereas the transformation from the hidden unit space to the output space is linear.

3. Dimension of each centre for a p input network is $p \times 1$.

2.3.1 Radial Basis Functions

The radial basis function in the hidden layer produces a significant non-zero response only when the input falls within a small localized region of the input space. Each hidden unit, known as radial centre, has its own 'receptive field' in the input space, i.e., each centre is representative of one or some of the input patterns. This is called 'local representation of inputs', and the network is also known as 'localized receptive field network'. Consider, for instance, an input vector x which lies in the receptive field for centre c_i. This would activate the hidden centre c_i and by a proper choice of weights, the target output is obtained. Suppose an input vector lies between two receptive field centres, then both those hidden units will be appreciably activated. The output will be a weighted average of the corresponding targets. In short, the inputs are clustered around the centres and the output is linear in terms of weights, w_i.

$$y = \sum_{i=1}^{h} \phi_i w_i \qquad (2.27)$$

The response of the ith radial centre in RBFN usually expressed by the following expression:

$$\phi_i = \phi(\|x - c_i\|) \qquad (2.28)$$

where $\|x - c_i\|$ is the Euclidean distance between x and c_i. $\phi(.)$ is a radial basis function. When this basis function is Gaussian, each node produces an identical output for inputs within a fixed radial distance from the centre, i.e., they are radially symmetric. This accounts for the nomenclature 'radial basis function'. Other activation functions include

$\phi(z) = e^{-z^2/2\sigma^2}$ Gaussian radial function
$\phi(z) = z^2 \log z$ thin plate spline
$\phi(z) = (z^2 + r^2)^{1/2}$ quadratic
$\phi(z) = \frac{1}{(z^2 + r^2)^{1/2}}$ inverse quadratic

2.3.2   Learning in RBFN

Training of an RBF network requires optimal selection of the centre c_i and weights, w_i, $i = 1$ to h. This is a two-fold problem, unlike in a multi-layer network. As the different layers of the network perform different tasks, both the layers are optimized using different techniques and in different time scales. Several strategies are applicable depending on the way in which the radial centres are specified. Some of these strategies are discussed below.

Pseudo-Inverse Technique

Pseudo-inverse approach computes the weight vector w in batch mode, i.e., the weight update is done after, the network computes the forward responses of all input patterns. Assume that the radial functions for the hidden units are Gaussian. The centres are chosen randomly from the training data set (provided the training set is representative). The Gaussian function is normalized, i.e., for any x, $\sum_i \phi_i = 1$. The width of the radial basis function is usually determined by an ad hoc choice:

$$\sigma = \frac{d}{\sqrt{2h}} \qquad (2.29)$$

where d is the spread of the centres, i.e., the maximum distance between two centres, and h is the number of centres.

Such a choice of σ helps to avoid extremas. Moreover, the performance of the network is not very sensitive to the precise values of σ.

Once the width of the basis function is fixed, the next task is to learn the weights. For a given input–output pattern (x^p, y^p), the output in terms of weight vector w can be written as

$$y^p = \phi^{p^T} w \qquad (2.30)$$

where ith element of ϕ^p is computed as $\phi_i^p = e^{(-\frac{h}{d^2}\|x^p - c_i\|^2)}$. Collection of such expressions over all patterns will lead to the following least square problem:

$$\Phi w = Y \qquad (2.31)$$

where ith row of Φ is ϕ^{i^T}. $Y = [y^1, \ldots, y^p, \ldots]^T$. The weight vector w can be solved using the following pseudo-inverse:

$$w = (\Phi^T \Phi)^{-1} \Phi^T Y \qquad (2.32)$$

Example 2.2 Consider the RBF network as shown in Figure 2.9 to solve an EX-NOR problem

x_1	x_2	y
0	0	1
0	1	0
1	0	0
1	1	1

The output of the RBF network is given by

$$y = w_1 \phi_1 + w_2 \phi_2 + \theta$$

Assuming Gaussian radial centres, $\phi x = e^{-x^2}$, find out the weights, w_1 and w_2, associated with the two inputs of the network and θ, the weight, associated with the output bias.

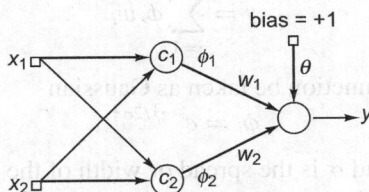

Figure 2.9 RBF network for solving the problem

Solution
Applying the four training patterns one after another, the following equations are obtained:

$$w_1 + w_2 e^{-2} + \theta = 1$$

$$w_1 e^{-1} + w_2 e^{-1} + \theta = 0$$

$$w_1 e^{-1} + w_2 e^{-1} + \theta = 0$$

$$w_1 e^{-2} + w_2 + \theta = 1$$

Thus, there are four equations to be solved for three unknowns w_1, w_2, and θ. In this case,

$$\Phi = \begin{bmatrix} 1 & 0.1353 & 1 \\ 0.3679 & 0.3679 & 1 \\ 0.3679 & 0.3679 & 1 \\ 0.1353 & 1 & 1 \end{bmatrix}, \; w = \begin{bmatrix} w_1 \\ w_2 \\ \theta \end{bmatrix}, \; Y = \begin{bmatrix} 1 \\ 0 \\ 0 \\ 1 \end{bmatrix}$$

Using Eqn (2.32), we get the desired weights as

$$w = \begin{bmatrix} 2.5031 \\ 2.5031 \\ -1.848 \end{bmatrix}$$

Gradient–Descent Algorithm

One of the most natural approaches to update centres and weights is supervised training by error correction learning. This is easily implemented by using a gradient–descent procedure. The update rule for centre learning is given below.

$$c_{ij}(t+1) = c_{ij}(t) - \eta_1 \frac{\partial E(t)}{\partial c_{ij}(t)}$$

for $i = 1$ to p and $j = 1$ to M.

Similarly the updates for weights are given as

$$w_i(t+1) = w_i(t) - \eta_2 \frac{\partial E(t)}{\partial w_i(t)}$$

where the cost function is $E = \frac{1}{2} \sum (y^d - y)^2$

The actual response of the RBF network as shown in Figure 2.8 is computed as

$$y = \sum_{i=1}^{h} \phi_i w_i$$

Let the radial basis function be taken as Gaussian

$$\phi_i = e^{-z_i^2 / 2\sigma^2}$$

where $z_i = \| x - c_i \|$ and σ is the spread or width of the Gaussian function.

Differentiating E with respect to w_i yields

$$\frac{\partial E}{\partial w_i} = \frac{\partial E}{\partial y} \times \frac{\partial y}{\partial w_i} = -(y^d - y)\phi_i \tag{2.33}$$

Differentiating E with respect to c_{ij} yields

$$\frac{\partial E}{\partial c_{ij}} = \frac{\partial E}{\partial y} \times \frac{\partial y}{\partial \phi_i} \times \frac{\partial \phi_i}{\partial c_{ij}} \tag{2.34}$$

$$= -(y^d - y) \times w_i \times \frac{\partial \phi_i}{\partial z_i} \times \frac{\partial z_i}{\partial c_{ij}} \tag{2.35}$$

$$\frac{\partial \phi_i}{\partial z_i} = -\frac{z_i}{\sigma^2} \phi_i \tag{2.36}$$

$$\frac{\partial z_i}{\partial c_{ij}} = \frac{\partial}{\partial c_{ij}} (\sum_j (x_j - c_{ij})^2)^{1/2} \tag{2.37}$$

$$= -(x_j - c_{ij})/z_i \tag{2.38}$$

After simplification, the update rule for the centres is

$$c_{ij}(t+1) = c_{ij}(t) + \eta_1 (y^d - y) w_i \frac{\phi_i}{\sigma^2} (x_j - c_{ij}) \tag{2.39}$$

The update rule for the weights

$$w_i(t+1) = w_i(t) + \eta_2 (y^d - y) \phi_i \tag{2.40}$$

The gradient–descent vector $\partial E / \partial c_{ij}$ exhibits a clustering effect. Note that for an RBF network, there is no back propagation of error unlike the supervised learning in a multi-layered network.

Example 2.3   Consider the dynamical system described by Eqn (2.25). Identify the system using a radial basis function network.

Solution
As mentioned in Example 2.1, since the system is dynamic while the RBF is a static network, the previous state is also input to the network along with the original system input during the identification process. The input is generated randomly between ± 1. A total of 1000 data pairs are generated with this input range using Eqn (2.25). The number of RBFN centres is taken as 100. The gradient descent algorithm is employed to update the weights as well as the centres of the network. The training results are shown in Figure 2.10 where (a) shows the input–output relationship of the training data as well as the network output after training. Figure (b) shows the desired and actual output of the network at every sampling instant.

The convergence of the mean square error over 1500 epochs is shown in Figure 2.11.

Once the training is over, 1000 new data points are generated to validate the identified model while taking the input to the system as $u(k) = \sin(0.1\,k)$. The

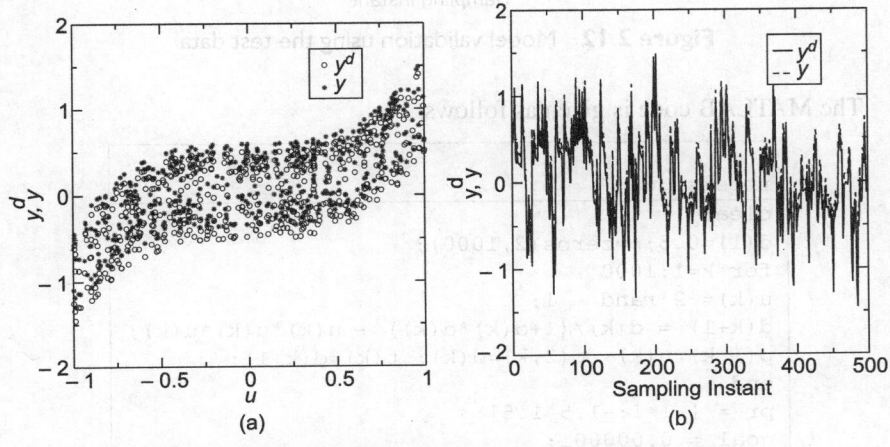

Figure 2.10   System identification results after training

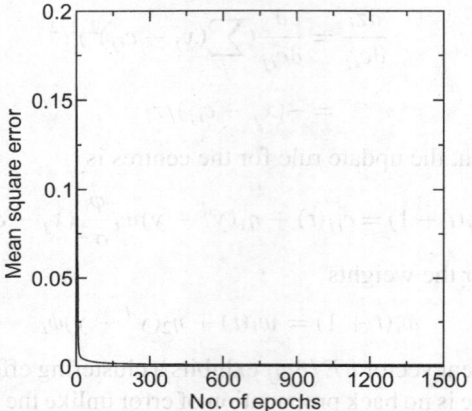

Figure 2.11 Epoch-wise convergence of the mean square error

result of model validation is shown in Figure 2.12. The RMS error for the training data set is found to be 0.026, while for the testing data set it is 0.04.

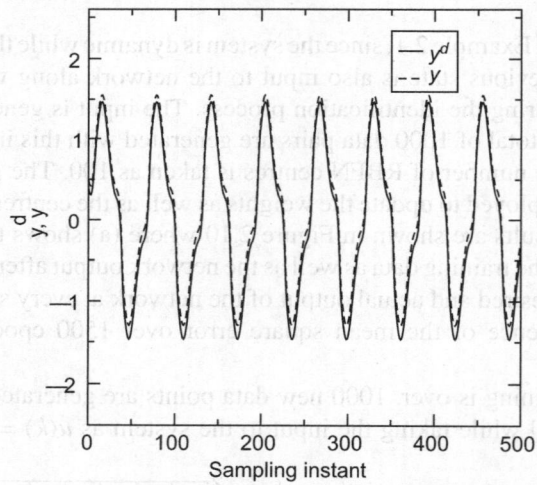

Figure 2.12 Model validation using the test data

The MATLAB code is given as follows.

```
rbfsiso.m
clear
d(1)=0.5;p=zeros(2,1000);
for k=1:1000
u(k)= 2*rand - 1;
d(k+1) = d(k)/(1+d(k)*d(k)) + u(k)*u(k)*u(k);
p(1,k)=u(k); p(2,k)=d(k); t(k)=d(k+1);
end
pr = [-1 1;-1.5 1.5];
goal = 0.000001;
```

```
net = newrb(p,t,goal);
y = sim(net,p);
size(y)
plot(u,t,'*',u,y,'o')

d(1)=0.5;p=zeros(2,1000);
for k=1:1000
u(k)= sin(0.1*k);
d(k+1) = d(k)/(1+d(k)*d(k)) + u(k)*u(k)*u(k);
p(1,k)=u(k); p(2,k)=d(k); t(k)=d(k+1);
end
y = sim(net,p);
figure
plot(t);
hold on;
plot(y,'r')
```

Hybrid Learning

In hybrid learning, the radial basis functions relocate their centres in a self-organized manner, while the weights of the output layers are computed using the supervised learning rule. Hence the name 'hybrid learning'. Due to self-organized learning, the radial centres are placed in those regions where significant input data is present. When a pattern is presented to the RBF network during training, either a new centre is grown if the pattern is sufficiently novel or the network parameters in both layers are updated using gradient–descent. The test for novelty relies on two criteria:

1. Is the Euclidean distance between the input pattern and the nearest existing centre greater than a threshold $\delta(t)$?

2. Is the mean square error at the output greater than a desired accuracy $\in$?

A new centre is allocated when both of the novelty criteria are satisfied.

K-mean clustering technique is employed for the self-organized selection of centres. It dynamically updates cluster centres for each iteration until they reach a stable state. The centre moves to the densest part of the cluster and hence the intra-cluster distance is minimized. The input vector x is classified according to the most frequently occurring label among those of the K nearest sample. The procedure of the centre update is as follows:

Find out a centre that is closest to x in terms the Euclidean distance. This particular centre is updated according to the following rule:

$$c_i(t+1) = c_i(t) + \alpha(x - c_i(t))$$

Thus, the centre moves closer to x. If $P(x)$ is the uniform probability of x, $\int ||x - c_i||^2 P(x)dx$, $\alpha(0 \to 1)$ is gradually decreased to 0.

While centres are updated using unsupervised scheme, the weights can be updated using any of the least square algorithms like gradient–descent or recursive least squares (RLS). The RLS [29] algorithm is presented below.

The ith output of the RBFN described earlier can be written as

$$x_i = \phi^T \theta_i \quad i = 1, \ldots, n$$

where $\phi \in \Re^l$ is the output vector of the hidden layer and $\theta_i \in \Re^l$ is the connection weight vector from the hidden units to the ith output unit. The weight update equations as per the RLS algorithm are described as

$$\theta_i(k+1) = \theta_i(k) + P(k+1)\phi(k)[x_i(k+1) - \phi(k)^T \theta_i(k)]$$
$$P(k+1) = P(k) - P(k)\phi(k)[1 + \phi(k)^T P(k)\phi(k)]^{-1}\phi(k)^T P(k)$$

where $P(k) \in \Re^{l \times l}$ is known as covariance matrix. The RLS is more accurate and fast compared to the least mean square (LMS) [30] algorithm.

The simplest approach is to update the centres using the gradient–descent algorithm and the weights can be updated using a simple LMS algorithm. Although, computational requirement increases by adjusting the centres, the number of centres can be substantially reduced by this approach. The generalization performance of such a network is much better as compared to the hybrid learning scheme where the centres are fixed or learned unsupervised and the weights are updated using the recursive least squares algorithm.

Example 2.4   Consider the dynamical system described by Eqn (2.25). Identify the system using a hybrid learning scheme. Show the initial and final distributions of the RBFN centres.

Solution
For the dynamics system (2.25), the input is generated randomly between ± 1. A total of 1000 data pairs are generated with this input range using Eqn (2.25). The number of RBFN centres is taken as 100. K-SOM clustering is employed to update the centres and the RLS algorithm is used to update the weights of the network. The initial covariance matrix P for the RLS is taken as a diagonal matrix with the diagonal element 100. Mapping of the input data and centres are shown in Figure 2.13, where (a) shows how the centres are initialized and (b) shows the spreading of the centres towards the nearest inputs after learning.

Identification results are found to be satisfactory with an RMS error of 0.018 for training and 0.027 for testing data.

2.4   ADAPTIVE LEARNING RATE

Faster convergence and function approximation accuracy are two key issues in choosing a training algorithm. The popular method for training a feed-forward network has been the back propagation. One of the main drawbacks of back propagation algorithm is its slow rate of convergence and its inability to ensure global convergence. Some heuristic methods like adding a momentum term to the original BP algorithm and standard numerical optimization techniques using quasi-Newton methods have been proposed to improve the convergence rate of the BP algorithm [31, 32]. The problem with quasi-Newton methods are that the storage and memory requirements go up as the square of the size of the network. The non-linear optimization techniques such as the Newton

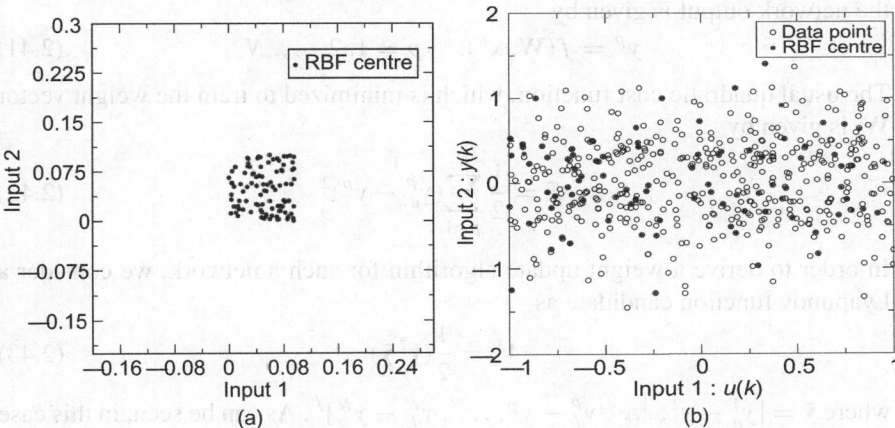

Figure 2.13 Centre training results for Example 2.4

method, conjugate-gradient, etc. [33, 34] have been used for training. Though the algorithm converges in fewer iterations than the BP algorithm, it requires too much computation per pattern. Other algorithms for faster convergence include extended Kalman filtering (EKF) [35], recursive least square (RLS) [36], and Levenberg–Marquardt (LM) [37, 38]. In order to overcome the computational complexity in these algorithms, a number of improvements have also been suggested [39, 40]. However, these improvements do not bring them closer to back-propagation algorithm as far as simplicity and ease of implementation is concerned.

In a recent work [41], two novel algorithms have been proposed on adaptive learning rate using the Lyapunov stability theory. The proposed algorithm has an exact parallel with the popular BP algorithm where the fixed learning rate in the BP algorithm is replaced by an adaptive learning rate. It is observed that this adaptive learning rate increases the speed of convergence. Although Yu et.al. [42] and Yu et. al. [43] have used Lyapunov function based weight update algorithms, none of them address the issue of computation of adaptive learning rate in a formal manner. The nature of convergence has also not been discussed in their work.

It is shown in Section 2.2.5 that a momentum term may be added to the BP algorithm in order to speed up its convergence rate. Such a term arises naturally in the Lyapunov function based algorithms, and thus, it provides a theoretical justification for such kind of modifications.

2.4.1 Lyapunov Function Based Adaptive Learning Rate

Consider a simple feed-forward neural network with a single output where $\phi(.)$ is the non-linear activation function for neurons. The network is parametrized in terms of its weights which is represented as a weight vector $\mathbf{W} \in R^m$. For a specific function approximation problem, the training data consists of N patterns, $\{\mathbf{x}^p, y^p\}$, $p = 1, 2, ..., N$. For a specific pattern p, if the input vector is $\mathbf{x}^p$, then the network output is given by

$$y^p = f(\mathbf{W}, \mathbf{x}^p), \quad p = 1, 2, \ldots, N \tag{2.41}$$

The usual quadratic cost function, which is minimized to train the weight vector $\mathbf{W}$, is given by

$$E = \frac{1}{2} \sum_{p=1}^{N} (y_d^p - y^p)^2 \tag{2.42}$$

In order to derive a weight update algorithm for such a network, we consider a Lyapunov function candidate as

$$V_1 = \frac{1}{2}(\tilde{\mathbf{y}}^T \tilde{\mathbf{y}}) \tag{2.43}$$

where $\tilde{\mathbf{y}} = [y_d^1 - y^1, \ldots, y_d^p - y^p, \ldots, y_d^N - y^N]^T$. As can be seen, in this case the Lyapunov function is the same as the usual quadratic cost function minimized during batch update using the back propagation learning algorithm. The time derivative of the Lyapunov function V_1 is given by

$$\dot{V}_1 = -\tilde{\mathbf{y}}^T \frac{\partial \mathbf{y}}{\partial W} \dot{\mathbf{W}} = -\tilde{\mathbf{y}}^T \mathbf{J} \dot{\mathbf{W}} \tag{2.44}$$

where

$$J = \frac{\partial \mathbf{y}}{\partial \mathbf{W}} \in R^{N \times m} \tag{2.45}$$

Theorem 2.1 If an arbitrary initial weight $\mathbf{W}(0)$ is updated by

$$\mathbf{W}(t') = \mathbf{W}(0) + \int_0^{t'} \dot{\mathbf{W}} dt \tag{2.46}$$

where

$$\dot{\mathbf{W}} = \frac{\| \tilde{\mathbf{y}} \|^2}{\| \mathbf{J}^T \tilde{\mathbf{y}} \|^2} \mathbf{J}^T \tilde{\mathbf{y}} \tag{2.47}$$

then $\tilde{\mathbf{y}}$ converges to zero under the condition that $\dot{\mathbf{W}}$ exists along the convergence trajectory.

Proof: Substitution of Eqn (2.47) into Eqn (4.38) gives

$$\dot{V}_1 = - \| \tilde{\mathbf{y}} \|^2 \leq 0 \tag{2.48}$$

where $\dot{V}_1 < 0$ for all $\tilde{\mathbf{y}} \neq 0$. If $\dot{V}_1$ is uniformly continuous and bounded, then according to *Barbalat's lemma* [44] as $t \to \infty$, $\dot{V}_1 \to 0$ and $\tilde{\mathbf{y}} \to 0$. The weight update law given in Eqn (2.47) is a batch update law. Analogous to the instantaneous gradient–descent (or BP) algorithm, the instantaneous LF I learning algorithm can be derived as

$$\dot{\mathbf{W}} = \frac{\| \tilde{\mathbf{y}} \|^2}{\| \mathbf{J_p}^T \tilde{\mathbf{y}} \|^2} \mathbf{J_p}^T \tilde{\mathbf{y}} \tag{2.49}$$

where $\tilde{\mathbf{y}} = y_d^p - y^p \in R$ and $\mathbf{J_p} = \frac{\partial y^p}{\partial \mathbf{W}} \in R^{1 \times m}$ is the instantaneous value of the Jacobian. The difference equation representation of the weight update algorithm based on Eqn (2.49) is given by

$$\mathbf{W}(t + 1) = \mathbf{W}(t) + \mu \dot{\mathbf{W}}(t) \tag{2.50}$$

$$= \mathbf{W}(t) + \left(\mu \frac{\| \tilde{y} \|^2}{\| \mathbf{J_p}^T \tilde{y} \|^2} \right) \mathbf{J_p}^T \tilde{y} \tag{2.51}$$

where μ is a constant which is selected heuristically. A very small constant ϵ can be added to the denominator of Eqn (2.49) to avoid numerical instability when error $\tilde{y}$ goes to zero. The Lyapunov function based algorithm is compared with the BP algorithm based on the gradient–descent principle.

In the instantaneous gradient–descent method, the expression has the form

$$\triangle \mathbf{W} = -\eta \left(\frac{\partial E}{\partial \mathbf{W}} \right)^T \tag{2.52}$$

$$= \eta \mathbf{J_p}^T \tilde{y} \tag{2.53}$$

$$\mathbf{W}(t + 1) = \mathbf{W}(t) + \eta \mathbf{J_p}^T \tilde{y} \tag{2.54}$$

where η is the learning rate. Comparing Eqn (2.54) with Eqn (2.50), we can see a very interesting similarity where the fixed learning rate η in the BP algorithm is replaced by its adaptive version η_a given by

$$\eta_a = \left(\mu \frac{\| \tilde{y} \|^2}{\| \mathbf{J_p}^T \tilde{y} \|^2} \right) \tag{2.55}$$

This is the most remarkable finding of [41]. Earlier, there have been many research papers concerning the adaptive learning rate [31, 45, 46, 47]. However, the computation of adaptive learning rate using the Lyapunov function approach is a key contribution to this field.

Theorem 4.1 states that the global convergence of the learning algorithm (2.47) is guaranteed provided $\dot{\mathbf{W}}$ exists and is non-zero along the convergence trajectory. This, in turn, necessitates $\| \frac{\partial V_1}{\partial \mathbf{W}} \| = \| J^T \tilde{y} \| \neq 0$. The condition $\| \frac{\partial V_1}{\partial \mathbf{W}} \| = 0$ represents local minima of the scalar function (2.43). Thus, theorem 4.1 says that the global minimum is reached only when local minima are avoided during training. Since instantaneous update rule introduces noise, it may be possible to reach a global minimum in some cases; however, the global convergence is not guaranteed.

When the algorithm is implemented to learn a XOR map, Figure 2.14 shows how the learning rate naturally evolves and converges to zero as learning is over. Readers are referred to [41] for more details.

2.5 FEEDBACK NETWORKS

The versatility of a neural network is considerably increased by the incorporation of feedback. Feed-forward networks as discussed earlier in this chapter learn the map from the input space to the output space. Once the weights are fixed, the state of each neuron is determined by the input vector, i.e., the neuron state is independent of the initial and past state of this neuron. Hence a feed-forward

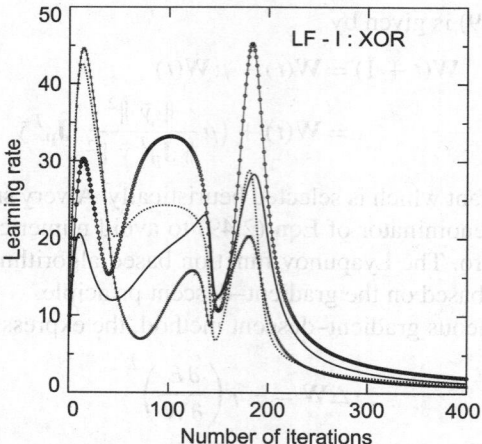

Figure 2.14 Adaptive learning rate while learning a XOR map

network is a static network. There are no dynamics involved. In contrast, a
feedback network allows feedback connection as shown in Figure 2.15. Such a

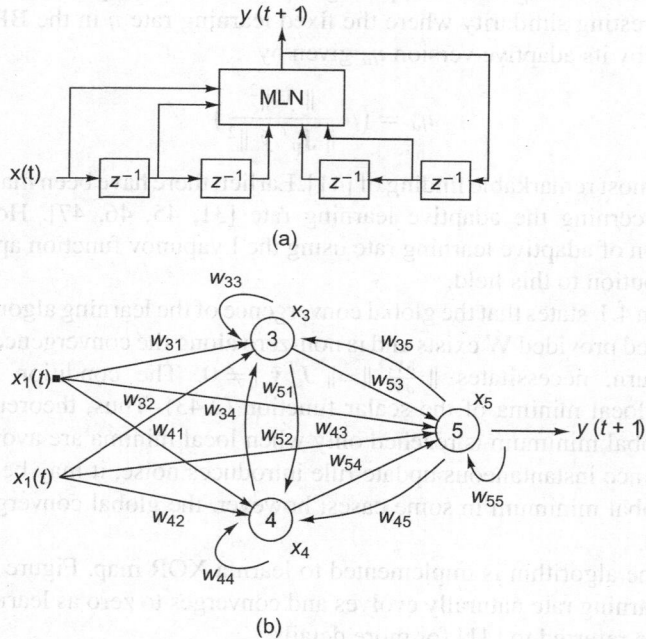

Figure 2.15 (a) Partially connected recurrent networks and (b) fully connected
recurrent networks

feedback network, often called recurrent network, becomes a non-linear dynamic
system and exhibits a highly non-linear dynamic behaviour, which makes them
highly interesting. Such a system has very rich temporal and spatial behaviours,
such as stable and unstable fixed points and limit cycles, and chaotic behaviors.
These behaviors can be utilized to model certain cognitive functions, such as
associative memory, unsupervised learning, self-organizing maps, and temporal

reasoning. In this section, the recurrent networks will be presented from a system identification perspective. Readers can refer to the following works [48, 49, 50, 51], based on which the subject matter of this section has been developed.

Consider a system expressed in discrete dynamics

$$y(t) = f(y(t-1)) + g(y(t-1))u(t-1) \qquad (2.56)$$

where $y(t)$ may be a scalar variable or a vector, $f(\cdot)$ and $g(\cdot)$ are two arbitrary non-linear functions of $y(t-1)$. The feed-forward network will learn this dynamics as

$$\hat{y}(t) = \hat{f}(y(t-1), u(t-1)) \qquad (2.57)$$

where $y(t-1)$ is the actual system output at the $(t-1)$ instant. The recurrent network learns this map as

$$\hat{y}(t) = \hat{f}(\hat{y}(t-1) \quad u(t-1)) \qquad (2.58)$$

which indicates that the network has internal memory, i.e., the present output state is a function of the previous output of the network. Figure 2.16 summarizes the difference between a feed-forward network and a feedback network.

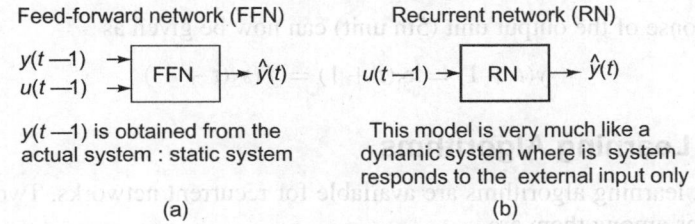

<table>
<tr><td>Feed-forward network (FFN)</td><td>Recurrent network (RN)</td></tr>
</table>

$y(t-1) \rightarrow$ | FFN | $\rightarrow \hat{y}(t)$
$u(t-1) \rightarrow$

$u(t-1) \rightarrow$ | RN | $\rightarrow \hat{y}(t)$

$y(t-1)$ is obtained from the actual system : static system

This model is very much like a dynamic system where is system responds to the external input only

(a)   (b)

Figure 2.16   (a) Feed-forward network and (b) recurrent network

2.5.1   Response of Recurrent Networks

Let us find the response $y(t+1)$ of the network shown in Figure 2.15 (b). The figure shows a fully connected network. The following convention will be used to describe a recurrent structure. The neurons are serially numbered (including the input neurons). The input to the ith neuron is denoted as s_i. The response of the ith neuron is denoted as x_i (this also includes the input units). The connection weight from the jth neuron to the ith neuron is represented as w_{ij}. Using the above notations, the input to the ith neuron can be computed as

$$s_i(t+1) = \sum_j w_{ij}x_j(t)$$

where t represents the temporal argument. The response of the above unit can be expressed as

$$x_i(t) = f(s_i(t+1))$$

where $f(.)$ represents the sigmoid activation function, i.e.,

$$f(h) = \frac{1}{1 + e^{-h}}$$

For the given network as shown in Figure 2.15(b), the response of the unit number 3 can be given as

$$s_3(t+1) = w_{32}x_2(t) + w_{33}x_3(t) + w_{34}x_4(t) + w_{35}x_5(t)$$

$$= \sum_{i=1}^{5} w_{3i}x_i(t)$$

$$x_3(t+1) = f(s_3(t+1))$$

Similarly, for the 4th unit, the response can be given as

$$x_4(t+1) = f(s_4(t+1))$$

$$s_4(t+1) = \sum_{i=1}^{5} w_{4i}x_i(t)$$

The response of the output unit (5th unit) can now be given as

$$y(t+1) = x_5(t+1) = f(s_5(t+1))$$

2.5.2   Learning Algorithms

Different learning algorithms are available for recurrent networks. Two popular algorithms among them are

1. Back propagation through time
2. Real time recurrent learning

2.5.3   Back Propagation Through Time

Back Propagation Through Time (BPTT) is an extension of the standard back propagation algorithm. It is characterized by *unfolding* the temporal operation of the recurrent network into a multi-layered feed-forward network, with a new layer added at every time step. Consider the recurrent network shown in Figure 2.17 (a). The output of the network is given as

$$y(t+1) = f(w\,x(t) + g\,y(t))$$

where f is the sigmoidal activation function, w is the connection weight between the input and the output neuron and g is the self-feed back connection weight between the neuron and itself. The unfolded network with $t = 5$ is shown in Figure 2.17 (b). The unfolded network is a feed-forward network where the basic network repeats every time step. For example, in the first time step, the output $y(1)$ is a function of the previous state and the input $y(0)$ and $u(0)$, respectively.

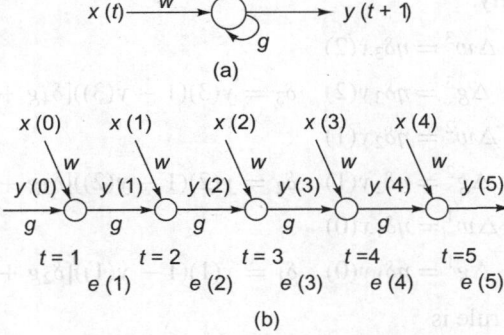

(a)

(b)

Figure 2.17 (a) Original network with a single neuron and (b) unfolded network

Example 2.5 Unfold the fully recurrent network given in Figure 2.15(b) in time.

Solution
The unfolded network until time step $t = 4$ is shown in Figure 2.18.

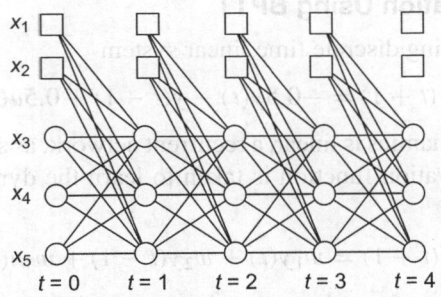

Figure 2.18 Unfolded version of the fully connected network

Example 2.6 Enumerate steps for training the unfolded network of the simple recurrent network given in Figure 2.17.

Solution
The steps to proceed further are as follows:

1. Compute the response of the sequence $y(1)$ to $y(5)$, given the sequence $x(0)$ to $x(4)$, and $y(0)$.

2. Compute the error $e(5) = y^d(5) - y(5)$
   $\Delta w^5 = \eta \delta_5 x(4)$
   $\Delta g^5 = \eta \delta_5 y(4)$, where $\delta_5 = y(5)(1 - y(5))e(5)$

3. Compute $e(4)$, $e(3)$, $e(2)$, and $e(1)$ and proceed the same way until the first node $t = 1$
   $\Delta w^4 = \eta \delta_4 x(3)$
   $\Delta g^4 = \eta \delta_4 y(3)$
   where $\delta_4 = y(4)(1 - y(4))[\delta_5 g + e(4)]$

Similarly,

$$\Delta w^3 = \eta \delta_3 x(2)$$

$$\Delta g^3 = \eta \delta_3 y(2) \quad \delta_3 = y(3)(1 - y(3))[\delta_4 g + e(3)]$$

$$\Delta w^2 = \eta \delta_2 x(1)$$

$$\Delta g^2 = \eta \delta_2 y(1) \quad \delta_2 = y(2)(1 - y(2))[\delta_3 g + e(2)]$$

$$\Delta w^1 = \eta \delta_1 x(0)$$

$$\Delta g^1 = \eta \delta_1 y(0) \quad \delta_1 = y(1)(1 - y(1))[\delta_2 g + e(1)]$$

The update rule is

$$w^{\text{new}} = w^{\text{old}} + \sum_{i=1}^{5} \Delta w^i$$

$$g^{\text{new}} = g^{\text{old}} + \sum_{i=1}^{5} \Delta g^i$$

System Identification Using BPTT

Consider the following discrete time linear system

$$y(t + 1) = -0.5y(t) - y(t - 1) + 0.5u(t) \tag{2.59}$$

Since the system dynamics is linear, a recurrent network, as shown in Figure 2.19, with the linear activation function is taken to learn the dynamics. The network response is given as

$$y(t + 1) = w_1 y(t) + w_2 y(t - 1) + w_3 u(t) \tag{2.60}$$

where w_1, w_2, and w_3 are unknown weights. A weight update algorithm using BPTT needs to be derived using the same principle of back propagation so that these parameters converge to the actual parameters associated with the dynamics (2.59).

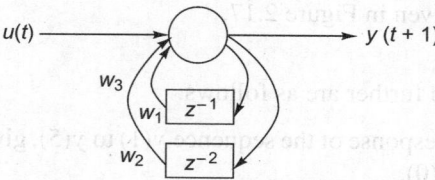

Figure 2.19 A recurrent network that can learn the linear dynamics given in (2.59)

The network in Figure 2.19 is unfolded and the unfolded network in time is shown in Figure 2.20. The response of the network is computed at different time steps as follows

$$y(2) = w_1 y(1) + w_2 y(0) + w_3 u(1)$$

$$y(3) = w_1 y(2) + w_2 y(1) + w_3 u(2)$$

$$\vdots$$

$$y(n+1) = w_1 y(n) + w_2 y(n-1) + w_3 u(n)$$

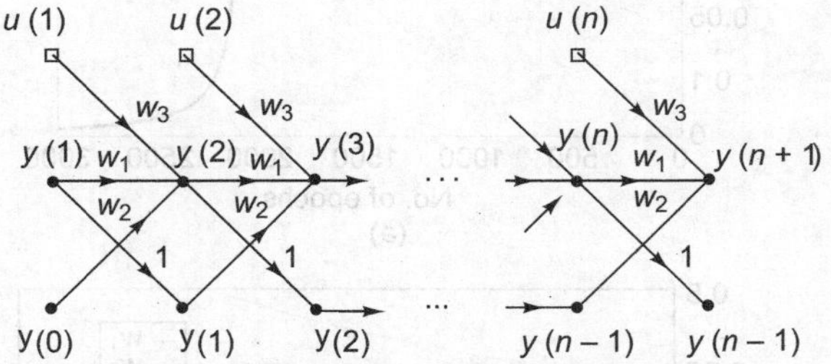

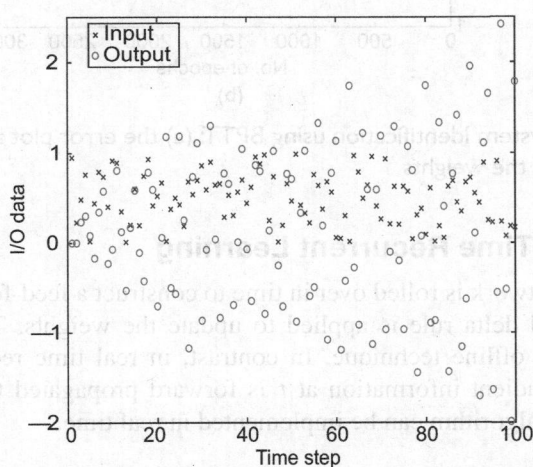

Figure 2.20 Unfolding of the network in Figure 2.19

The network has to be trained using the generalized delta rule as explained in the previous example. A total of 100 data points have been generated using the dynamic model of the system (2.59) where input $u(t)$ is generated randomly between 0 and 1. The training data are shown in Figure 2.21. From this figure, it is shown that the output range is between -2 and $+2$.

Figure 2.21 Training data for learning the dynamics (2.59)

The network is trained using BPTT. The RMS error versus the number of epochs is shown in Figure 2.22(a). The weights are found to be

$$w_1 = -0.4999, \quad w_2 = -1.0002, \quad \text{and} \quad w_3 = 0.4995 \qquad (2.61)$$

The evolution of the weights during training is shown in Figure 2.22(b).

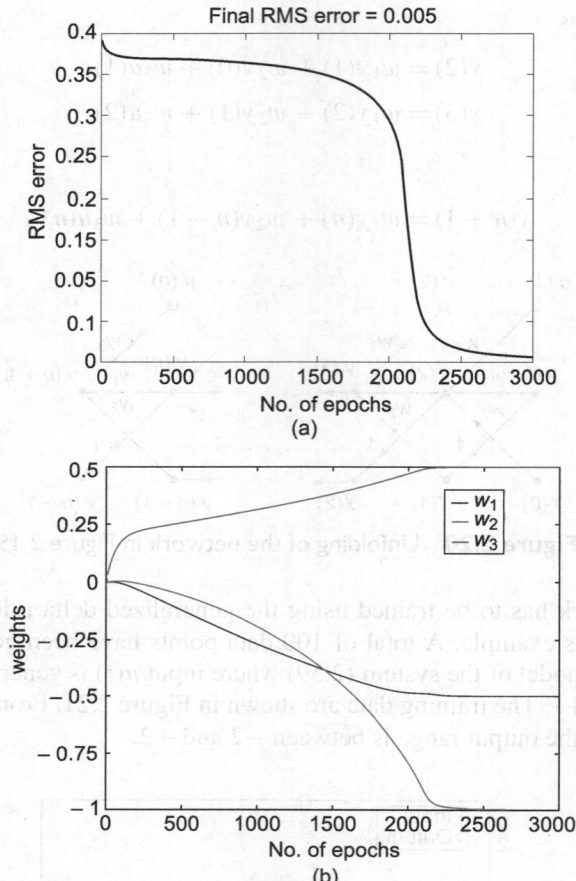

Figure 2.22 System Identification using BPTT: (a) the error plot and (b) evolution of the weights

2.5.4 Real Time Recurrent Learning

In BPTT, the network is rolled over in time to construct a feed-forward network. The generalized delta rule is applied to update the weights. Thus, the BPTT algorithm is an offline technique. In contrast, in real time recurrent learning (RTRL), the gradient information at t is forward propagated to the next time step $t + 1$. The algorithm can be implemented in real time.

RTRL: A Simple Example

Consider the recurrent network with a single neuron as shown in Figure 2.17 where $x(t)$ is the input vector at time t and $y(t + 1)$ is the output vector at time $t + 1$. The forward response of the network is given as

$$y(t + 1) = f(s(t + 1))$$

where $s(t + 1) = wx(t) + gy(t)$ and $f(h) = \frac{1}{1+e^{-h}}$

The cost function to be minimized is defined as

$$E(t + 1) = \frac{1}{2}(y^d(t + 1) - y(t + 1))^2 \qquad (2.62)$$

The network has two parametric weights w and g. Using the gradient–descent algorithm we can write

$$w(t + 1) = w(t) - \eta \frac{\partial E(t + 1)}{\partial w}$$

$$g(t + 1) = g(t) - \eta \frac{\partial E(t + 1)}{\partial g}$$

Differentiating E with respect to the synaptic weight vector w

$$\frac{\partial E(t + 1)}{\partial w} = -[y^d(t + 1) - y(t + 1)] \frac{\partial y(t + 1)}{\partial w} \qquad (2.63)$$

Let us define two variables

$$P_w(t + 1) = \frac{\partial y(t + 1)}{\partial w} \quad \text{with} \quad P_w(0) = 0 \quad \text{and}$$

$$P_g(t + 1) = \frac{\partial y(t + 1)}{\partial g} \quad \text{with} \quad P_g(0) = 0$$

The objective is to derive a recursive relation for $P_w(t + 1)$ and $P_g(t + 1)$ in terms of $P_w(t)$ and $P_g(t)$

Compute $\frac{\partial y(t+1)}{\partial w}$ as

$$\frac{\partial y(t + 1)}{\partial w} = \frac{\partial y(t + 1)}{\partial s(t + 1)} \times \frac{\partial s(t + 1)}{\partial w}$$

Since $y(t + 1) = \frac{1}{1+e^{-s(t+1)}}$, we can write

$$\frac{\partial y(t + 1)}{\partial s(t + 1)} = y(t + 1)(1 - y(t + 1)) \qquad (2.64)$$

Similarly, $\frac{\partial s(t+1)}{\partial w}$ can be computed as

$$\frac{\partial s(t + 1)}{\partial w} = \frac{\partial}{\partial w}[wx(t) + gy(t)]$$

$$= x(t) + g \frac{\partial y(t)}{\partial w} = x(t) + g P_w(t)$$

Therefore, $\dfrac{\partial y(t + 1)}{\partial w} = y(t + 1)(1 - y(t + 1))[x(t) + g P_w(t)]$

In a similar fashion we can compute

$$\frac{\partial y(t + 1)}{\partial g} = y(t + 1)(1 - y(t + 1))[x(t) + g P_g(t)]$$

which implies

$$\frac{\partial E(t+1)}{\partial w} = -[y^d(t+1) - y(t+1)]y(t+1)$$

$$(1 - y(t+1))[x(t) + g P_w(t)]$$

$$\frac{\partial E(t+1)}{\partial g} = -[y^d(t+1) - y(t+1)]y(t+1)$$

$$(1 - y(t+1))[x(t) + g P_g(t)]$$

The weight update laws can thus be written as

$$w(t+1) = w(t) + \eta[y^d(t+1) - y(t+1)]P_w(t+1)$$

$$g(t+1) = g(t) + \eta[y^d(t+1) - y(t+1)]P_g(t+1)$$

where the values of $P_w(t+1)$ and $P_g(t+1)$ are computed using the following set of relations:

$$P_w(t+1) = y(t+1)(1 - y(t+1))[x(t) + g P_w(t)]$$

$$P_g(t+1) = y(t+1)(1 - y(t+1))[x(t) + g P_g(t)]$$

$$\text{where} \quad P_w(t) = \frac{\partial y(t)}{\partial w} \quad \text{and} \quad P_g(t) = \frac{\partial y(t)}{\partial g}$$

Thus, $P_w(1) = y(1)(1 - y(1))[x(0) + g P_w(0)]$, where $P_w(0) = 0$ and so on. Similarly for $P_g(.)$.

Example 2.7 Derive the RTRL algorithm for the network shown in Figure 2.23.

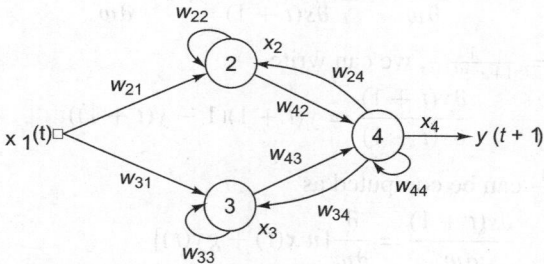

Figure 2.23 A real time recurrent network

Solution

The forward response of the network is computed as follows:

$$s_2(t+1) = w_{21}x_1(t) + w_{22}x_2(t) + w_{24}x_4(t)$$

$$s_3(t+1) = w_{31}x_1(t) + w_{33}x_3(t) + w_{34}x_4(t)$$

$$s_4(t+1) = w_{42}x_2(t) + w_{43}x_3(t) + w_{44}x_4(t)$$

$$x_i(t+1) = f(s_i(t+1)), \quad \text{for} \quad i = 2, 3, 4$$

$$y(t+1) = x_4(t+1), \quad f(z) = \frac{1}{1 + e^{-z}}$$

The weight update law is given by

$$w_{jk}(t+1) = w_{jk}(t) - \eta \frac{\partial E(t+1)}{\partial w_{jk}}$$

where $E(t+1) = \frac{1}{2}(y^d(t+1) - y(t+1))^2$

$$\frac{\partial E(t+1)}{\partial w_{jk}} = -(y^d(t+1) - y(t+1)) \frac{\partial y(t+1)}{\partial w_{jk}}$$

Let us define, $\dfrac{\partial y(t+1)}{\partial w_{jk}} = \dfrac{\partial x_4(t+1)}{\partial w_{jk}} = P^4_{jk}(t+1)$. We can write

$$P^4_{jk}(t+1) = f'(s_4(t+1))[w_{42} P^2_{jk}(t) + w_{43} P^3_{jk}(t) + w_{44} P^4_{jk}(t) + \delta_{jk} x_k(t)]$$

where $\delta_{jk} = 1$ if $j = 4$ and $f'(s_4(t+1)) = y(t+1)(1 - y(t+1))$. f' denotes the derivative of f with respect to its argument. Similarly, we can write

$$P^3_{jk}(t+1) = f'(s_3(t+1))[w_{33} P^3_{jk}(t) + w_{34} P^4_{jk}(t) + \delta_{jk} x_k(t)]$$

$$P^2_{jk}(t+1) = f'(s_2(t+1))[w_{22} P^2_{jk}(t) + w_{24} P^4_{jk}(t) + \delta_{jk} x_k(t)]$$

Again

$$P^4_{pq}(t+1) = f'(s_4(t+1)) = [w_{42} P^2_{pq}(t) + w_{43} P^3_{pq}(t) + w_{44} P^4_{pq}(t) + \delta_{P4} x_q(t)]$$

when $P = 4$ and $\delta_{P4} = 1$, and

$$P^3_{pq}(t+1) = f'(s_3(t+1)) = [w_{33} P^3_{pq}(t) + w_{34} P^4_{pq}(t) + \delta_{3P} x_q(t)]$$

$$P^2_{pq}(t+1) = f'(s_2(t+1)) = [w_{22} P^3_{pq}(t) + w_{24} P^4_{pq}(t) + \delta_{2P} x_q(t)]$$

For example, let us compute $w_{31}(t+1)$.

$$w_{31}(t+1) = w_{31}(t) - \eta \frac{\partial E(t+1)}{\partial w_{31}}$$

$$\frac{\partial E(t+1)}{\partial w_{31}} = -[y^d(t+1) - y(t+1)] \frac{\partial y(t+1)}{\partial w_{31}}$$

$$= -[y^d(t+1) - y(t+1)] P^4_{31}(t+1)$$

Therefore,

$$w_{31}(t+1) = w_{31}(t) + \eta[y^d(t+1) - y(t+1)] P^4_{31}(t+1)$$

where

$$P^4_{31}(t+1) = y(t+1)[1 - y(t+1)][w_{42} P^2_{31}(t) + w_{43} P^3_{31}(t) + w_{44} P^4_{31}(t)]$$

Similarly, one can proceed for $P^i_{31}(t) = 0$ for $i = 2, 3, 4$.

System Identification Using RTRL

Consider the discrete time linear system as described by Eqn (2.59). A recurrent network with a linear activation function is chosen to identify the system dynamics. The response of the network is given as

$$y(t+1) = w_1 y(t) + w_2 y(t-1) + w_3 u(t) \tag{2.65}$$

where w_1, w_2, and w_3 are unknown weights. A weight update algorithm is to be found using RTRL so that these parameters converge to the actual system parameters associated with the dynamics (2.59). The cost function is taken as $E(t+1) = \frac{1}{2}(y^d(t+1) - y(t+1))^2$. The weight update law is given as

$$w_i(t+1) = w_i(t) - \eta \frac{\partial E(t+1)}{\partial w_i} \quad \text{for} \quad i = 1, 2, 3$$

The partial derivatives of the cost function with respect to the weights are computed as follows:

$$\frac{\partial E(t+1)}{\partial w_1} = -(y^d(t+1) - y(t+1))P_{w_1}(t+1) \tag{2.66}$$

where

$$P_{w_1}(t+1) = \frac{\partial y(t+1)}{\partial w_1} = y(t) + w_1 P_{w_1}(t) + w_2 P_{w_1}(t-1)$$

Similarly

$$\frac{\partial E(t+1)}{\partial w_2} = -(y^d(t+1) - y(t+1))P_{w_2}(t+1)$$

$$P_{w_2}(t+1) = \frac{\partial y(t+1)}{\partial w_2} = y(t-1) + w_1 P_{w_2}(t) + w_2 P_{w_2}(t-1)$$

$$\frac{\partial E(t+1)}{\partial w_3} = -(y^d(t+1) - y(t+1))P_{w_3}(t+1)$$

$$P_{w_3}(t+1) = \frac{\partial y(t+1)}{\partial w_3} = u(t) + w_1 P_{w_3}(t) + w_2 P_{w_3}(t-1)$$

A total of 100 input–output data points, as generated for the same example using BPTT, have been used to train the network. The training data are shown in Figure 2.21. Using RTRL for one sequence of time, the final weights are found to be

$$w_1 = -0.5000, \quad w_2 = -0.9999, \quad \text{and} \quad w_3 = 0.5010 \tag{2.67}$$

Figure 2.22(a) shows the RMS error versus number of epochs, while (b) shows evolution of the weights during training.

2.6 KOHONEN SELF-ORGANIZING MAP

Kohonen [52] proposed an unsupervised learning algorithm that can form clusters for a given data set while preserving topology. A simple configuration of the Kohonen self-organizing feature map is illustrated in Fig. 2.25. The most prominent feature is the concept of excitatory learning with a neighbourhood around the winning neuron. The size of the neighbourhood slowly decreases with each iteration, see for example Figure 2.25. A more detailed description is provided below:

The first step is to initialize the synaptic weights in the network. There are three essential processes involved in the formation of the SOM.

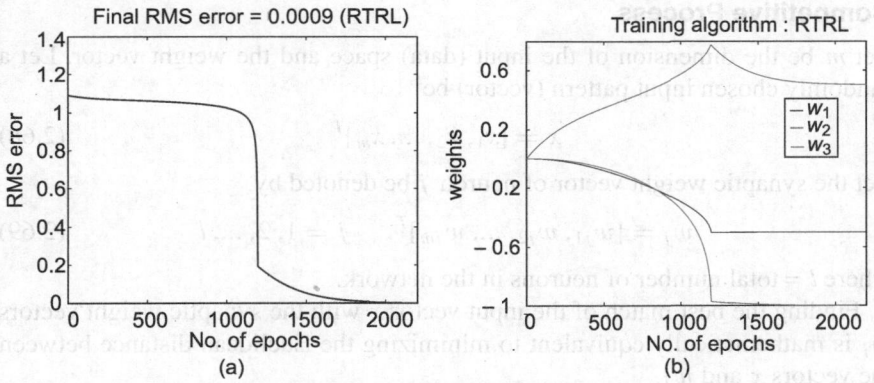

Figure 2.24 System Identification using RTRL: (a) the error plot and (b) evolution of the weights

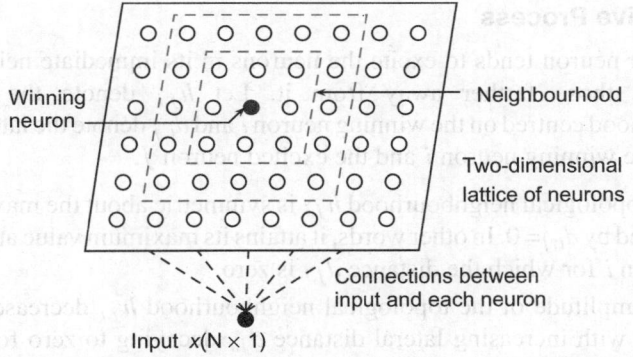

Figure 2.25 A two-dimensional self-organizing feature map (By updating all the weight connecting to a neighbourhood of the target neurons, it enables the neighbouring neuron to become more responsive to the same input pattern. Consequently, the correlation between neighbouring nodes can be enhanced. Once such a correlation is established, the size of a neighbourhood can be decreased gradually, based on the desire of having a stronger identity of individual nodes.)

- *Competition:* For each input pattern, the neurons in the network compute their respective values of a discriminant function. The neuron with the largest value of that function is declared winner.
- *Cooperation:* The winning neuron determines the spatial location of a topological neighbourhood of excited neurons, i.e., cooperative neighbouring neurons.
- *Synaptic adaptation:* This enables the excited neurons to adjust their synaptic weights in relation to the input pattern.

Competitive Process

Let m be the dimension of the input (data) space and the weight vector. Let a randomly chosen input pattern (vector) be

$$x = [x_1, x_2, ..., x_m]^T \tag{2.68}$$

Let the synaptic weight vector of neuron j be denoted by

$$w_j = [w_{j1}, w_{j2}, ..., w_{jm}]^T, \quad j = 1, 2, ..., l \tag{2.69}$$

where l = total number of neurons in the network.

Finding the best match of the input vector x with the synaptic weight vectors w_j is mathematically equivalent to minimizing the Euclidean distance between the vectors x and w_j.

Let $i(x)$ = index to identify the neuron that best matches x,

$$i(x) = \arg \min_j ||x - w_j||, \quad j = 1, 2, ..., l \tag{2.70}$$

Cooperative Process

The winner neuron tends to excite the neurons in its immediate neighbourhood more than those farther away from it. Let $h_{j,i}$ denote the topological neighbourhood centred on the winning neuron i and $d_{i,j}$ denote the lateral distance between the winning neuron i and the excited neuron j.

- The topological neighbourhood $h_{j,i}$ is symmetric about the maximum point defined by $d_{i,j} = 0$. In other words, it attains its maximum value at the winning neuron i for which the distance $d_{j,i}$ is zero.

- The amplitude of the topological neighbourhood $h_{j,i}$ decreases monotonically with increasing lateral distance $d_{j,i}$ decaying to zero for $d_{j,i} \to \infty$ which is a necessary condition for convergence.

A typical choice of $h_{j,i}$ that satisfies these requirements is the Gaussian function as shown in Figure 2.6. The expression of a Gaussian neighbourhood function is given as

$$h_{j,i(x)}(n) = \exp \left(-\frac{d_{j,i}^2}{2\sigma^2(n)} \right), \quad n = 0, 1, 2, ... \tag{2.71}$$

where $\sigma(n)$ is the width of neighbourhood function. During the training process, the width can vary in the following manner:

$$\sigma(n) = \sigma_0 \exp \left(-\frac{n}{\tau_1} \right) \tag{2.72}$$

where $\qquad \sigma_0$ = initial value of σ

$\qquad\qquad \tau_1$ = time constant of width function

$\qquad\qquad n$ = number of training steps

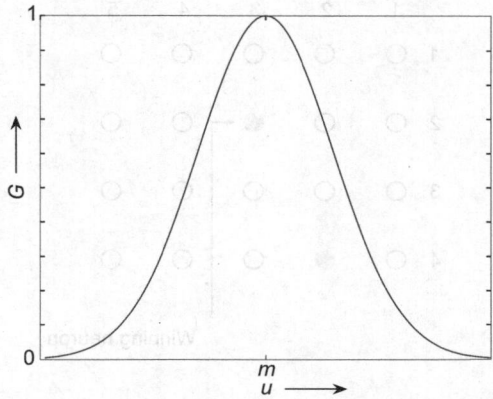

Figure 2.26 Gaussian neighbourhood function

One-dimensional Lattice

For a one-dimensional lattice, $d_{j,i}$ is measured as

$$d_{j,i} = |j - i| \tag{2.73}$$

$$d_{j,i} = |j - i|$$

Consider the above one-dimensional Kohonen lattice. Let index of winning neuron i be 3 and index of two neighbourhood neurons are 2 and 5. Thus,

$$d_{8,5} = |2 - 3| = 1$$
$$d_{3,5} = |5 - 3| = 2$$

Two-dimensional Lattice

For a two-dimensional lattice, $d_{j,i}^2$ is measured as

$$d_{j,i}^2 = ||r_j - r_i||^2 \tag{2.74}$$

where 　　　　　r_i = position of winning neuron$i$

　　　　　　　　r_j = position of neighborhood neuron j

Let the index of the winning neuron i be {2, 3} and the index of a neighbourhood neuron j is {4, 2}

$$d_{j,i} = (4 - 2)^2 + (2 - 3)^2 = 5$$

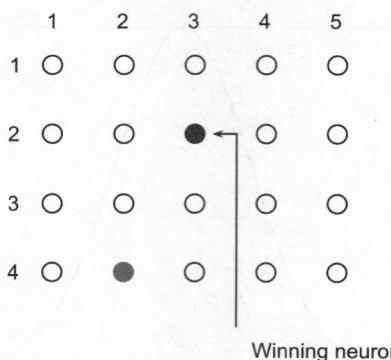

Winning neuron

Adaptive Process

Weights associated with the winning neuron and its neighbours are updated as per a neighbourhood index. The winning neuron is allowed to be maximally benefited from this weight update, while the neuron that is farthest from the winner is minimally benefited. The change in weight in each training step is given by

$$\Delta w_j(n) = \eta(n)h_{j,i(x)}(n)(x - w_j(n))$$

$$\eta(n) = \eta_0 \exp\left(-\frac{n}{\tau_2}\right) \quad n = 0, 1, 2, \dots$$

where

$\eta(n)$ = learning rate parameter

η_0 = initial value of the learning rate parameter

τ_2 = time constant of the learning rate

The updated weight vector $w_j(n + 1)$ at time $(n + 1)$ is defined by

$$w_j(n + 1) = w_j(n) + \eta(n)h_{j,i(x)}(n)(x - w_j(n)) \tag{2.75}$$

This equation will move the weight of winning neuron w_i towards the input vector x.

Example 2.8 $1 - D$ SOM learns $2 - D$ topology

In the simulation, a neural network is chosen with 100 neuron organized in a one-dimensional lattice. The network is trained with a two dimensional input vector x.

- Consider the input data coming randomly from a 2-D topology.
- Since each data point is two-dimensional, $x = [x_1 \ x_2]^T$, where
  x_1 represents the x-coordinate
  and x_2 represents the y-coordinate
- Here w_i associated with each neuron is also two-dimensional.

The training is done for 6000 iterations.

$$m = 2$$
$$x = [x_1, x_2]^T$$
$$w_j = [w_{j,1}, w_{j,2}]^T \quad j = 1, 2, \ldots, 100$$

The weights are initialized from a random set

$$(-0.4 < w_{j,1} < 0.4 \text{ and } -0.4 < w_{j,2} < 0.4)$$

The input x is uniformly distributed in the region ($0 < x_1 < 1$ and $0 < x_2 < 1$). Figure 2.27 shows the input data space. Figure 2.28 shows the network weights before training and Figure 2.29 shows that the weights of the network preserve the topology of the input space.

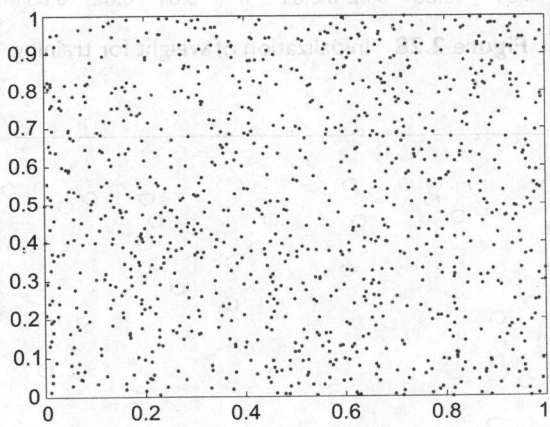

Figure 2.27 Input space for training

Next we will consider the case where the input data are coming randomly from an 'L' shaped input space.

The weights are initialized from a random set
$(-0.4 < w_{j,1} < 0.4 \text{ and } -0.4 < w_{j,2} < 0.4)$

The input x is uniformly distributed in the region
$(0 < x_1 < 1 \text{ and } 0 < x_2 < 0.3)$.
$(0 < x_1 < 0.3 \text{ and } 0.3 < x_2 < 1)$. The training is done for 6000 iterations. Figure 2.30 shows the input data space. Figure 2.28 shows the network weights before training and Figure 2.31 shows that the weights of the network preserve the topology of the input space.

Example 2.9 2-D SOM learns 2-D topology

In the simulation, a neural network is chosen with 100 neuron organized in a two-dimensional lattice with 10 rows and 10 column. The network is trained with

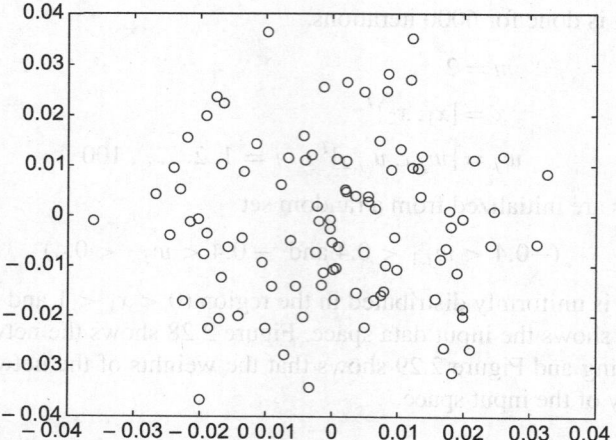

Figure 2.28 Initialization of weight for training

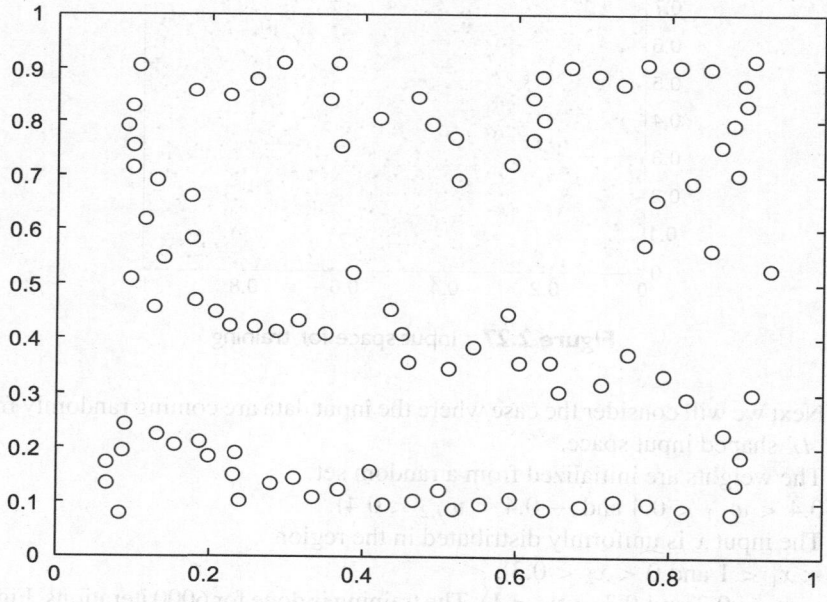

Figure 2.29 Weights after the completion of training

a two-dimensional input vector x. Thus,

$$m = 2$$

$$x = [x_1, x_2]^T$$

$$w_{i,j} = [w_{i,j,1}, w_{i,j,2}]^T \quad i = 1, 2, \ldots, 10 \quad j = 1, 2, \ldots, 10$$

The weights are initialized from a random set ($-0.08 < w_{i,j,1} < 0.08$ and $-0.08 < w_{i,j,2} < 0.08$). The input x is uniformly distributed in the region ($0 <$

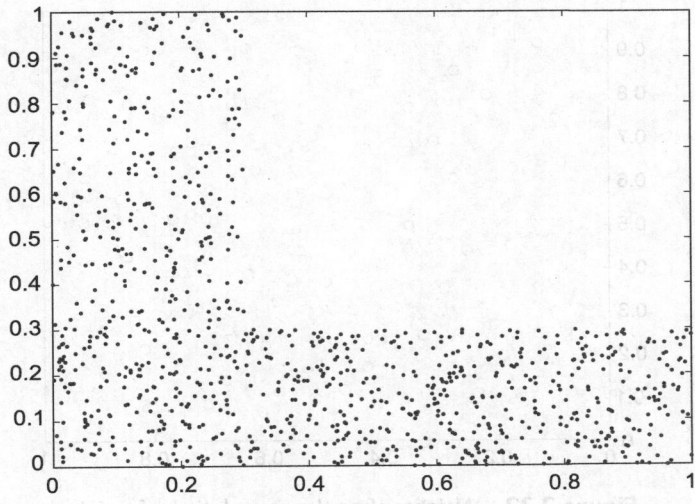

Figure 2.30 Input space for training

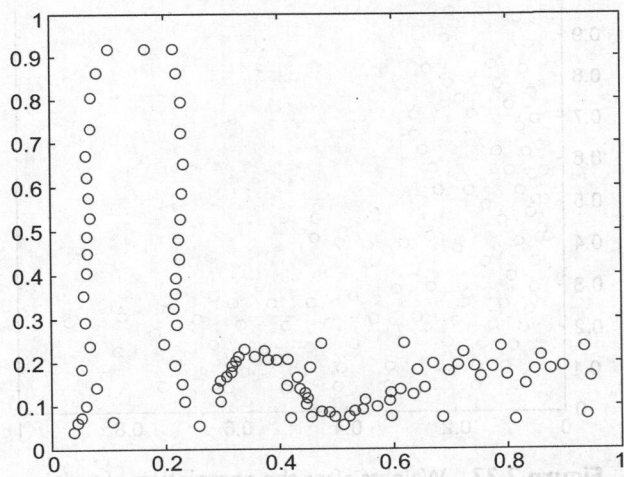

Figure 2.31 Weights after the completion of training

$x_1 < 1$ and $0 < x_2 < 1$). Figure 2.27 shows the input data space. Figure 2.28 shows the network weights before training and Figure 2.32 shows that the weights of the network preserve the topology of the input space.

Next we will show the topology preservation of an L shaped input space with the same two-dimensional lattice. Figure 2.30 shows the input data space. Figure 2.28 shows the network weights before training and Figure 2.33 shows that the weights of the network preserve the topology of the input space.

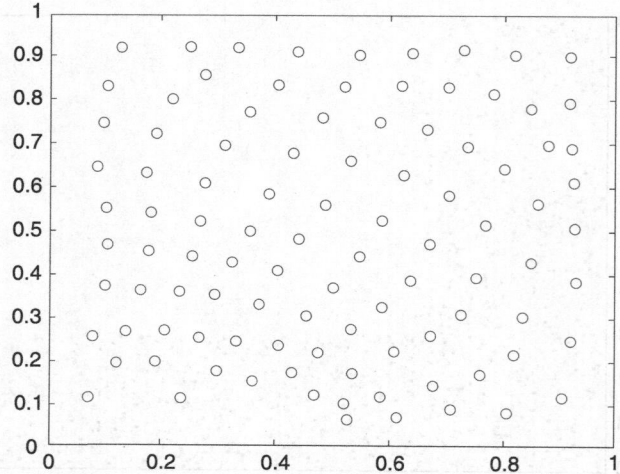

Figure 2.32   Weights after the completion of training

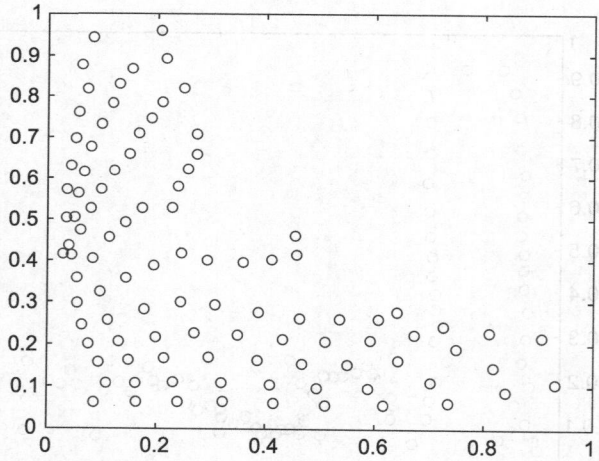

Figure 2.33   Weights after the completion of training

2.7   SYSTEM IDENTIFICATION USING NEURAL NETWORKS

In control, the mathematical model of the plant dynamics is derived using the knowledge of physics, chemistry, or biology that governs the plant dynamics. The model needs to be close enough to the actual dynamics so that the controller designed based on the mathematical model can guarantee stability of the actual system and can provide good performance. However, it is difficult to derive accurate mathematical models for systems such as chemical processes, a robot operating in an unstructured environment, moving aircraft being subjected to uncertain forces, and visually guided manipulators. In such situations, the designer takes help of the experimental data directly conducted to observe the input–output response of the plant. The process of deriving models from experimental data is called *system identification*. With the coming of fast and powerful computing processors, system identification of non-linear plants has become easier.

System model representations have been discussed in Chapter 1. It is convenient to model systems using discrete time representation using feed-forward networks. Although various forms of discrete time models are available [53], non-affine and input affine forms are two typical representations. The non-affine form is given as

$$x(k + 1) = f(x(k), u(k)) \tag{2.76}$$

where $u(\cdot) \in R^m$ and $x(\cdot) \in R^n$ are the input and state vectors, respectively. When the function f in (2.76) is unknown, the problem of system identification can be formally stated as

Suppose that the dynamics of a causal, time invariant discrete time plant is described by Eqn (2.76), where the input $u(\cdot)$ is a uniformly bounded signal. The plant is assumed to be stable. Then a feed-forward network $N(\cdot)$ approximately represents the function f if $N(\cdot)$ predicts an output $\hat{x}(k + 1)$, given the input sequence $u(k)$ and previous system state $x(k)$ such that $\sum_{k=1}^{P} \|x(k + 1) - \hat{x}(k + 1)\| \leq \epsilon$ for a small desired $\epsilon > 0$ where P is the number of discrete time instants.

Figure 2.34 shows a learning framework for generating a neural system model of (2.76). We can select either an MLN or an RBF network for neural representation.

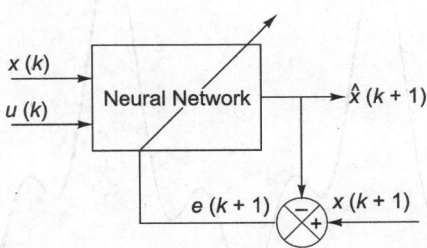

Figure 2.34 Non-affine system model using feed-forward networks

When the system is in the input affine form, the governing equation becomes

$$x(k + 1) = f(x(k)) + g(x(k))u(k) \tag{2.77}$$

The system in (2.77) can be approximated using two neural networks as shown in Figure 2.35.

Example 2.10 The plant is described by the following difference equation:

$$x(k + 1) = 0.3x(k) + 0.6x(k - 1) + g[u(k)] \tag{2.78}$$

where g has a form $g(u) = 0.6 \sin(\iota u) + 0.3 \sin(3\pi u) + 0.1 \sin(5\pi u)$. The unforced linear system is asymptotically stable. Derive a neural model using a series–parallel representation

$$\hat{x}(k + 1) = 0.3x(k) + 0.6x(k - 1) + N[u(k)] \tag{2.79}$$

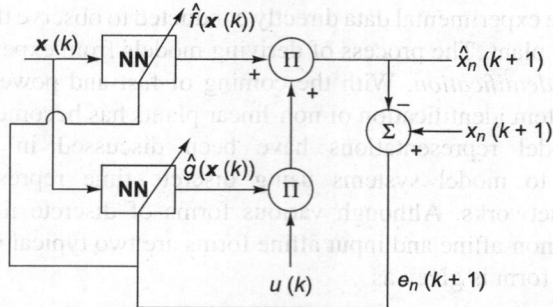

Figure 2.35 Affine system model using feed-forward networks

where the unknown function *g* is approximated by a neural network N.

Solution

The neural network is a three layered network with two hidden layers of 20 and 10 neurons, respectively. The weights of the neural network N are adjusted using the instantaneous back propagation algorithm with a learning rate of 0.25. The input is randomly chosen from the uniform interval $[-1, \ 1]$. The identification is carried out for 500 time steps and the result is shown in Figure 2.36. The RMS error between the output of the model and the plant is found to be 0.003.

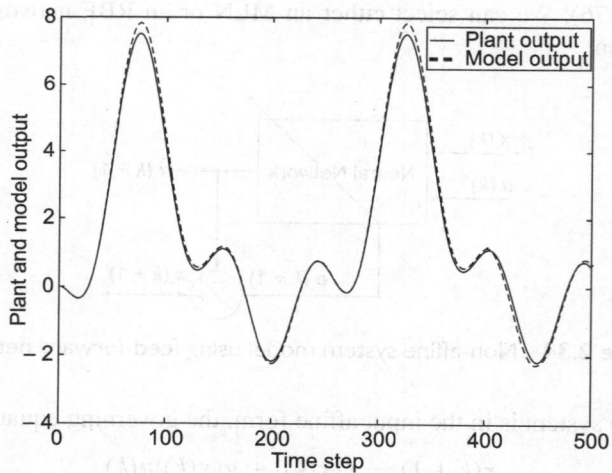

Figure 2.36 Response of the plant and the neural model after training

Example 2.11 Consider a MIMO plant which is governed by the following difference equation:

$$
\begin{bmatrix} x_1(k+1) \\ x_2(k+1) \end{bmatrix} = \begin{bmatrix} \frac{x_1(k)}{1+x_2^2(k)} \\ \frac{x_1(k)x_2(k)}{1+x_2^2(k)} \end{bmatrix} + \begin{bmatrix} u_1(k) \\ u_2(k) \end{bmatrix}
\tag{2.80}
$$

Derive a neural model using the following series–parallel representation:

$$\begin{bmatrix} \hat{x}_1(k+1) \\ \hat{x}_2(k+1) \end{bmatrix} = \begin{bmatrix} N^1[x_1(k), x_2(k)] \\ N^2[x_1(k), x_2(k)] \end{bmatrix} + \begin{bmatrix} u_1(k) \\ u_2(k) \end{bmatrix} \tag{2.81}$$

Solution
The identification is carried out for random inputs u_1 and u_2 which are uniformly distributed in the interval $[-1, 1]$. N^1 and N^2 correspond to two three-layered networks with hidden units of 15 and 10, respectively. The learning rate is taken as 0.1. Figure 2.37 shows the identification results.

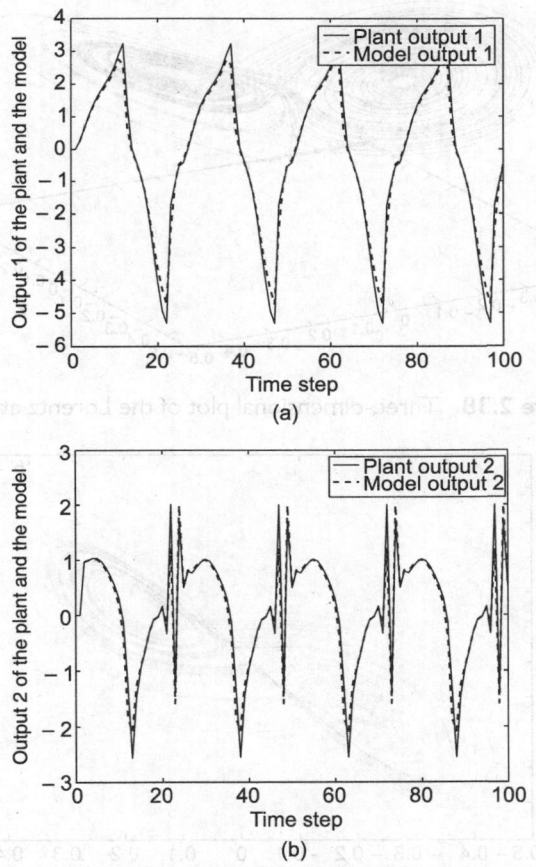

Figure 2.37 Responses of the plant and the identification model

Example 2.12 Lorentz attractor problem
The Lorentz attractor refers to the solutions of the system of differential equations as proposed by Lorentz for modelling the motion of convective fluids.

The differential equations are described as follows:

$$\left.\begin{array}{l} \dfrac{\partial x}{\partial t} = \sigma(y - x) \\[2mm] \dfrac{\partial y}{\partial t} = x(r - z) - y \\[2mm] \dfrac{\partial z}{\partial t} = xy - bz \end{array}\right\} \tag{2.82}$$

The identification is carried out for random initial value and w which are uniformly distributed in the interval $[-1, 1]$. A^1 and A^2 correspond to two three-layered networks with hidden units of 15 and 10, respectively. The learning rate is taken as 0.1. Figure 2.37 shows the identification results.

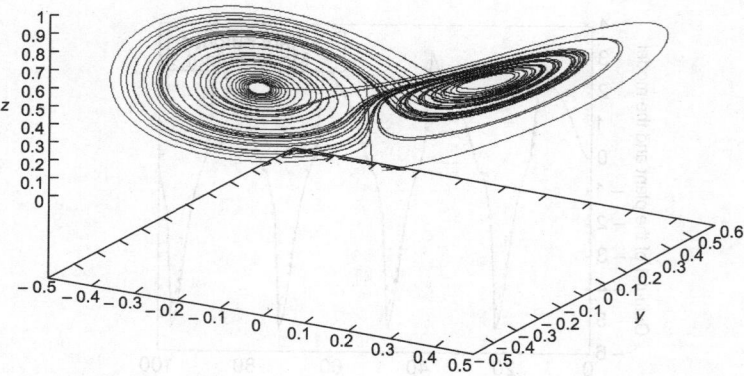

Figure 2.38 Three-dimensional plot of the Lorentz attractor

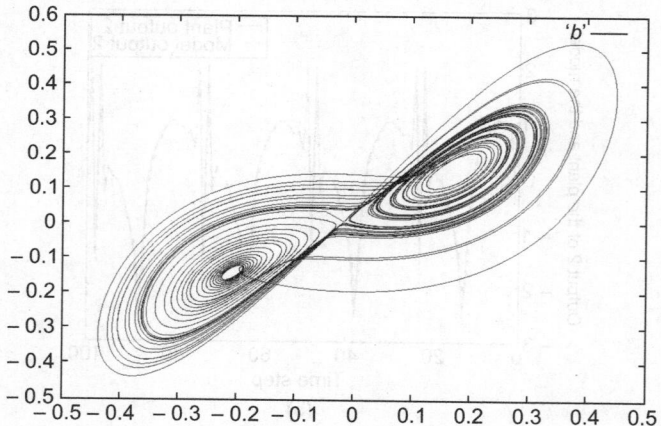

Figure 2.39 The x-y projection of the Lorentz attractor

The parameters of this system of equations are σ , r, and b which are fixed for a specific problem. For our study, the values of the parameters are taken as $\sigma = 16$, $r = 45.92$, and $b = 4$. The solutions of this system of differential equations cannot be found by methods of the analytic integration and hence the solutions are found by the method of Runge–Kutta fourth order ODE solver algorithm which numerically solves for the solution with a good amount of accuracy.

The solution of these differential equations traces out a path in the three-dimensional phase space. The three dimensional plot of the Lorentz attractor set solved by the Runge–Kutta method is shown in Figure 2.38. A plot of the projection on the x-y plane is shown in Figure 2.39. Similarly, a plot of the Lorentz attractor set on the x-axis is also shown in Figure 2.40. It shows the variation of the x-signal with time. All the three plots are obtained for a set of 30, 000 data points, numerically calculated using the fourth-order Runge–Kutta differential solving model.

In the following sections, we shall model this attractor using modified forms of the radial basis function network. This network is more or less completely similar to the usual radial basis function network described earlier in this chapter except for the learning rule used for the centres. The centre learning in this network is modified to accommodate the clustering algorithm, inspired from the k-means clustering algorithm based on the philosophy 'winner takes all'. Here a similar method is applied, but a 'winner takes most' paradigm is used. Another change from the usual clustering algorithm is that here the clustering is done in a higher dimensional space. The use of the input–output clubbed for clustering is not a new idea. It has been previously researched and its benefits in the RBF networks were reported in [54]. The input space and the output space are clubbed together and the learning algorithm is applied in this higher dimensional space and later the projection of the learned vectors on the input space is taken as the centres. Therefore, the locations of the centres are not only influenced by the input vectors but also by the output sample deviations. The learning in this network is done in two stages. In the first stage, only centre learning is done. We start by placing some random centres in the clubbed input–output space. Then a clustering algorithm is used to modify their positions such that the input–output vectors are *equally clustered* among the centres of the neurons. After this the projection of the clubbed input–output vectors is taken on the input space, which provides the network with the required centres. Then a gradient–descent algorithm is applied to learn the weight vectors.

For testing the performance of the radial basis function network, a three-input and three-output neural network was used to first train and then predict the Lorentz attractor. This network consisted of 700 centres. The values of σ (the neighbourhood width) and η (the learning rate) were taken as 0.3 and 0.25, respectively. The training set consisted of 10, 000 data points which were randomly given to the network. After training, the network was used to predict 300, 000 data points of the Lorentz attractor. The results of this are shown in figures here. Figure 2.41 denotes the x-signal with time. This figure can be compared with Figure 2.40 to make a qualitative judgment of the effectiveness of the network. The mean square error produced in the x-signal estimation is 0.0791083, for the y-signal it is 0.062171, and for the z-signal it is 0.0588676, while the total error is 0.116514 . It is seen that this network architecture has been successful in capturing the dynamics of the Lorentz attractor to some extent, but

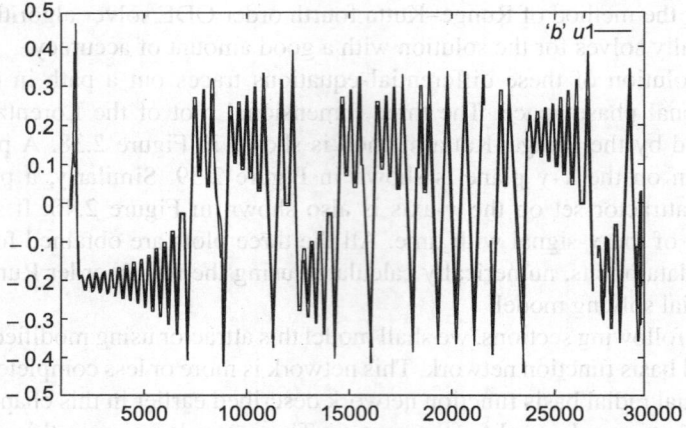

Figure 2.40 Plot of the x-signal of the actual Lorentz attractor with time

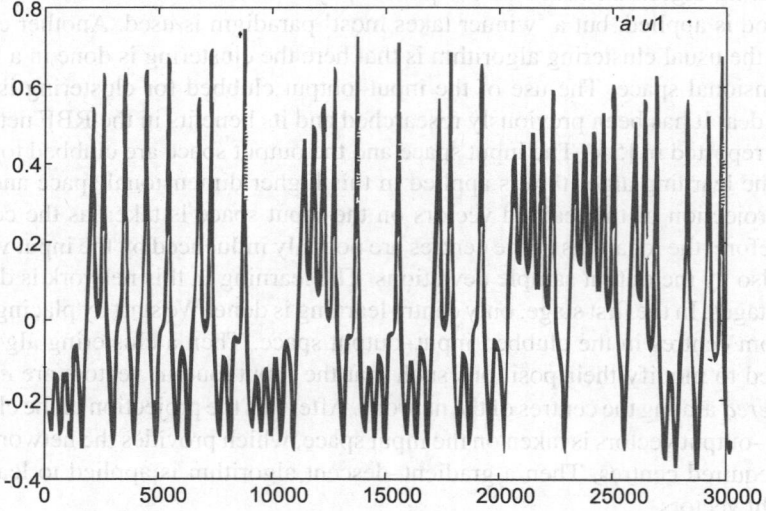

Figure 2.41 Plot of the x-signal coming from the trained RBFN

certainly it does leave scope for improvement. This network demonstrates the power of the general class of radial basis neural networks.

The RBFN performance can be improved if the clustering is done in a higher dimensional space. The use of the input–output clubbed for clustering has been previously researched and its benefits in the RBF networks were reported in [54]. The input space are the output space and clubbed together and the learning algorithm is applied in this higher dimensional space and later the projection of the learnt vectors on the input space is taken as the centres. Therefore, the locations of the centres are not only influenced by the input-vectors but also by the output sample deviations. The learning in this network is done in two stages. In the first stage, only centre learning is done. We start by placing some random centres in the clubbed input–output space. Then a clustering algorithm

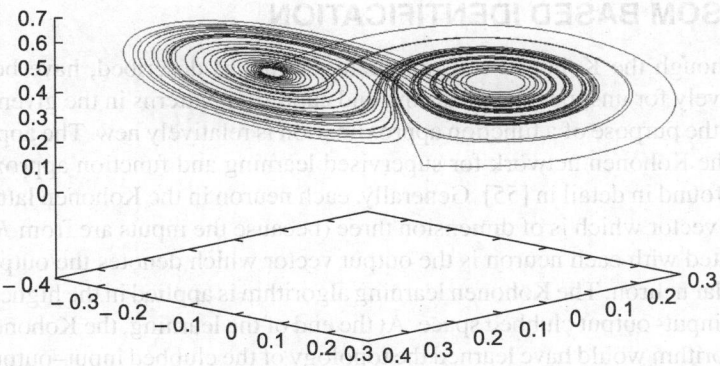

Figure 2.42 Three-dimensional plot of the predicted Lorentz system using RBFN with higher order clustering

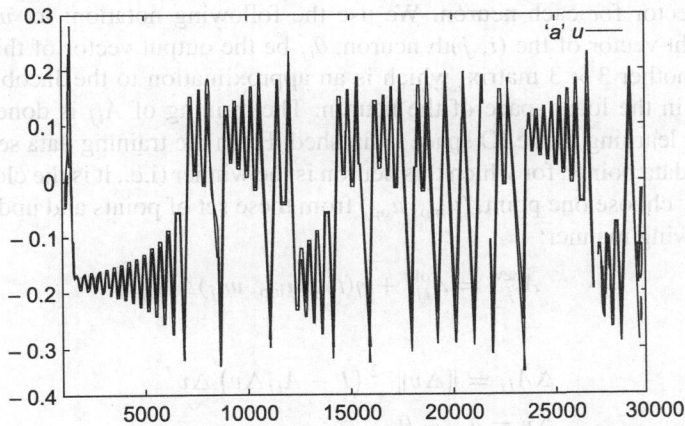

Figure 2.43 The x-signal plot of the predicted Lorentz system using RBFN with higher order clustering

is used to modify their positions such that the input–output vectors are *equally clustered* among the centres of the neurons. After this the projection of the clubbed input–output vectors is taken in the input space, which provides the network with the required centres. Then a gradient–descent algorithm is applied to learn the weight vectors. The number of centres in the network was 500 and the values of σ (the neighbourhood width) and η (the learning rate) were taken as 1.5 and 0.30, respectively. Here also a training set of 10,000 data points was given to the network randomly. After training, the network was used to predict 300, 000 data points of the Lorentz attractor given the same initial points. The results of this are shown in the plots. Figure 2.42 shows the three-dimensional plot of the predicted Lorentz attractor after the training was complete. Similarly Figure 2.43 shows the behaviour of the x-signal obtained from the RBFN using higher order clustering. The mean square errors(MSE) for this network were as follows: MSE in the x-signal = 0.0632505, MSE in the y-signal = 0.0818714, MSE in z-signal = 0.043263, while the total MSE = 0.0112162.

2.8 SOM BASED IDENTIFICATION

Even though the Kohonen self organizing maps, as described, have been used extensively for unsupervised learning and capturing patterns in the given data, its use for the purpose of a function approximation is relatively new. The approach of using the Kohonen network for supervised learning and function approximation can be found in detail in [55]. Generally, each neuron in the Kohonen lattice has a weight vector which is of dimension three (because the inputs are from R^3). Also associated with each neuron is the output vector which denotes the output of the particular neuron. The Kohonen learning algorithm is applied in the higher dimensional, input–output clubbed space. At the end of the learning, the Kohonen learning algorithm would have learned the topology of the clubbed input–output space.

Thus, we can now define an output vector for each neuron in the final Kohonen lattice [55]. The projection of the clubbed vectors on the input space gives the weight vector, while the projection on the output space will give the associated output vector for each neuron. We use the following notation: let w_{ij} denote the weight vector of the (i, j)th neuron, θ_{ij} be the output vector of this neuron, A_{ij} be another 3×3 matrix, which is an approximation to the Jacobian of the function in the local space of the neuron. The training of A_{ij} is done after the Kohonen learning in the IO space is finished. From the training data set, we find all those data points, for which this neuron is the winner (i.e., it is the closest). We randomly choose one point, $(u_{\text{inp}}, u_{\text{op}})$ from these set of points and update A_{ij} in the following manner:

$$A_{ij}^{\text{new}} = A_{ij}^{\text{old}} + \eta(t)g(u_{\text{inp}}, w_{ij})\Delta A_{ij} \tag{2.83}$$

where

$$\Delta A_{ij} = \|\Delta v\|^{-2} \left(I - A_{ij}\Delta v\right)\Delta v^T$$

$$\Delta v = u_{\text{op}} - \theta_{ij}$$

$$\eta = \text{learning rate}$$

$$g(u_{\text{inp}}, w_{ij}) = \exp\left(-\frac{\|u_{\text{inp}} - w_{ij}\|^2}{2\sigma^2}\right)$$

$$I = 3 \times 3 \text{ Identity matrix}$$

Therefore, for any given input u, the corresponding output u_{ij} from each neuron is given by

$$u_{ij} = \theta_{ij} + A_{ij}\left(u - w_{ij}\right) \tag{2.84}$$

We find the final output of the whole network by doing a weighted average over all the outputs from various neurons. The weight depends the distance of the neuron to the 'winner' neuron (corresponding to this input). Thus, we get

$$u_{\text{out}} = \frac{\sum_{i,j} h(i, j, i^0, j^0)w_{ij}}{\sum_{i,j} h(i, j, i^0, j^0)} \tag{2.85}$$

where

$$h(i, i^0, t) = \exp(\frac{-||(i, j) - (i^0, j^0)||^2}{2\sigma(t)^2})$$

Here, (i^0, j^0) represent the 'winner' neuron to the input u. In this particular problem $\sigma(t)$ and $\eta(t)$ are both high initially and then they are decreased to very low positive value by Eqn 2.86.

$$\epsilon(t) = \epsilon_{initial} \left(\frac{\epsilon_{final}}{\epsilon_{initial}} \right)^{\frac{t}{t_{max}}} \tag{2.86}$$

where $\epsilon \in \{\sigma, \eta\}$.

Two-dimensional Kohonen Lattice

The use of the hybrid structure of the Kohonen SOM as a function approximation technique is described in Section 2.8. We will use this approach to model the Lorentz attractor as given in Example 2.12.

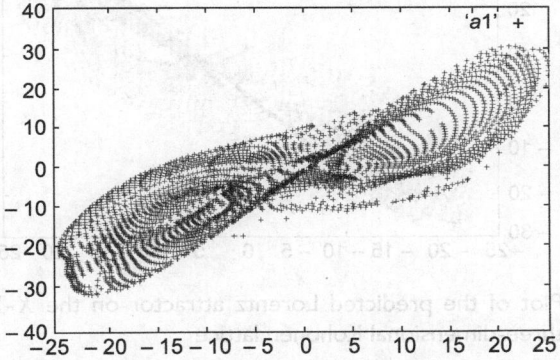

Figure 2.44 Plot of the predicted Lorentz attractor on the X-Y plane using two-dimensional Kohonen lattice

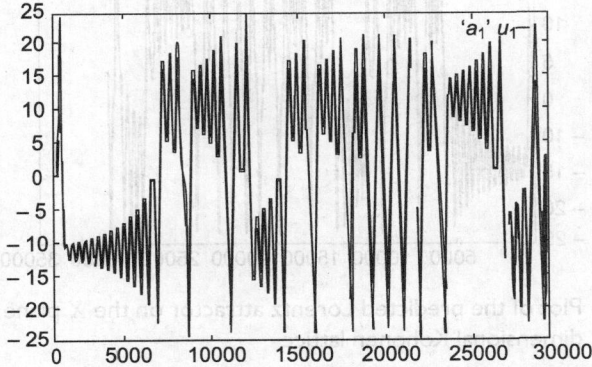

Figure 2.45 Plot of the x-signal of the predicted Lorentz attractor using two-dimensional Kohonen lattice

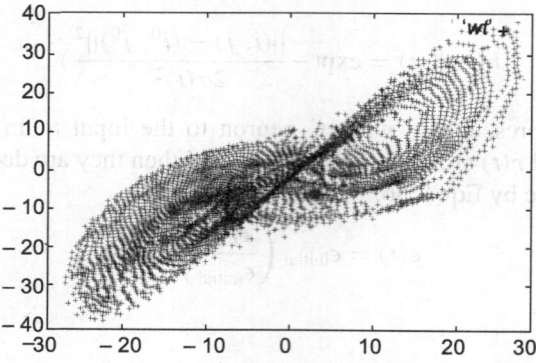

Figure 2.46 Plot of the weight vectors in the input space after learning

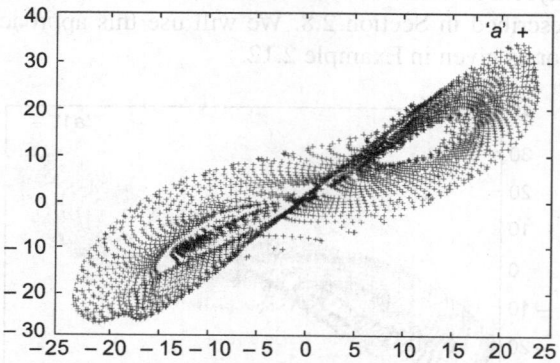

Figure 2.47 Plot of the predicted Lorentz attractor on the X-Y plane using a three-dimensional Kohonen lattice

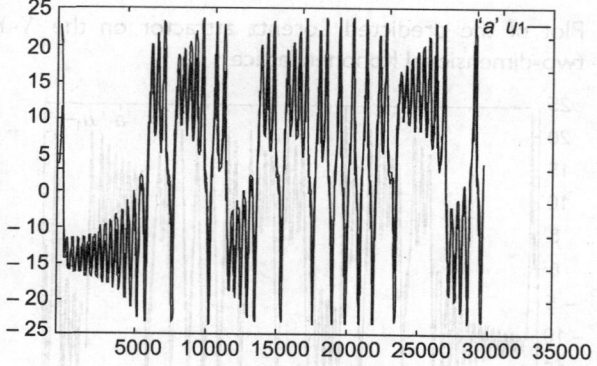

Figure 2.48 Plot of the predicted Lorentz attractor on the X plane using a three-dimensional Kohonen lattice

A two-dimensional network is considered for the purpose which consists of a 70×70 lattice and the values of σ_{initial}, σ_{final}, η_{initial}, and η_{final} are taken as 40, 0.01, 0.90, and 0.01, respectively. The network is trained over a training set of 10, 000 data points. Then this network is used to predict the Lorentz attractor for 30, 000 data points. The results obtained for this network are shown in Figures 2.43 and 2.44. From these plots the topological preserving nature of the Kohonen lattice is very well seen, in spite of the fact that this learning was done in the higher dimensional space after clubbing the input–output vectors. The next plot (Figure 2.46) shows the positioning of the weight vectors of the neurons of the two-dimensional Kohonen lattice in the input space after learning. The topology preservation is very clearly seen which indicates that the network has efficiently learnt the data and has effectively captured the dynamics of the time series.

Three-dimensional Kohonen Lattice

This network is very similar to that of the network described in Section 2.8. The

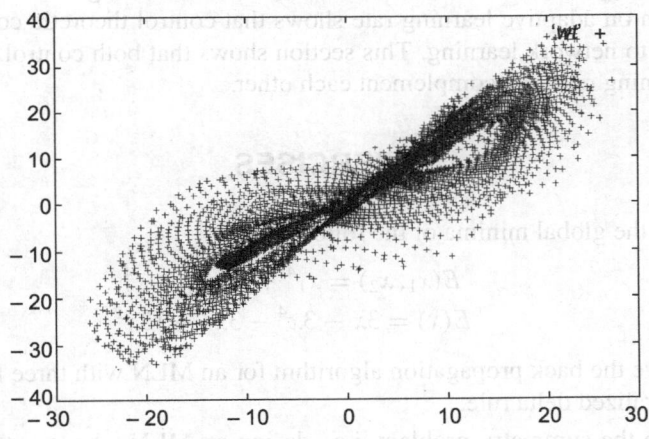

Figure 2.49 Plot of the weight vectors in the $X-Y$ plane using a three-dimensional Kohonen lattice

only difference between this network and the one above is in the dimension of the Kohonen lattice. In this case a three-dimensional lattice is used instead of a two-dimensional one. The output vectors are also learned by the use of a higher order clustering algorithm, i.e., the input and output vectors are clubbed and then the Kohonen learning rule is applied to these vectors as described earlier. The network consists of a $15 \times 12 \times 12$ lattice and the values of σ_{initial}, σ_{final}, η_{initial}, and η_{final} are taken as 20, 0.01, 0.90, and 0.005, respectively. Once the training is over, the projection of these vectors on the input space becomes the weight vectors of the neurons, while the projected vectors on the output space form the output of the respective neurons. The network is trained over a training set of 10, 000 data points. Then this network is used to predict the Lorentz attractor for 30, 000 data points. It is intuitively felt that a three-dimensional Kohonen lattice should perform better in predicting and capturing the dynamics of the Lorentz attractor. This is indeed the case as demonstrated by the results obtained. The plots of the obtained results are shown in Figures 2.47, 2.48, and 2.49. Readers should be able to appreciate the use of an SOM network in system identification as these results show that even chaotic attractors can be modelled using such networks.

SUMMARY

This chapter on neural networks is self-sufficient for the readers to grasp the intelligent control concepts using neural networks covered in this book. Network architectures and the associated learning algorithms for multi-layered networks, radial basis function networks, recurrent networks, and SOM networks have been presented in a tutorial manner in this chapter. The use of these networks for function approximations has been described through many illustrative examples. The section on system identification allows readers to understand the process of function approximation for non-linear systems including a chaotic system. The section on adaptive learning rate shows that control theoretic concepts can be applied to network learning. This section shows that both control theory and neural learning research complement each other.

EXERCISES

1. Find the global minima of the following:

$$E(x_1, x_2) = x_1{}^4 + 2x_1^2 x_2^2 + x_2^4$$
$$E(x) = 3x - 3x^4 - 3x^3 + 13x^2$$

2. Derive the back propagation algorithm for an MLN with three layers. Use generalized delta rule.

3. Solve the symmetry problem, i.e., design an MLN whose output will be true (+1) if the input pattern is symmetric about the centre and false (−1) otherwise. Use a two-layered network with six input and two hidden units.

4. Show that the network shown in Figure 2.50 solves the XOR problem by constructing decision regions and a truth table for the network.

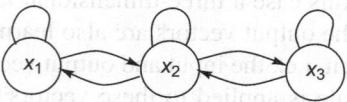

Figure 2.50 Multi-layered networks

5. Can the back propagation learning using a sigmoidal non-linearity be used to achieve the following one-to-one mapping?

   (a) $f(x) = \frac{1}{x}$ $1 \leq x \leq 100$
   (b) $f(x) = \log_{10} x$ $1 \leq x \leq 10$

(c) $f(x) = \sin x$ $1 \le x \le \frac{\pi}{2}$

   Set up two sets of data, one for training and another for testing. Use the training data set to compute the synaptic weights of the network. Evaluate the computation accuracy of the network by using the test data. Use a single hidden layer but a variable number of hidden neurons. Note the effect of the change in the size of the hidden layer on the network performance.

6. The MLN in Figure 2.51 uses a single hidden layer with a sigmoidal activation function.

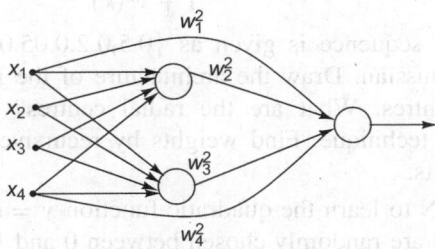

Figure 2.51   Multi-layered networks

   (a) Find out the response of the network using forward pass.
   (b) Find out the update law for the weights w_1^2 and w_4^2.

7. Construct an RBFN with four hidden units, with each radial basis function centre being determined by each piece of the input data, to obtain an exact solution to the XOR problem. The four possible input patterns are defined by (0,0), (0.1), (1,1), and (1,0), which represent the cyclically ordered corners of a square.

   (a) Construct the interpolation matrix for the resulting RBFN. Hence, compute the inverse matrix Φ^{-1}.
   (b) Calculate the linear weights of the output layer of the network.

8. Design an RBFN for the following set of training examples $(x \rightarrow y)$:

$$x^1 = [-7 \; +2]^T \quad y^1 = +1$$
$$x^2 = [-4 \; -2]^T \quad y^2 = -1$$

   Use the radial basis function $\Phi(r) = (r^2 + 9)^{0.5}$.

   (Hint: Fix the number of centres and the corresponding centroid. No centre updating is required. Only find the value of the weight vector.)

9. For an RBFN, find the update rule for the centres using the gradient–descent technique while the basis function is assumed to be inverse quadratic, i.e.,

$$\Phi(r) = \frac{1}{(r^2 + c^2)^{\frac{1}{2}}}$$

10. Realize an XOR function using a radial basis function network whose basis function is given by $\Psi(r) = r^2 log(r)$.

11. An RBFN has five radial centres. It has one input u and one output y. The weight vector is denoted by W. The centre of each unit is denoted by c_i. The radial basis function is Gaussian.
    (a) Draw the architecture of the network.
    (b) Write the response y of the network in terms of u, c, and W.
    (c) Derive the expression dy/du for the network.

12. Generate five sets of training data using the following model:

$$y(k+1) = \frac{y(k)}{1+y^2(k)}u^3(k)$$

    The input sequence is given as $\{0.5, 0.2, 0.05, 0.8, 0.35\}$. The radial function is Gaussian. Draw the architecture of the RBFN network with five radial centres. What are the radial centres? Find weights using peudo-inverse technique. Find weights by recursive gradient algorithm. Compare results.

13. Train an RBFN to learn the quadratic function $y = u^2$. The network has 50 centres and are randomly chosen between 0 and 1. The input range is [0, 1]. The radial function used is Gaussian type $\Phi(r) = e^{-r^2}$
    (a) Find the response of the network when the input takes a value from the set [0.2, 0.5, 0.6, 0.9]. Does the result indicate that the network has learned above mapping.
    (b) One of the ways to understand the proper training is to test for both forward mapping (response to a given input) and inverse mapping (predict input for a given response). For inverse mapping, the iterative algorithm is given as

$$u(t+1) = u(t) - \eta\frac{\partial E(t)}{\partial u(t)}$$

    (i) Find the explicit expression $\partial E/\partial u$ for the network given.
    (ii) Using the iterative inversion algorithm predict the input when the desired response takes a value from the set [0.09, 0.25, 0.36, 0.64]. Select the initial value randomly from [0, 1].
    (iii) Repeat above steps while updating both centres and weights.

14. Figure 2.52 shows a recurrent network consisting of three computation nodes with self-feedback applied to all of them. Construct a multi-layer feed forward network by unfolding the temporal behaviour of this network.

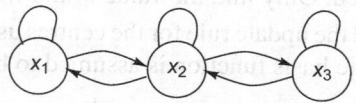

Figure 2.52 A recurrent network

15. Generate a set of random input vectors $u(k)$, $k = 1, 2, \ldots, 200$. Obtain the training data set defined by the relation

$$y(k+1) = \frac{1}{1+y^2(k)}u(k)$$

Given $y(0) = 0$. The network is shown in Figure 2.53. Unfold the network in time up to $t = 200$.

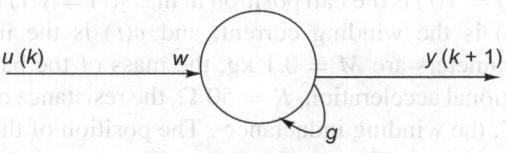

Figure 2.53

16. Unfold the network shown in Figure 2.54 in time.

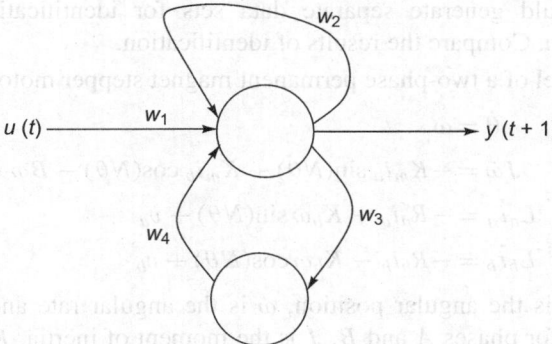

Figure 2.54

17. Consider the dynamics of a single link manipulator

$$ml^2\ddot{\theta} + mgl\cos\theta = \tau$$

where θ is the manipulator angle, $\dot{\theta}$ is the angular velocity, and τ is the joint angle torque. Taking the system parameters as $m = 1$ kg, $l = 1$ m, $g = 10$ m/s^2, discretize the state–space model of the system using the Euler method. Take the sampling interval T as 0.001 s. Generate 3000 pairs of the input–output data and identify the system using a
   • Multi-layered network
   • Radial basis function network
Since the system is open-loop unstable, a PD controller with random sinusoidal trajectories as the desired trajectories can be used for data generation. Various dither signals like noise, impulse, and step can be added to the PD controller output so that the generated data span almost the entire workspace of the system.
Generate 1000 pairs of new test data and verify the identification result.

18. The state–space model of a magnetic ball suspension is given by

$$\frac{dx_1(t)}{dt} = x_2(t)$$

$$\frac{dx_2(t)}{dt} = g - \frac{x_3^2(t)}{Mx_1(t)}$$

$$\frac{dx_3(t)}{dt} = -\frac{R}{L}x_3(t) + \frac{1}{L}v(t)$$

where $x_1(t) = y(t)$ is the ball position in m, $x_2(t) = \dot{y}(t)$ is the ball velocity, $x_3(t) = i(t)$ is the winding current, and $v(t)$ is the input voltage. The system parameters are $M = 0.1$ kg, the mass of the ball, $g = 9.81$ m/s^2, the gravitational acceleration, $R = 50\ \Omega$, the resistance of the winding, and $L = 0.5\ H$, the winding inductance . The position of the ball is measured by a position sensor. Express the system in a discrete dynamical form and identify the dynamics of the system using

- Multi-layered network
- Radial basis function network

You should generate separate data sets for identification and model validation. Compare the results of identification.

19. The model of a two-phase permanent magnet stepper motor is given by

$$\dot{\theta} = \omega$$

$$J\dot{\omega} = -K_m i_a \sin(N\theta) + K_m i_b \cos(N\theta) - B\omega - T_L$$

$$L_a \dot{i}_a = -R_a i_a + K_b \omega \sin(N\theta) + v_a$$

$$L_b \dot{i}_b = -R_b i_b - K_b \omega \cos(N\theta) + v_b$$

where θ is the angular position, ω is the angular rate and i_a, i_b are the currents for phases A and B, J is the moment of inertia, K_m is the motor torque constant, N is the number of teeth on the rotor, B is the viscous friction coefficient, L_a, L_b, and R_a, R_b are the inductances and resistances of phases A and B, K_b is the back emf constant, and T_L is the load torque. Design a control scheme for voltages v_a and v_b such that θ follows a desired trajectory $r(t)$. Model the load torque for a mechanical link of length l and mass m as $mgl\sin(\theta)$ where m is the mass of the shaft.

The system parameters are below:

$J = 0.0733$ kg-m^2, $B = 0.002$ N m s/rad

$m = 0.4014$ kg, $N = 50$, $l = 1$ m

$L_a = L_b = 0.7$ H, $R_a = 0.9\ \Omega$, $R_b = 1.2\ \Omega$

$K_m = 0.25$ N m/A, $K_b = 0.025$ V-sec/rad, $g = 9.81$ m/s^2

Find out the discrete time representation of the system and identify the dynamics using a radial basis function network.

20. Consider the following discrete time representation of a surge tank system:

$$h(k+1) = h(k) + T\frac{-\sqrt{2gh(k)}}{A(h(k))} + \frac{1}{A(h(k))}u(k)$$

where $u(k)$ is the input flow, $h(k)$ is the liquid level of the tank (output of the system), and $A(h(k)) = \sqrt{ah^2(k) + b}$ is the cross-sectional area of the tank at the kth instant. The input $u(k)$ can be both positive or negative, i.e., it can pull liquid out of the tank as well as put it in the tank. The parameters a and b are given as $a = 1$ and $b = 2$. Considering $T = 0.1$, identify the system dynamics using a recurrent network. Employ both BPTT and RTRL for learning the network.

21. The discrete dynamical representation of a Henon map is given as follows:

$$x_1(k + 1) = -1.4\, x_1^2(k) + x_2(k) + u(k) + 1$$
$$x_2(k + 1) = 0.3\, x_2(k)$$

Identify the above dynamics using a self-organizing map.

21. The discrete dynamics

CHAPTER

3

Fuzzy Logic

Fuzzy logic [56, 57] is an extension of the propositional logic from two perspectives. First, instead of the binary valuation space (truth/falsehood) of the propositional logic, the fuzzy logic provides a multi-valued truth space in [0,1]. Secondly, the propositional logic generates inferences based on complete matching of the antecedent clauses with the available data, whereas the fuzzy logic is capable of generating inferences even when a partial matching occurs. The fuzzy logic provides a simple way to arrive at a definite conclusion based on the vague, ambiguous, imprecise, noisy, or missing input information.

3.1 CLASSICAL SETS

Classical sets are characterized by definite or crisp boundaries, i.e., there is no uncertainty involved in the location of the boundaries for these sets. On the other hand, a fuzzy set is defined by its vague and ambiguous properties and hence the boundaries are specified ambiguously. The crisp sets are the sets without ambiguity in their membership. Let us revisit a few definitions from the classical set theory.

Definition 3.1 Set: A set is a collection of the objects that have one or more common characteristics.

Definition 3.2 Members/elements: The objects belonging to a set are called the members or elements of that set and are represented as $x \in A$, where A is a set.

Figure 3.1 shows two sets of natural numbers and real numbers.

Definition 3.3 Subset: A set B is said to be a subset of set A if and only if $y \in B \Rightarrow y \in A, \forall y$. This is represented as $B \subseteq A$.

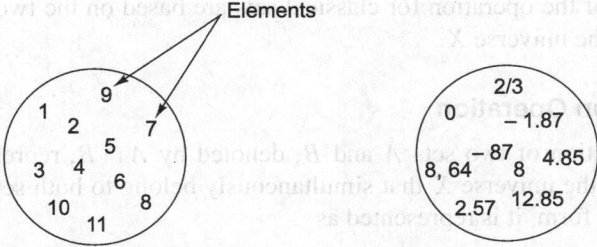

Figure 3.1   Pictorial representation of a set

Definition 3.4 Proper subset: A set B is said to be a proper subset of set A if and only if B is a subset of A and $\exists x \in A$ such that $x \notin B$.

Figure 3.2 shows two sets A and B, where B is a proper subset of A.

Figure 3.2   Proper subset

Definition 3.5 Equal sets: Two sets A and B are said to be equal if and only if $\forall x \in A$ and $\forall y \in B$, $x = y$.

Figure 3.3 shows two sets A and B, which are equal.

Figure 3.3   Equal sets

Definition 3.6 Null set: If a set ϕ does not contain any element in it, it is called a null set.

Definition 3.7 Universal set: A universal set X is a set that has all possible members of a particular domain.

3.1.1   Operations on Classical Sets

There are various operations that can be performed on the classical or crisp sets. The results of the operations performed on the classical sets will be definite. The definitions of the operation for classical sets are based on the two sets A and B defined on the universe X.

Intersection Operation

The intersection of two sets A and B, denoted by $A \cap B$, represents all those elements in the universe X that simultaneously belong to both sets A and B. In set theoretic form, it is represented as

$$A \cap B = x \mid x \in A \text{ and } x \in B$$

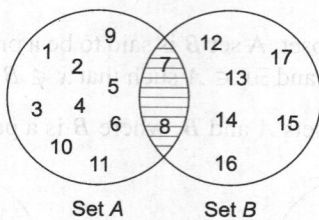

Set A   Set B

Figure 3.4   Intersection of two sets

Figure 3.4 shows the intersection of two sets A and B, which contains the elements 7 and 8.

Union Operation

The union of two sets A and B, denoted by $A \cup B$, represents all those elements in the universe X that belong to either of the sets A and B. It is represented as

$$A \cup B = x \mid x \in A \text{ or } x \in B$$

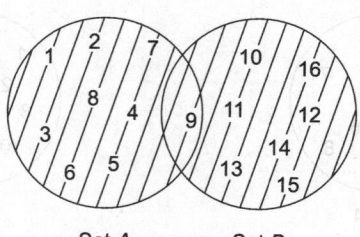

Set A   Set B

Figure 3.5   Union of two sets

Figure 3.5 shows the union of two sets A and B, which contains all the elements of A and B.

Complement

The complement of a set A, denoted by $\overline{A}$, is defined as the collection of all those elements in the universe X that do not belong to the set A. It is represented as

$$\overline{A} = x \mid x \notin A, x \in X$$

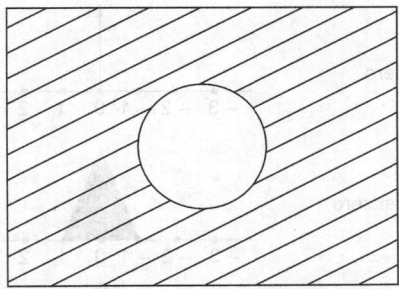

Figure 3.6   Complement of a set

Figure 3.6 shows the pictorial representation of the complement of a set.

Difference

The difference of a set A with respect to another set B, denoted by $A|B$, is defined as the collection of all those elements in the universe X that belong to A but do not belong to B simultaneously. In set theoretic form, it is represented as

$$A|B = x| \; x \in A \text{ and } x \notin B$$

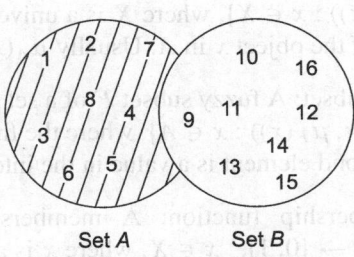

Set A Set B

Figure 3.7   Difference $A|B$

Figure 3.7 shows the difference of two sets A and B.

3.2   FUZZY SETS

The concept of fuzzy sets arose as an answer to the problems of paradoxes, uncertainties, and the absence of precision found in real-world data, which could not be accommodated using crisp sets. The fuzzy set theory is an extension of the classical set theory. The degree of belongingness in the fuzzy sets is defined by the term 'membership function'. The operations on fuzzy sets are defined using the membership functions.

3.2.1   Concept of a Fuzzy Number

A fuzzy number is a quantity whose value is imprecise, rather than exact, as in the case of ordinary numbers. Figure 3.8 shows a pictorial representation of fuzzy numbers where a fuzzy number can be defined as 'almost zero' or 'near zero' depending on the closeness of the number to 0.

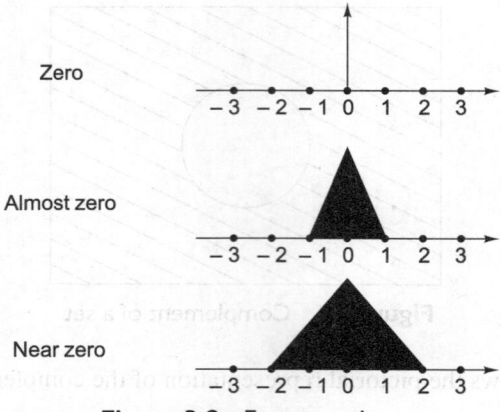

Figure 3.8 Fuzzy number

Every member x of a fuzzy set A is assigned a fuzzy index $\mu_A(x)$ in the interval of $[0, 1]$, which is often called the grade of membership of x in A. In a classical set, the membership grade $\mu_A(x)$ is either 0 or 1. Formally one can define the fuzzy set, subset, and membership functions as follows.

Definition 3.8 Fuzzy set: A fuzzy set can be defined as a set of ordered pairs given by $A = \{(x, \mu_A(x)) : x \in X\}$, where X is a universal set and $\mu_A(x)$ is the grade of membership of the object x in A. Usually $\mu_A(x)$ lies in $[0, 1]$.

Definition 3.9 Fuzzy subset: A fuzzy subset F of a set A can be defined as a set of ordered pairs $F = \{(x, \mu_F(x)) : x \in A\}$, where the first element is an element of the set A and the second element is a value in the interval $[0, 1]$.

Definition 3.10 Membership function: A membership function $\mu_A(x)$ is characterized by $\mu_A : x \to [0, 1], \quad x \in X$, where x is a real number describing an object or its attribute, X is the universe of discourse, and $A \subset X$.

For example, consider a universal set T that denotes temperature. Cold, Normal, and Hot are the subsets of the universal set T. In the classical approach, one can define these subsets as

$$\text{Cold} = \{\text{temp} \in T : 5^\circ \, \text{C} \le \text{temp} < 15^\circ \text{C}\}$$

$$\text{Normal} = \{\text{temp} \in T : 15^\circ \, \text{C} \le \text{temp} < 25^\circ \text{C}\}$$

$$\text{Hot} = \{\text{temp} \in T : 25^\circ \, \text{C} \le \text{temp} < 35^\circ \text{C}\}$$

One should note that 14.9°C is cold while 15.1°C is normal implying that the classical sets have rigid boundaries. In contrast, the fuzzy sets have soft boundaries. One approach to define fuzzy subsets for the universal set T is shown in Figure 3.9. The temperature 15°C is a member of the two fuzzy sets, *Cold* and *Normal* with a membership grade $\mu_T(\text{Cold}) = \mu_T(\text{Normal}) = 0.5$.

Let the elements of a set X be $x_1, x_2, \ldots, x_n$. Then the fuzzy set $A \subseteq X$ is denoted by any of the following types of nomenclature:

1. $A = \{(x_1, \mu_A(x_1)), (x_2, \mu_A(x_2)), \ldots, (x_n, \mu_A(x_n))\}$

2. $A = \left\{ \frac{x_1}{\mu_A(x_1)}, \frac{x_2}{\mu_A(x_2)}, \cdots, \frac{x_n}{\mu_A(x_n)} \right\}$

3. $A = \left\{ \frac{\mu_A(x_1)}{x_1}, \frac{\mu_A(x_2)}{x_2}, \ldots, \frac{\mu_A(x_n)}{x_n} \right\}$

where $\frac{x_1}{\mu_A(x_1)}$ or $\frac{\mu_A(x_1)}{x_1}$ represents a tuple, not a division.

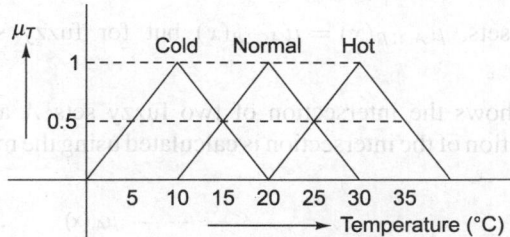

Figure 3.9 Fuzzy membership

3.2.2 Operations on Fuzzy Sets

The operations on fuzzy sets are usually described with reference to the membership functions. Following are the common operations defined on fuzzy sets:

1. Fuzzy complement
2. Fuzzy intersection
3. Fuzzy union

Fuzzy Complement

The membership function $\mu_{\overline{A}}(x)$ of the complement of A (denoted by $\overline{A}$) is defined by

$$\mu_{\overline{A}}(x) = 1 - \mu_A(x) \; \forall x \in X$$

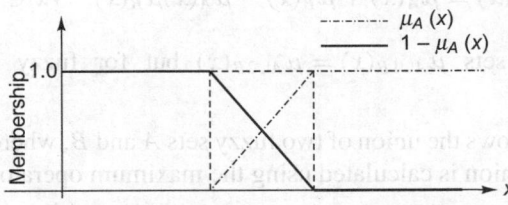

Figure 3.10 Fuzzy complement

Figure 3.10 shows the membership grade μ_A of a set A, where the solid line represents the membership grade $\mu_{\overline{A}}$ of its complement, i.e., $\overline{A}$.

Fuzzy Intersection

The intersection of a fuzzy set A with another fuzzy set B is defined by

$$A \cap B = x | x \in A \wedge x \in B$$

Two most important operators for the intersection are given below.

1. Minimum operator

$$\mu_{A \cap 1 B}(x) = \mu_A(x) \wedge \mu_B(x) = \min\{\mu_A(x), \mu_B(x)\}, \quad \forall x \in X$$

2. Product operator

$$\mu_{A \cap 2 B}(x) = \mu_A(x)\mu_B(x), \quad \forall x \in X$$

For crisp sets, $\mu_{A \cap 1 B}(x) = \mu_{A \cap 2 B}(x)$ but for fuzzy sets, $\mu_{A \cap 1 B}(x) \geq \mu_{A \cap 2 B}(x)$.

Figure 3.11 shows the intersection of two fuzzy sets A and B, where the membership function of the intersection is calculated using the minimum operator.

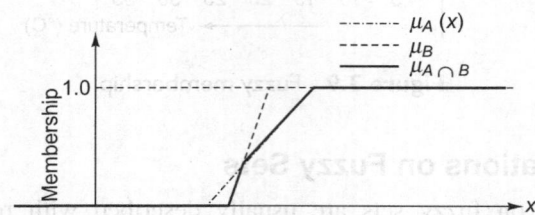

Figure 3.11   Fuzzy intersection: extreme operator

Fuzzy Union

The union of a set A with another set B is defined by

$$A \cup B = x | x \in A \vee x \in B, \quad \forall x \in X$$

Two most important operators for the union are given below:

1. Maximum operator

$$\mu_{A \cup 1 B}(x) = \mu_A(x) \vee \mu_B(x) = \max\{\mu_A(x), \mu_B(x)\}, \quad \forall x \in X$$

2. Sum operator

$$\mu_{A \cup 2 B}(x) = \mu_A(x) + \mu_B(x) - \mu_A(x)\mu_B(x), \quad \forall x \in X$$

For crisp sets $\mu_{A \cup 1 B}(x) = \mu_{A \cup 2 B}(x)$ but for fuzzy sets $\mu_{A \cup 1 B}(x) \leq \mu_{A \cup 2 B}(x)$

Figure 3.12 shows the union of two fuzzy sets A and B, where the membership function of the union is calculated using the maximum operator.

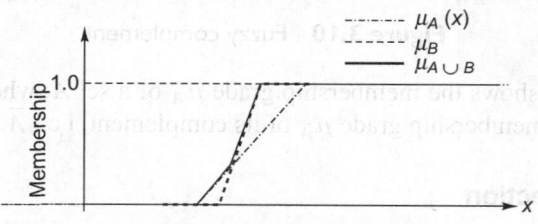

Figure 3.12   Fuzzy union

3.2.3   Other Fuzzy Operations

1. De Morgan's law

$$\overline{A \cap B} = \overline{A} \cup \overline{B}$$
$$\overline{A \cup B} = \overline{A} \cap \overline{B}$$

2. Difference

$$A|B = A \cap \overline{B}$$
$$B|A = B \cap \overline{A}$$

$$\mu_{\overline{A \cup B}} = \min\{1 - \mu_A, 1 - \mu_B\} \quad \text{and} \quad \mu_{A \cap \overline{B}} = \min\{\mu_A, 1 - \mu_B\}$$

3.2.4   Properties of Fuzzy Sets

1. Commutativity: $A \cup B = B \cup A$ and $A \cap B = B \cap A$
2. Associativity: $A \cup (B \cup C) = (A \cup B) \cup C$ and $A \cap (B \cap C)$ $= (A \cap B) \cap C$
3. Distributivity: $A \cup (B \cap C) = (A \cup B) \cap (A \cup C)$ and $A \cap (B \cup C)$ $= (A \cap B) \cup (A \cap C)$
4. Idempotency: $A \cup A = A$ and $A \cap A = A$
5. Identity: $A \cup \emptyset = A$ and $A \cap X = A$, $A \cap \emptyset = \emptyset$ and $A \cup X = X$

3.2.5   Some Typical Membership Functions

The γ-function: The γ-function is defined as follows:

$$\gamma(x; \alpha, \beta) = 0, \qquad x \le \alpha$$
$$= (x - \alpha)/(\beta - \alpha) \qquad \alpha < x \le \beta$$
$$= 1 \qquad x > \beta$$

Pictorial representation of the γ-function is shown in Figure 3.13.

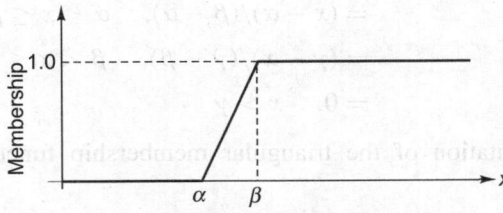

Figure 3.13   The γ-function

s-function: The s-function is defined as follows:

$$s(x; \alpha, \beta, \gamma) = 0, \quad x \le \alpha$$
$$= 2[(x - \alpha)/(\gamma - \alpha)]^2, \quad \alpha < x \le \beta$$
$$= 1 - 2[(x - \gamma)/(\gamma - \alpha)]^2, \quad \beta < x \le \gamma$$
$$= 1, \quad x > \gamma$$

Pictorial representation of the *s*-function is shown in Figure 3.14.

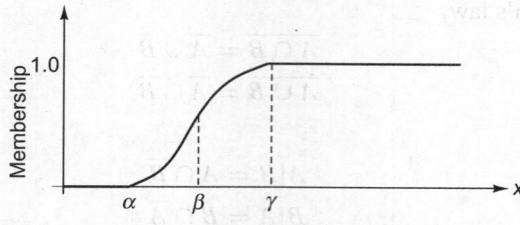

Figure 3.14 The *s*-function

L-function: The *L*-function is defined as follows:

$$L(x; \alpha, \beta) = 1, \quad x < \alpha$$
$$= (\alpha - x)/(\beta - \alpha), \quad \alpha \le x \le \beta$$
$$= 0, \quad x > \beta$$

Pictorial representation of the *L*-function is shown in Figure 3.15.

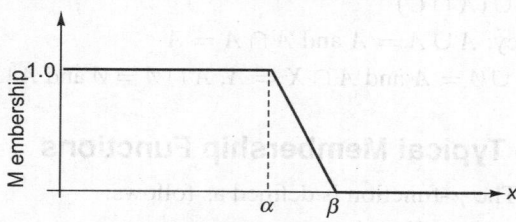

Figure 3.15 *L*-function

Triangular membership function: A triangular membership function is defined as follows:

$$\Lambda(x; \alpha, \beta, \gamma) = 0, \quad x \le \alpha$$
$$= (x - \alpha)/(\beta - \alpha), \quad \alpha < x \le \beta$$
$$= (\gamma - x)/(\gamma - \beta), \quad \beta < x \le \gamma$$
$$= 0, \quad x > \gamma$$

Pictorial representation of the triangular membership function is shown in Figure 3.16.

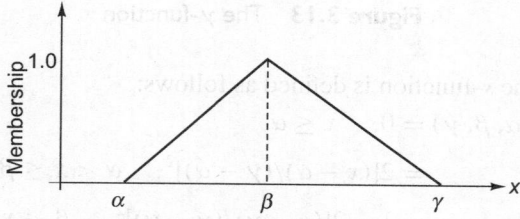

Figure 3.16 A triangular membership function

$\prod$-function: The $\prod$-function is defined as follows:

$$\prod(x; \alpha, \beta, \delta) = 0, \quad x \leq \alpha$$
$$= (x - \alpha)/(\beta - \alpha), \quad \alpha < x \leq \beta$$
$$= 1, \quad \beta < x \leq \gamma$$
$$= (\delta - x)/(\delta - \gamma), \quad \gamma < x \leq \delta$$
$$= 0, \quad x > \delta$$

Pictorial representation of the $\prod$-function is shown in Figure 3.17.

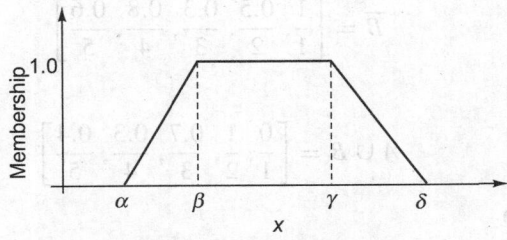

Figure 3.17 The $\prod$-function

Gaussian membership function: The Gaussian membership function is defined as follows:

$$G(x; c, \sigma) = e^{-\frac{(x-c)^2}{\sqrt{(2\sigma)}}}$$

where the parameters c and σ control the centre and width of the membership function. Pictorial representation of the Gaussian membership function is shown in Figure 3.18.

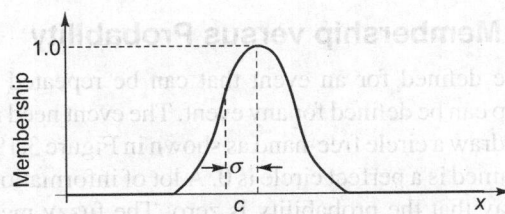

Figure 3.18 The Gaussian function

Discrete Fuzzy Sets

A discrete fuzzy set can be represented as

$$A = \left\{ \frac{\mu_A(x_1)}{x_1}, \frac{\mu_A(x_2)}{x_2}, \ldots, \frac{\mu_A(x_n)}{x_n} \right\}$$

All the discussions made so far are also applicable to the discrete fuzzy sets. Let us consider two fuzzy sets

$$A = \left[\frac{0}{1}, \frac{1}{2}, \frac{0.5}{3}, \frac{0.3}{4}, \frac{0.2}{5} \right]$$

and

$$B = \begin{bmatrix} \dfrac{0}{1}, \dfrac{0.5}{2}, \dfrac{0.7}{3}, \dfrac{0.2}{4}, \dfrac{0.4}{5} \end{bmatrix}$$

We can now evaluate different fuzzy operations as follows:

1. Complement

$$\overline{A} = \begin{bmatrix} \dfrac{1}{1}, \dfrac{0}{2}, \dfrac{0.5}{3}, \dfrac{0.7}{4}, \dfrac{0.8}{5} \end{bmatrix}$$

and

$$\overline{B} = \begin{bmatrix} \dfrac{1}{1}, \dfrac{0.5}{2}, \dfrac{0.3}{3}, \dfrac{0.8}{4}, \dfrac{0.6}{5} \end{bmatrix}$$

2. Union

$$A \cup B = \begin{bmatrix} \dfrac{0}{1}, \dfrac{1}{2}, \dfrac{0.7}{3}, \dfrac{0.3}{4}, \dfrac{0.4}{5} \end{bmatrix}$$

3. Intersection

$$A \cap B = \begin{bmatrix} \dfrac{0}{1}, \dfrac{0.5}{2}, \dfrac{0.5}{3}, \dfrac{0.2}{4}, \dfrac{0.2}{5} \end{bmatrix}$$

4. Difference

$$A|B = A \cap \overline{B} = \begin{bmatrix} \dfrac{0}{1}, \dfrac{0.5}{2}, \dfrac{0.3}{3}, \dfrac{0.3}{4}, \dfrac{0.2}{5} \end{bmatrix}$$

and

$$B|A = B \cap \overline{A} = \begin{bmatrix} \dfrac{0}{1}, \dfrac{0}{2}, \dfrac{0.5}{3}, \dfrac{0.2}{4}, \dfrac{0.4}{5} \end{bmatrix}$$

3.2.6 Fuzzy Membership versus Probability

Probability can be defined for an event that can be repeated again and again. Fuzzy membership can be defined for any event. The event need not be repeatable. Suppose we try to draw a circle free-hand as shown in Figure 3.19. The probability that the figure obtained is a perfect circle is 0. A lot of information about the figure is lost when we say that the probability is zero. The fuzzy membership, on the

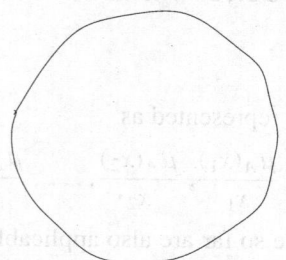

Figure 3.19 Freehand circle

other hand, by assigning a value, say, 0.8, tells us that the figure is somewhat near to a perfect circle.

3.2.7 Extension Principle of Fuzzy Sets

Here we will discuss one-to-one and many-to-one mapping.

One-to-One Mapping

Let $f(.)$ be a mapping from a fuzzy universal set X to another fuzzy universal set Y, and A and B are subsets of X and Y, respectively. Let the fuzzy set A be given by

$$A = \left\{ \frac{\mu_A(x_1)}{x_1}, \frac{\mu_A(x_2)}{x_2}, \cdots, \frac{\mu_A(x_n)}{x_n} \right\}$$

If there is a one-to-one mapping from x_i to $y_i = f(x_i)$, then B is given by

$$B = \left\{ \frac{\mu_B(y_1)}{y_1}, \frac{\mu_B(y_2)}{y_2}, \cdots, \frac{\mu_B(y_n)}{y_n} \right\}$$

where $\mu_B(y_i) = \mu_A(x_i)$. The one-to-one mapping is illustrated in Figure 3.20.

One-to-one mapping

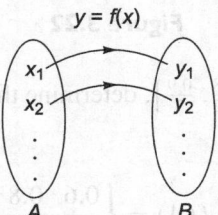

Figure 3.20 Illustrating X- and Y- projections of a fuzzy relation

Many-to-One Mapping

If a many-to-one mapping exists from a universal set X to another set Y then a maximum of the memberships of $f(x_i)$, $f(x_j)$, $\ldots$, $f(x_k)$, where $f(x_i) = f(x_j) = f(x_k)$ should be taken. Formally, for a many-to-one mapping from the set X to Y,

$$\text{Given } A = \left\{ \frac{\mu_A(x_1)}{x_1}, \frac{\mu_A(x_2)}{x_2}, \ldots, \frac{\mu_A(x_n)}{x_n} \right\},$$

$$B = \left\{ \frac{\mu_B(y_1)}{y_1}, \frac{\mu_B(y_2)}{y_2}, \ldots, \frac{\mu_B(y_n)}{y_n} \right\}$$

where $\mu_B(y_i) = \max[\mu_A(x_j) : x_j \in f^{-1}(y_i)]$. The many-to-one mapping is illustrated in Figure 3.21.

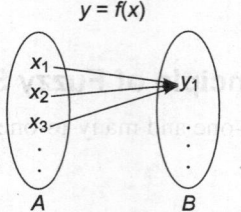

Many-to-one mapping

$$y = f(x)$$

Figure 3.21 Illustrating X- and Y- projections of a fuzzy relation

Example 3.1 Consider two fuzzy sets A and B and a many-to-one mapping between them ($A \to B$) as shown in Figure 3.22.

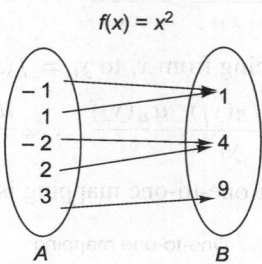

$$f(x) = x^2$$

Figure 3.22

Given $A = \left\{ \frac{0.2}{-1}, \frac{0.4}{-2}, \frac{0.6}{1}, \frac{0.8}{0.2}, \frac{0.9}{3} \right\}$, determine the fuzzy set B.

Solution

We have

$$B = f(A) = \left\{ \frac{0.6}{1}, \frac{0.8}{4}, \frac{0.9}{9} \right\}$$

So far, we have discussed a mapping from a one-dimensional space X to another one-dimensional space Y. In general, the mapping can be defined from an n-dimensional product space $X_1 \times X_2 \times \cdots \times X_n$ to a single universal Y. Here $X_1, X_2, \ldots, X_n$ denote fuzzy universes. The mapping in the present context is denoted by $f(x_1, x_2, \ldots, x_n)$, where $x_i \in X_i$. If $A_1, A_2, \ldots, A_n$ are n fuzzy sets in X_1, then

$$B = \mu_B(y)/y, \quad \text{where } y = f(x_1, x_2, \ldots, x_n)$$

and $\mu_B(y) = \max[\min\{\mu_{A1}(x_1), \mu_{A2}(x_2), \ldots, \mu_{An}(x_n)\}], \quad \text{if } f^{-1}(y) \neq \emptyset$

$\qquad\qquad = 0, \quad \text{otherwise}$

3.2.8   Crisp Relation

The Cartesian product of two universal sets is determined as

$$X \times Y = \{(x, y) \mid x \in X, y \in Y\}$$

The crisp relation $\mu_R(x, y)$ is defined as

$$\mu_R(x, y) = \begin{cases} 1, & (x, y) \in X \times Y \\ 0, & (x, y) \notin X \times Y \end{cases}$$

where 1 implies complete relation and 0 implies no relation. When the sets are finite, the relation is represented by a matrix, R, called relation matrix. For example, let us consider two sets X and Y as

$$X = (1, 2, 3)$$
$$Y = (a, b, c)$$

For the above sets, two relation matrices R_1 and R_2 are defined as follows:

$$R_1 = \begin{array}{c|ccc} & a & b & c \\ \hline 1 & 1 & 1 & 1 \\ 2 & 1 & 1 & 1 \\ 3 & 1 & 1 & 1 \end{array} \quad \text{and} \quad R_2 = \begin{array}{c|ccc} & a & b & c \\ \hline 1 & 0 & 0 & 1 \\ 2 & 1 & 0 & 0 \\ 3 & 0 & 1 & 0 \end{array}$$

The relation matrix R_1 implies that each element in X is completely related to each element in Y, whereas the relation matrix R_2 implies that only the pairs $(1, c)$, $(2, a)$, and $(3, b)$ are related.

Composition

Consider three universal sets X, Y, and Z.

1. Let R be a relation that relates elements from X to Y.
2. Let S be a relation that relates elements from Y to Z.
3. Let T be a relation that relates the same elements in X that R contains to the same elements in Z that S contains.

Given R and S, T is determined using the principle of composition: $T = R \circ S$.

1. If it is max–min composition, then

$$\mu_T(x, z) = \max_{y \in Y} (\min[\mu_R(x, y), \mu_S(y, z)])$$

2. If it is max–product composition, then

$$\mu_T(x, z) = \max_{y \in Y} \mu_R(x, y).\mu_s(y, z)$$

Example 3.2 Given $R = \begin{array}{c|cccc} & y_1 & y_2 & y_3 & y_4 \\ \hline x_1 & 1 & 0 & 1 & 0 \\ x_2 & 0 & 0 & 0 & 1 \\ x_3 & 0 & 0 & 0 & 0 \end{array}$ and $S = \begin{array}{c|cc} & z_1 & z_2 \\ \hline y_1 & 0 & 1 \\ y_2 & 0 & 0 \\ y_3 & 0 & 1 \\ y_4 & 0 & 0 \end{array}$ find $T = R \circ S$.

Solution
We get

$$T = \begin{array}{c|cc} & z_1 & z_2 \\ \hline x_1 & 0 & 1 \\ x_2 & 0 & 0 \\ x_3 & 0 & 0 \end{array}$$

One should verify that in the case of crisp relations, max–min or max–product will yield the same result.

3.2.9 Fuzzy Relations

If X and Y are two universal sets, the fuzzy relation $R(x, y)$ is given by

$$R(x, y) = \left\{ \frac{\mu_R(x, y)}{(x, y)} \middle| (x, y) \in X \times Y \right\}$$

Example 3.3 Let $X = \{1, 2, 3\}$ and $Y = \{1, 2\}$. If the membership function associated with each ordered pair (x, y) is given by $\mu_R(x, y) = e^{-(x-y)^2}$, then derive the fuzzy relation $R(x, y)$.

Solution

The fuzzy relation $R(x, y)$ can be defined in two ways. The one, which uses the standard nomenclature, is as follows:

$$R(x, y) = \left\{ \frac{e^{-(1-1)^2}}{(1, 1)}, \frac{e^{-(1-2)^2}}{(1, 2)}, \frac{e^{-(2-1)^2}}{(2, 1)}, \frac{e^{-(2-2)^2}}{(2, 2)}, \frac{e^{-(3-1)^2}}{(3, 1)}, \frac{e^{-(3-2)^2}}{(3, 2)} \right\}$$

$$= \left\{ \frac{1.0}{(1, 1)}, \frac{0.43}{(1, 2)}, \frac{0.43}{(2, 1)}, \frac{1.0}{(2, 2)}, \frac{0.16}{(3, 1)}, \frac{0.43}{(3, 2)} \right\}$$

The second method of defining the relation is through the relational matrix given by

X \ Y	1	2
1	1.0	0.43
2	0.43	1.0
3	0.16	0.43

$R(x, y)$

Since the membership function describes the closeness between sets X and Y, it is obvious that smaller value imply stronger relations.

Fuzzy Relations: Formal Definition

Definition 3.11 A fuzzy relation is a fuzzy set defined in the Cartesian product of crisp sets $X_1, X_2, \ldots, X_n$. A fuzzy relation $R(x_1, x_2, \ldots, x_n)$ is thus defined as

$$R(x_1, \ldots, x_n) = \left\{ \frac{\mu_R(x_1, \ldots, x_n)}{(x_1, \ldots, x_n)} \middle| (x_1, \ldots, x_n) \in X_1 \times \ldots \times X_n \right\}$$

where $\mu_R : X_1 \times X_2 \times \cdots \times X_n \to [0, 1]$

3.2.10 Projection of Fuzzy Relations

A fuzzy relation R is usually defined in the Cartesian space $X \times Y$. Often, a projection of this relation on any of the sets X or Y may become useful for further information processing.

Definition 3.12 Projection of a fuzzy relation $R(x_1, x_2, \ldots, x_n)$ onto $X_i \times X_j \times \cdots \times X_k$ for any i, j, and k in $[1, n]$ is defined as a fuzzy relation R_p, where

$$R_p(x_i, x_j, \ldots, x_k) = \left\{ \max_{x_i \in X_i, \ldots, x_k \in X_k} \frac{\mu_{Rp}(x_i, x_j, \ldots, x_k)}{(x_i, x_j, \ldots, x_k)} \right\}$$

1. The projection of $R(x, y)$ on X, denoted by $R1$ is given by $\mu_{R1}(x) = \max_{y \in Y}[\mu_R(x, y)]$.

2. The projection of $R(x, y)$ on Y, denoted by $R2$ is given by $\mu_{R2}(y) = \max_{x \in X}[\mu_R(x, y)]$.

Consider $R(x, y)$ from Example 3.3. Figure 3.23 illustrates the X- and Y-projections of a fuzzy relation. For the X-projection, the maximum value in each row is retained, while the maximum value in each column is retained for the Y-projection.

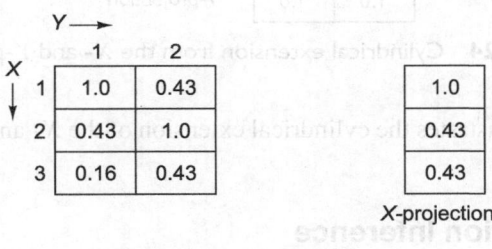

Figure 3.23 Projection of fuzzy relations

3.2.11 Cylindrical Extension of Fuzzy Relations

Cylindrical extension from the X-projection means filling all the columns of the relational matrix by the X-projection. Similarly, cylindrical extension from the Y-projection means filling all the rows of the relational matrix by the Y-projection. A cylindrical extension of a fuzzy relation $R_1(x)$ can thus be defined on $X \times Y$ as a binary fuzzy relation given by

$$R_{1_c}(x, y) = \left\{ \frac{\mu_{R_1}(x)}{(x, y)} \right\}$$

Cylindrical Extension of Projection: Formal Definition

Let R_1 be a fuzzy relation in $X_i \times X_j \times \cdots \times X_k$ for any $(i, j, \ldots, k)$ in $[1, n]$. The cylindrical extension of R_1 to $X_1 \times X_2 \times \cdots \times X_n$ is a fuzzy relation R_{1_c} given by

$$R_{1_c}(x_1, x_2, \ldots, x_n) = \left\{ \frac{\mu_{R_1}(x_1, x_2, \ldots, x_n)}{(x_1, x_2, \ldots, x_n)} \right\}$$

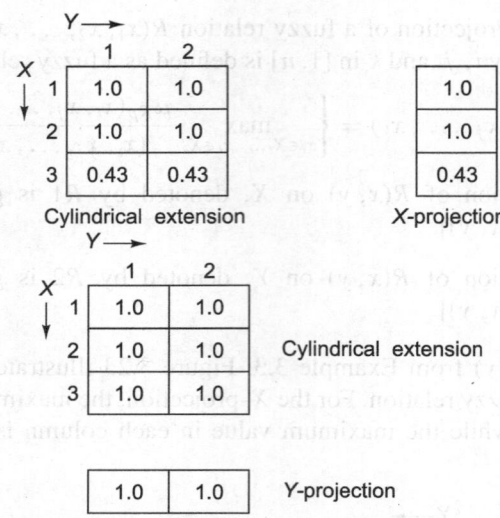

Figure 3.24 Cylindrical extension from the X- and Y-projections

Figure 3.24 illustrates the cylindrical extension of the X- and Y-projections of a fuzzy relation.

3.2.12   Relation Inference

Fuzzy relation	Cartesian space
R_1	$X \times Y$
R_2	$Y \times Z$
R_3	$X \times Z$

If R_1 and R_2 are given, then R_3 is inferred using either of the following:

1. Fuzzy max–min composition operation
2. Fuzzy max–product composition operation

Fuzzy Max–Min Composition Operation

Let us consider two fuzzy relations R_1 and R_2 defined on Cartesian spaces $X \times Y$ and $Y \times Z$, respectively. The max–min composition of R_1 and R_2 is a fuzzy set defined on the Cartesian space $Y \times Z$ as

$$R_3 = R_1 \circ R_2 = \left\{ \frac{\mu_{R3}(x, z)}{(x, z)} \right\}$$

where

$$\mu_{R3}(x, z) = \max_{y}\{\min(\mu_{R1}(x, y), \mu_{R2}(y, z))\}|x \in X, y \in Y, z \in Z\}$$

Properties of Max–Min Composition

Let P, Q, and R be the three relational matrices defined on $X \times Y$, $Y \times Z$, and $Z \times W$, respectively. The max–min operation satisfies the following properties:

1. Associative

   $P \circ (Q \circ R) = (P \circ Q) \circ R$

2. Distributive over union

   $P \circ (Q \cup R) = (P \circ Q) \cup (P \circ R)$

3. Weakly distributive over intersection

   $P \circ (Q \cap R) \subseteq (P \circ Q) \cup (P \circ R)$

4. Monotonic

   $Q \subseteq R \Rightarrow (P \circ Q) \subseteq (P \circ R)$

Figure 3.25 illustrates the fuzzy max–min composition operation.

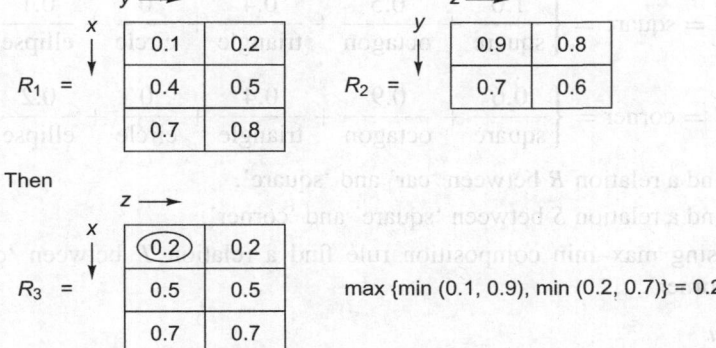

Figure 3.25 Max–min composition operation

Fuzzy Max–Product Composition Operation

The max–product composition of R_1 and R_2 is a fuzzy set defined by

$$R_3 = R_1 * R_2 = \left\{ \frac{\mu_{R3}(x, z)}{(x, z)} \right\}$$

where $\mu_{R3}(x, z) = \max_y [(\mu_{R1}(x, y) * \mu_{R2}(y, z))]$.

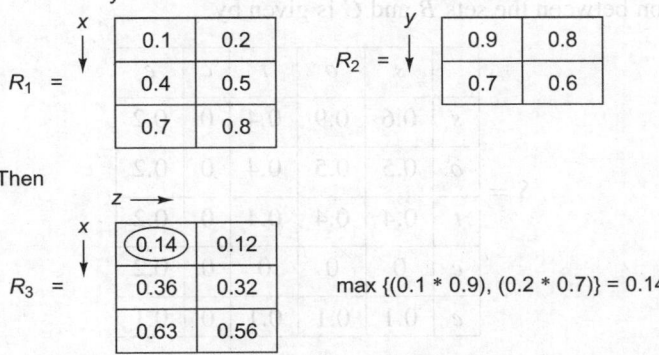

Figure 3.26 Max–product composition operation

Example 3.4 Let X be a universe of general, well-known objects, such as

$$X = \{\text{car, boat, house, bike, tree, mountain}\}$$

and let Y be a universe of simple geometric shapes, such as

$$Y = \{\text{square, octagon, triangle, circle, ellipse}\}$$

Define a simple fuzzy sets of objects, such as 'car', 'square' and 'corner' as given below:

$$A = \text{car} = \left\{ \frac{1.0}{\text{car}} + \frac{0.4}{\text{boat}} + \frac{0.1}{\text{house}} + \frac{0.6}{\text{bike}} + \frac{0.1}{\text{tree}} + \frac{0}{\text{mountain}} \right\}$$

$$B = \text{square} = \left\{ \frac{1.0}{\text{square}} + \frac{0.5}{\text{octagon}} + \frac{0.4}{\text{triangle}} + \frac{0}{\text{circle}} + \frac{0.1}{\text{ellipse}} \right\}$$

$$C = \text{corner} = \left\{ \frac{0.6}{\text{square}} + \frac{0.9}{\text{octagon}} + \frac{0.4}{\text{triangle}} + \frac{0}{\text{circle}} + \frac{0.2}{\text{ellipse}} \right\}$$

1. Find a relation R between 'car' and 'square'.
2. Find a relation S between 'square' and 'corner'.
3. Using max–min composition rule find a relation T between 'car' and 'corner'.

Solution

The relation matrix R between the fuzzy sets A and B is given by

$$R =$$

	s	o	t	c	e
c	1	0.5	0.4	0	0.1
b	0.4	0.4	0.4	0	0.1
h	0.1	0.1	0.1	0	0.1
b	0.6	0.5	0.4	0	0.1
t	0.1	0.1	0.1	0	0.1
m	0	0	0	0	0

The relation between the sets B and C is given by

$$S =$$

	s	o	t	c	e
s	0.6	0.9	0.4	0	0.2
o	0.5	0.5	0.4	0	0.2
t	0.4	0.4	0.4	0	0.2
c	0	0	0	0	0.2
e	0.1	0.1	0.1	0	0.1

The relation $T = R \circ S$ between the fuzzy sets A and C using the max–min rule is computed as

$$T = \begin{array}{c|ccccc} & s & o & t & c & e \\ \hline c & 0.6 & 0.9 & 0.4 & 0 & 0.2 \\ b & 0.4 & 0.4 & 0.4 & 0 & 0.2 \\ h & 0.1 & 0.1 & 0.1 & 0 & 0.1 \\ b & 0.6 & 0.6 & 0.4 & 0 & 0.2 \\ t & 0.1 & 0.1 & 0.1 & 0 & 0.1 \\ m & 0 & 0 & 0 & 0 & 0 \end{array}$$

3.3 FUZZY RULE BASE AND APPROXIMATE REASONING

In this section, we will discuss about the linguistic variables, fuzzy rule base, fuzzy implication relations, fuzzy compositional rules, approximate reasoning for discrete, and continuous fuzzy sets using graphical representations.

3.3.1 Fuzzy Linguistic Variables

Algebraic variables take numbers as values, while linguistic variables take words or sentences as values. Let x be a linguistic variable with label 'temperature'. The fuzzy set temperature denoted by T can be written as

$$T = \{\text{'very cold', 'cold', 'normal', 'hot', 'very hot'}\}$$

Here temperature is the base variable, which is also called the universe of discourse. Each item in this fuzzy set is a fuzzy linguistic value for the variable x.

3.3.2 Linguistic Modifier

Let us consider another example:

Linguistic variable	Age	Height
Linguistic values	Young, old, very old	Short, medium, tall
Dynamic range	[15–100] years	[3–7] feet
Membership	μ_{young}, μ_{old}, $\mu_{very\ old}$	μ_{short}, μ_{tall}, etc.

Consider a fuzzy set

$$\text{Young} = \left[\frac{0.8}{20}, \frac{0.6}{30}, \frac{0.2}{40}, \frac{0}{60}\right]$$

The linguistic variable 'very young', where 'very' is a modifier, is a set

$$\text{Young}^2 = \left[\frac{0.64}{20}, \frac{0.36}{30}, \frac{0.04}{40}, \frac{0}{60}\right]$$

Similarly, the linguistic variable 'very very young' can be written by induction as

$$\text{Young}^4 = \left[\frac{0.4096}{20}, \frac{0.1296}{30}, \frac{0.0016}{40}, \frac{0}{60}\right]$$

Similarly, if a is a fuzzy set,

Extremely $a = a^3$ very $a = a^2$ more or less $a = a^{\frac{1}{2}}$,   slightly $a = a^{\frac{1}{3}}$

3.3.3 Rule-base Systems

Worldly knowledge is very conveniently expressed in natural language. The *rule base* is one of the ways to represent knowledge using natural language.

A generic form of a rule base is as follows:

IF premise (antedecent), THEN conclusion (consequent)

1. The above form is commonly referred to as the IF-THEN rule-based form.
2. It typically expresses an inference such that if we know a fact we can infer or derive another fact.

3.3.4 Fuzzy Rule Base

Fuzzy information can be represented in form of a rule base, which consists of a set of rules in the conventional antecedent–consequent form such as

Rule 1: IF x is A, THEN y is B, where A and B represent fuzzy propositions (sets).

Now suppose we introduce a new antecedent, say A', and we consider the following rule:

Rule 2: IF x is A', THEN y is B'

From information derived from Rule 1, is it possible to derive the consequent in Rule 2, B'?

The answer is YES. The consequent B' can be found from the composition operation $B' = A' o R$ (R : relational matrix).

3.3.5 Fuzzy Implication Relations

A fuzzy implication relation for a given rule 'IF x is A_i THEN y is B_i is formally denoted by

$$R_i(x, y) = \{\tfrac{\mu_{Ri}(x,y)}{(x,y)}\}$$

where the membership function $\mu_{Ri}(x, y)$ is constructed using implication rules *if p then q* ($p \rightarrow q$), where both p and q are fuzzy propositions. Following are various fuzzy implication relations.

1. **Dienes–Rescher implication:** *If p then q* states that *p is true but q is false is impossible*, i.e., $p \wedge \neg q$ is false. Using De Morgan's law, $p \wedge \neg q = \neg p \vee q$. Thus, the relational matrix can be computed as

$$\mu_{Ri}(x, y) = \max[1 - \mu_{Ai}(x), \mu_{Bi}(y)]$$

2. **Mamdani implication:** When fuzzy IF-THEN rules are locally true, then using the Mamdani implication $p \rightarrow q$ implies $p \wedge q$ is true. Thus, the

relational matrix can be computed using any of the following expressions:

$$\mu_{Ri}(x, y) = \min[\mu_{Ai}(x), \mu_{Bi}(y)]$$

$$\mu_{Ri}(x, y) = \mu_{Ai}(x)\mu_{Bi}(y)$$

The Mamdani implication rule is widely used in fuzzy systems and fuzzy control engineering. For example,

If the temperature is HOT, then the fan should run FAST. This rule does not imply if temperature is COLD, then the fan should run SLOW.

3. **Zadeh implication:** In the Zadeh implication $p \rightarrow q$ may imply that *either p and q are true or p is false.* Thus, $p \rightarrow q = (p \wedge q) \vee (\neg p)$. The relational matrix can be computed as follows:

$$\mu_{Ri}(x, y) = \max[\min(\mu_{Ai}(x), \mu_{Bi}(y)), 1 - \mu_{Ai}(x)]$$

Given a set of rules, we learnt various schemes by which we can construct a relational matrix between the antecedent and the consequent. The next step would be to utilize this relational matrix for inference. This method is commonly known as *compositional rule of inference.*

3.3.6 Fuzzy Compositional Rules

Following are the different rules for the fuzzy composition operation $B = AoR$:

$$\text{Max--min} : \mu_B(y) = \max_{x \in X} \{\min[\mu_A(x), \mu_R(x, y)]\}$$
$$\text{Max--product} : \mu_B(y) = \max_{x \in X} \{\mu_A(x) \cdot \mu_R(x, y)\}$$
$$\text{Min--max} : \mu_B(y) = \min_{x \in X} \{\max[\mu_A(x), \mu_R(x, y)]\}$$
$$\text{Max--max} : \mu_B(y) = \max_{x \in X} \{\max[\mu_A(x), \mu_R(x, y)]\}$$
$$\text{Min--min} : \mu_B(y) = \min_{x \in X} \{\min[\mu_A(x), \mu_R(x, y)]\}$$

Example 3.5 IF x is A THEN y is B, where $A = \left\{\frac{0.2}{1}, \frac{0.5}{2}, \frac{0.7}{3}\right\}$ and $B = \left\{\frac{0.6}{5}, \frac{0.8}{7}, \frac{0.4}{9}\right\}$, infer B' for the following rule:

IF x is A' THEN y is B', where $A' = \left\{\frac{0.5}{1}, \frac{0.9}{2}, \frac{0.3}{3}\right\}$.

Solution
Using the Mamdani implication relation

$$R = \begin{array}{c|ccc} & 5 & 7 & 9 \\ \hline 1 & 0.2 & 0.2 & 0.2 \\ 2 & 0.5 & 0.5 & 0.4 \\ 3 & 0.6 & 0.7 & 0.4 \end{array}$$

Using the max–min composition relation $B' = A' \circ R = \left\{\frac{0.5}{5}, \frac{0.5}{7}\frac{0.4}{9}\right\}$.

3.3.7 Inference Mechanism Compared

In the previous example, B' can be computed using various inference mechanisms, such as

$$\text{Max--min} : B' = \left\{\frac{0.5}{5}, \frac{0.5}{7}, \frac{0.4}{9}\right\}$$

$$\text{Max–product} : B' = \left\{ \frac{0.45}{5}, \frac{0.45}{7}, \frac{0.36}{9} \right\}$$

$$\text{Min–max} : B' = \left\{ \frac{0.5}{5}, \frac{0.5}{7}, \frac{0.4}{9} \right\}$$

$$\text{Max–max} : B' = \left\{ \frac{0.9}{5}, \frac{0.9}{7}, \frac{0.9}{9} \right\}$$

$$\text{Min–min} : B' = \left\{ \frac{0.2}{5}, \frac{0.2}{7}, \frac{0.2}{9} \right\}$$

3.3.8   Approximate Reasoning

Given a rule R and given a condition A, the inference B is obtained using the compositional rule of inference $B = A \circ R$. The fuzzy sets associated with each rule base may be discrete or continuous. A rule base may contain a single rule or multiple rules. Various inference mechanisms for a single rule are enumerated. The inference mechanism for multiple rules will be illustrated later in this section.

Example 3.6   Single rule with discrete fuzzy set
   Rule 1: IF temperature is hot, THEN fan should run fast
   Rule 2: IF temperature is moderately hot, THEN fan should run moderately fast
The temperature is expressed in °F and the speed is expressed in 1000 rpm. Given

$$H = \text{'hot'} = \left\{ \frac{0.4}{70}, \frac{0.6}{80}, \frac{0.8}{90}, \frac{0.9}{100} \right\}$$

$$F = \text{'fast'} = \left\{ \frac{0.3}{1}, \frac{0.5}{2}, \frac{0.7}{3}, \frac{0.9}{4} \right\}$$

$$H' = \text{'moderately hot'} = \left\{ \frac{0.2}{70}, \frac{0.4}{80}, \frac{0.6}{90}, \frac{0.8}{100} \right\}$$

Given the above rule base, find F'.

Solution
The relational matrix R that connects the fuzzy sets H and F is computed using the Mamdani implication

$$R = \begin{array}{c|cccc} & 1 & 2 & 3 & 4 \\ \hline 70 & 0.3 & 0.4 & 0.4 & 0.4 \\ 80 & 0.3 & 0.5 & 0.6 & 0.6 \\ 90 & 0.3 & 0.5 & 0.7 & 0.8 \\ 100 & 0.3 & 0.5 & 0.7 & 0.9 \end{array}$$

$$F' = H' \circ R = \left\{ \frac{0.3}{1}, \frac{0.5}{2}, \frac{0.7}{3}, \frac{0.8}{4} \right\}$$

Example 3.7   Multiple rules with discrete fuzzy sets

   Rule 1: IF height is tall, THEN speed is high
   Rule 2: IF height is medium, THEN speed is moderate

The fuzzy sets for height (in feet) and speed (in m/s) are as follows:

$$H_1 = \text{'Tall'} = \left\{ \frac{0.5}{5}, \frac{0.8}{6}, \frac{1}{7} \right\}$$

$$H_2 = \text{'Medium'} = \left\{ \frac{0.6}{5}, \frac{0.7}{6}, \frac{0.6}{7} \right\}$$

$$S_1 = \text{'High'} = \left\{ \frac{0.4}{5}, \frac{0.7}{7}, \frac{0.9}{9} \right\}$$

$$S_2 = \text{'Moderate'} = \left\{ \frac{0.6}{5}, \frac{0.8}{7}, \frac{0.7}{9} \right\}$$

Given $H' = \text{'ABOVE AVERAGE'} = \left\{ \frac{0.5}{5}, \frac{0.9}{6}, \frac{0.8}{7} \right\}$, find $S' = \text{'ABOVE NORMAL'}$.

Solution

The fuzzy relational matrices for rules 1 and 2 are computed using the Mamdani implication rule

$$R_1 = \begin{array}{c|ccc} & 5 & 7 & 9 \\ \hline 5 & 0.4 & 0.5 & 0.5 \\ 6 & 0.4 & 0.8 & 0.8 \\ 7 & 0.4 & 0.8 & 0.9 \end{array} \qquad R_2 = \begin{array}{c|ccc} & 5 & 7 & 9 \\ \hline 5 & 0.6 & 0.6 & 0.6 \\ 6 & 0.6 & 0.7 & 0.7 \\ 7 & 0.6 & 0.6 & 0.6 \end{array}$$

$$[S' = \text{'above-normal'} = \max[H' \circ R_1, H' \circ R_2] = H' \circ \max[R_1, R_2]$$

$$= \begin{bmatrix} 0.5 & 0.9 & 0.8 \end{bmatrix} \circ \begin{bmatrix} 0.6 & 0.6 & 0.6 \\ 0.6 & 0.8 & 0.8 \\ 0.6 & 0.8 & 0.9 \end{bmatrix} = \left\{ \frac{0.6}{5}, \frac{0.8}{7}, \frac{0.8}{9} \right\}$$

Multiple Rules with Continuous Fuzzy Sets

A continuous fuzzy system with two non-interactive inputs x_1 and x_2 (antecedents) and a single output y (consequent) is described by a collection of r linguistic IF-THEN rules

IF x_1 is A_1^k and x_2 is A_2^k, THEN y is B^k for $k = 1, 2, \ldots, r$

where A_1^k and A_2^k are the fuzzy sets representing the kth antecedent pairs and B^k are the fuzzy sets representing the kth consequent. Let us consider the case of a two-input system as shown in Figure 3.27(a). The crisp inputs to the system are x_1 and x_2 and the crisp output of the system is y. Two rules of the system are

Rule 1: IF x_1 is A_1^1 and x_2 is A_2^1 THEN y is B^1

Rule 2: IF x_1 is A_1^2 and x_2 is A_2^2 THEN y is B^2

The crisp value of x_1 in Rule 1 is A_1^1, while the crisp value of x_2 is A_2^1. Both inputs have a different membership value depending on the membership distributions. The minimum of the two is taken as the resultant membership of Rule 1 whose corresponding output is B_1. Similarly, the minimum of the membership values corresponding to the crisp values of A_1^2 and A_2^2 is taken as the resultant membership of Rule 2. The inferred output membership function is shown in Figure 3.27(b). The crisp output can be obtained from the output membership function using various methods. One of the most common methods is the centre of gravity method, where the output is calculated using the following formula:

$$y^* = \frac{\int y\mu(y)\, dy}{\int \mu(y)\, dy}$$

Figure 3.27 pictorially illustrates the complete procedure.

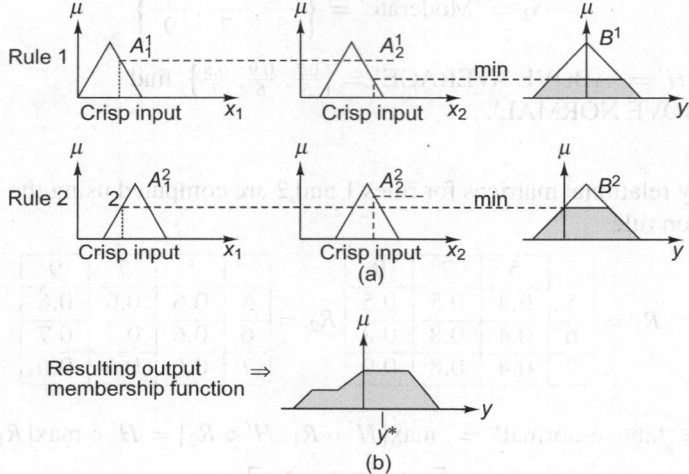

Figure 3.27 Defuzzification: centre of gravity method

3.4   FUZZY LOGIC CONTROL

Fuzzy logic control (FLC) [58, 59, 60] is a technique to embody human-like thinking into a control system. A fuzzy controller [61] can be designed to emulate human deductive thinking to infer conclusions from the past experience. The traditional control approach requires formal modelling of the physical reality. Fuzzy control incorporates ambiguous human logic into computer programs. It suits the control problems that cannot be easily represented by mathematical models. Following are some of the difficulties involved in a control system design where fuzzy control can be applied:

- Inaccurate model
- Parameter variation problem
- Unavailable or incomplete data
- Very complex plants
- Good qualitative understanding of plant or process operation

Design of such controllers leads to faster development or implementation cycles. Two typical fuzzy control systems are popularly known as Mamdani type and Takagi–Sugeno (T–S) type fuzzy systems. The Mamdani type fuzzy logic controller [62] is direct adaptive type where the controllers are designed directly based on the fuzzy rule base, so that the control objective is satisfied. Explicit system identification is not done in this case. The T–S type fuzzy logic controller [63] is indirect adaptive FLCs where the system to be controlled is identified in terms of a T–S fuzzy model, and the controller is designed based on the identified model.

3.4.1 Mamdani Model

The principal design components of a Mamdani-type FLC, as shown in Figure 3.28, are discussed below.

1. *Fuzzifier:* This decides fuzzification strategies and interpretation of a fuzzification operator, which basically involves discretization or normalization of a universe of discourse, fuzzy partition of the input and output spaces, and choice of the membership function of a primary fuzzy set. In a typical FLC, the input variables are chosen as error e and change in error Δe, which are converted into fuzzy sets using triangular membership functions as shown in Figure 3.29.

2. *Rule base:* The rule base is a collection of rules that describe the control strategy which depends on the choice of process state (input) variables and control (output) variables.

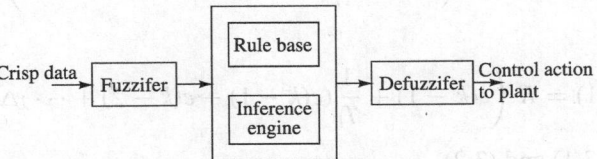

Figure 3.28   Architecture of an FLC: Mamdani type

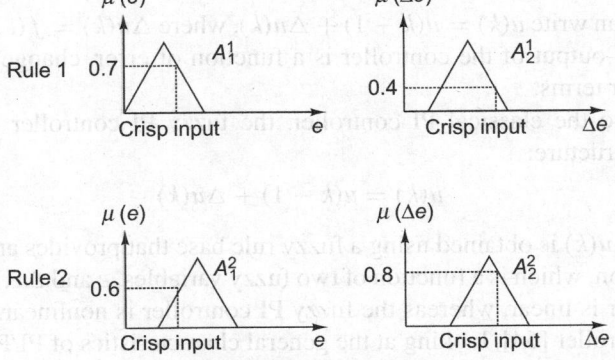

Figure 3.29   Fuzzification

3. *Fuzzy inference mechanism:* This establishes a logical connection between the input and output fuzzy sets.

4. *Defuzzifier:* This decides the defuzzification strategy to convert the output fuzzy variable into a corresponding crisp value. Among various defuzzification methods, the most popular is the centre of gravity (COG) method, which has already been described in the previous section. For an FLC, the output variable is the control input to the system. Thus, y in Figure 3.27 will be replaced either by the control input u in the case of a PD controller or by the change in the input Δu in the case of a PI controller.

A typical rule in a Mamdani model is as follows:

Rule$_j$: If x_1 is A_{1_j} and x_2 is A_{2_j} and ... x_n is A_{n_j} Then y is B_j, $j = 1, 2, \ldots, M$

Fuzzy PI Controller

We will now show how a typical fuzzy PI controller works. A classical PI control law is given as

$$u(t) = K\left(e(t) + \frac{1}{T_i}\int_0^t e(t')dt'\right)$$

Discrete version of the above controller can be written as

$$u(k) = K\left(e(k) + \frac{1}{T_i}(e(k) + e(k-1) + \cdots)\Delta T\right) \tag{3.1}$$

Similarly,

$$u(k-1) = K\left(e(k-1) + \frac{1}{T_i}(e(k-1) + e(k-2) + \cdots)\Delta T\right) \tag{3.2}$$

From Eqns (3.1) and (3.2),

$$u(k) - u(k-1) = K\left(e(k) - e(k-1) + \frac{1}{T_i}e(k)\Delta T\right)$$

Thus, one can write $u(k) = u(k-1) + \Delta u(k)$, where $\Delta u(k) = f(e, \Delta e)$, i.e., the incremental output of the controller is a function of error, change in error, and higher order terms.

Similar to the classical PI controller, the fuzzy PI controller can have the following structure:

$$u(k) = u(k-1) + \Delta u(k)$$

However, $\Delta u(k)$ is obtained using a fuzzy rule base that provides an incremental control action, which is a function of two fuzzy variables, e and Δe. The classical PI controller is linear, whereas the fuzzy PI controller is nonlinear. In the fuzzy PI/PID controller [64], looking at the general characteristics of PI/PID response, the rule base is formed. A typical response of the classical PI controller is shown in Figure 3.30, where PS, ZE, NS, PM, PL denote positive small, zero error, negative small, positive medium and positive large, respectively. The *error* is defined as the difference between the desired steady state output and the current output. Looking at the output response curve in Figure 3.30 one can argue that when the error e and change in error Δe are ZE, then there should not be any change in the control action, i.e., Δu should be ZE. Similarly, if e is PS and Δe is ZE, then Δu is PS. Continuing with the same logic, the following rules can be formed:

- IF e is ZE and $\dot{e}$ is ZE, THEN Δu is ZE
- IF e is PS and $\dot{e}$ is ZE, THEN Δu is PS
- IF e is NS and $\dot{e}$ is ZE, THEN Δu is NS
- IF e is PS and $\dot{e}$ is PS, THEN Δu is PM
- IF e is PL and $\dot{e}$ is PS, THEN Δu is PM
- IF e is PM and $\dot{e}$ is PM, THEN Δu is PL
- IF e is PS and $\dot{e}$ is NS, THEN Δu is ZE

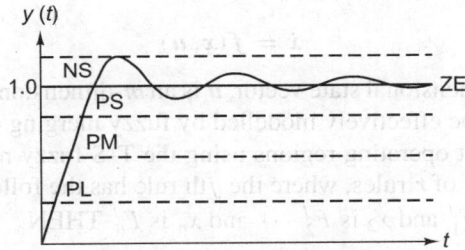

Figure 3.30   Response of a classical PI controller

Other rules can also be formed for all combinations of e and Δe.

Fuzzy PD Controller

A classical PD controller has the following structure:

$$u(k) = Ke(k) + T_d \frac{e(k) - e(k-1)}{\Delta T}$$

Thus, the control action is a function of the error $e(k)$ and change in error $\Delta e(k)$. Similar to that, a fuzzy PD controller can also be expressed as a function of the error and change in error which form the rule base of the fuzzy control system, i.e., the output of a fuzzy PD controller is directly computed from e and Δe. An example of the rule base of a fuzzy PD controller for a level control system is as follows:

- Rule 1: IF error e is positive high and change in error Δe is positive high THEN input flow u is positive high
- Rule 2: IF error e is positive low and change in error Δe is positive low THEN input flow u is positive low

In the above rules, the error, the change in error, and the input flow are three linguistic variables and positive high and positive low are subsets of the universe of discourse. The positive high error indicates that the output level is much below the desired output. The positive high change in error indicates that the level is still decreasing and thus the input flow must be positive high to increase the level. It is clear that such set of fuzzy rules drastically reduces the necessity of a large number of crisp rules.

3.4.2 Takagi–Sugeno Fuzzy Model

If a non-linear system is effectively represented as a fuzzy cluster of linear systems defined locally, then the systemic understanding can become more profound as linear systems are well understood. Such a powerful fuzzy model representation has been introduced by Takagi–Sugeno (T–S) [63], which has been referred to as T–S fuzzy model. Numerous research works have been carried out using the T–S fuzzy model as such a representation can provide a better systemic understanding.

Let us consider a general non-linear dynamical systems in continuous time, described by

$$\dot{x} = f(x, u)$$

where x is an n-dimensional state vector, u is an m-dimensional input vector. The above system can be effectively modelled by fuzzy merging of equivalent linear systems in different operating regions using the T–S fuzzy model. A T–S fuzzy model is composed of r rules, where the jth rule has the following form:

Rule$_j$: IF x_1 is F_1^j and x_2 is F_2^j $\cdots$ and x_n is F_n^j THEN

$$\dot{x} = A_j x + B_j u$$

where $x = [x_1, x_2, \ldots, x_n]^T$, $j = 1, \ldots, r$. Each rule represents a fuzzy zone in the state space. For an n-dimensional state-space system, the number of such fuzzy zones is $r = q^n$ if each state is fuzzified into q fuzzy regions. Thus, the non-linear system is represented by such r rules where each rule is associated with a local linear model.

Given a current state vector x and an input vector u, the T–S fuzzy model infers the system dynamics as

$$\dot{x} = \frac{1}{\sum_{j=1}^{r} \mu_j} \sum_{j=1}^{r} \mu_j \left(A_j x + B_j u \right) \tag{3.3}$$

where

$$\mu_j = \prod_{i=1}^{n} \mu_j^i(x_i) \tag{3.4}$$

Here $\mu_j^i(x_i)$ is the membership grade of the fuzzy term F_i^j, $j = 1, 2, \ldots, r$.

By defining the normalized membership grade associated with the jth rule as

$$\sigma_j = \frac{\mu_j}{\sum_{j=1}^{r} \mu_j} \tag{3.5}$$

the T–S fuzzy model representation of any non-linear continuous time system is expressed as:

$$\dot{x} = \sum_{j=1}^{r} \sigma_j \left(A_j x + B_j u \right) \tag{3.6}$$

Similarly the T–S model representation of a discrete time non-linear system is expressed as

$$x(k+1) = \sum_{j=1}^{r} \sigma_j \left(A_j x(k) + B_j u(k) \right) \tag{3.7}$$

One should note that summation of normalized memembership grades will always be 1, i.e.,

$$\sum_{j=1}^{r} \sigma_j = 1 \tag{3.8}$$

For an autonomous system, the T–S representation will have the following expression: $x(k+1) = \sum_{j=1}^{r} \sigma_j A_j x(k)$.

Example 3.8   Let us consider the dynamics of a single link manipulator as given by

$$\dot{x}_1 = x_2$$
$$\dot{x}_2 = -\frac{g}{l} \sin(x_1) + \frac{1}{ml^2} \tau$$

where $x_1 = \theta$ is the angle of the manipulator from the vertical, $x_2 = \dot{\theta}$ is the angular velocity of the manipulator, and τ is the control torque. The parameters are given as $m = 1$ kg, $l = 1$ m, and $g = 10$ m/s^2. Find out a T–S fuzzy model of the above system.

Solution
The fuzzy set of the variable x_1 is $M_1 = $ [PL, PS, ZE, NS, NL] and the fuzzy set of the variable x_2 is $M_2 = $ [PL, PS, ZE, NS, NL]. This will constitute 25 rules. Two of the rules are given as

Rule 1: IF $x_1(t)$ is PL and $x_2(t)$ is ZE, THEN

$$\dot{x} = \begin{bmatrix} 0 & 1 \\ -1.7 & 0 \end{bmatrix} x + \begin{bmatrix} 0 \\ 1 \end{bmatrix} \tau$$

Rule 2: IF $x_1(t)$ is PS and $x_2(t)$ is ZE, THEN

$$\dot{x} = \begin{bmatrix} 0 & 1 \\ -9.8 & 0 \end{bmatrix} x + \begin{bmatrix} 0 \\ 1 \end{bmatrix} \tau,$$

where PL stands for positive large, PS stands for positive small, ZE stands for zero, NL stands for negative large, and NS stands for negative small, respectively. The T–S fuzzy model of the above system will be a fuzzy weighted average of all the 25 rules.

Example 3.9   Let us consider the following two rules of a T–S fuzzy model.

Rule 1: IF x_1 is large and x_2 is moderate

THEN $x(k+1) = \begin{bmatrix} 0.5 & 0.4 \\ 2.0 & 1.0 \end{bmatrix} x(k)$

Rule 2: IF x_1 is large and x_2 is large

THEN $x(k+1) = \begin{bmatrix} 0.6 & 0.8 \\ 1.0 & 2.0 \end{bmatrix} x(k)$

where $x(k) = [x_1(k)\ x_2(k)]^T$. The measured values of x_1 and x_2 at $k = 1$ are $x_1(1) = 40$ units and $x_2(1) = 20$ units. Evaluate the actual system states at $k = 2$.

Solution

The membership functions are given as

$$\mu^1_{large}(x_1) = 0.3, \quad \mu^1_{moderate}(x_2) = 0.3;$$

$$\mu^2_{small}(x_1) = 0.7, \quad \mu^2_{large}(x_2) = 0.8;$$

According to the T–S model, $x(k+1) = \sum_{j=1}^{2} \sigma_j(k) A_j x(k)$

$$\mu_1(1) = \mu^1_{large}(x_1)\mu^1_{moderate}(x_2) = 0.3 \times 0.4 = 0.12$$

$$\mu_2(1) = \mu^2_{SMALL}(x_1)\mu^2_{LARGE}(x_2) = 0.7 \times 0.8 = 0.56$$

$$\sigma_1(1) = \frac{\mu_1(1)}{[\mu_1(1) + \mu_2(1)]} \qquad \sigma_2(1) = \frac{\mu_2(1)}{[\mu_1(1) + \mu_2(1)]}$$

$$= \frac{0.12}{[0.12 + 0.56]} = \frac{12}{68} \qquad = \frac{0.56}{[0.12 + 0.56]} = \frac{56}{68}$$

Thus the state vector at $k = 2$ can be evaluated as

$$x(2) = \sum_{j=1}^{2} \sigma_j(1) A_j x(1)$$

$$= \sigma_1(1) A_1 x(1) + \sigma_2(1) A_2 x(1)$$

$$= \frac{12}{68} \begin{bmatrix} 0.5 & 0.4 \\ 2.0 & 1.0 \end{bmatrix} \begin{bmatrix} 40 \\ 20 \end{bmatrix} + \frac{56}{68} \begin{bmatrix} 0.6 & 0.8 \\ 1.0 & 2.0 \end{bmatrix} \begin{bmatrix} 40 \\ 20 \end{bmatrix}$$

$$= [37.88\quad 83.52]^T$$

Once the system is identified in terms of a T–S fuzzy model, the controller can be designed using any existing design technique based on the identified model.

3.5 SYSTEM IDENTIFICATION USING T–S FUZZY MODELS

The T–S fuzzy model representation of a continuous time non-linear system has the following form:

$$\dot{x} = \sum_{j=1}^{r} \sigma_j \left(A_j x + B_j u \right) \tag{3.9}$$

Similarly, T–S fuzzy model representation of a discrete time non-linear system has the form:

$$x(k+1) = \sum_{j=1}^{r} \sigma_j (A_j x(k) + B_j u(k)) \tag{3.10}$$

where σ_j is defined in Eqn (3.5).

Given these representations, the system identification goal is to estimate parameters associated with local models A_j and B_j. Given real-time system data, these parameters can be estimated using popular gradient descent algorithms. However, if the non-linear model of a system is exactly known, then the T–S fuzzy representation can be derived using linearization techniques.

3.5.1 The T–S Model from Input–Output Data

The linear model parameters A_js and B_js can be found from the input–output data set using a fuzzy neural network [65, 66]. A typical fuzzy neural network is shown in Figure 3.31 where $x_i(k+1)$ is the individual state $i = 1, \ldots, n$. The parameters of the T–S fuzzy model, i.e., the elements of the matrices A_js and B_js in Eqn (3.10) are considered as the weights of the fuzzy neural network (FNN). The system input as well as the system states is input to the FNN. The fuzzy neural network describes the plant for each rule j as a linear combination of its weights. For a particular input data (from the input–output data set), the output of the network is taken as the weighted average of the output of each rule layer. This is compared with the desired output of the system (from the input–output data set), and depending on the error the weights of the network are updated using a gradient descent algorithm. A more detailed description of the identification method, including the weight update law, can be found in [66].

Example 3.10 (Rotational translational proof mass actuator) Consider the example of a rotational translational proof-mass actuator (RTAC) as shown in Figure 3.32. The RTAC system combines a translational oscillator with a rotational proof mass actuator. The oscillator consists of a cart connected to a fixed wall by a linear spring. The cart is constrained to move in the x-direction only. The proof mass actuator is attached to the cart and is controlled by an applied torque N. F is a disturbance force that perturbs the cart. Table 3.1 defines different parameter values of the system. The cart position and the actuator angle is x_c and

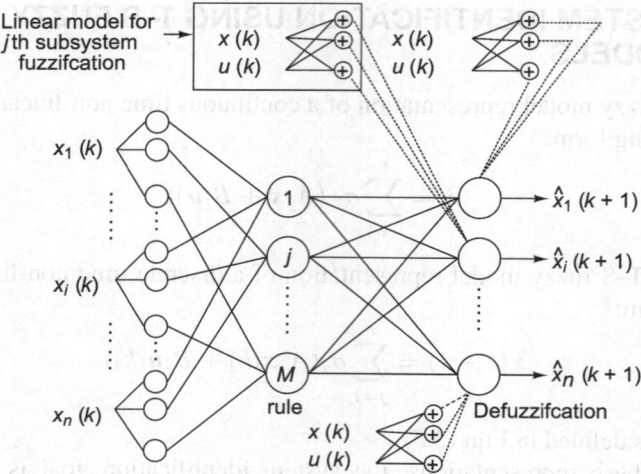

Figure 3.31 Fuzzy neural network model of a non-linear system with n states

θ, respectively. Define the normalized variable as

$$\xi = x_c \sqrt{\frac{M + m}{I + mL^2}}, \qquad \tau = t\sqrt{\frac{k}{M + m}}$$

$$u = N\frac{M + m}{k(I + mL^2)}, \qquad w = \frac{1}{k}\sqrt{\frac{M + m}{I + mL^2}}F$$

Table 3.1 RTAC parameters

Mass of the cart	$M = 1.36$ kg
Proof mass	$m = 0.096$ kg
Distance	$L = 0.059$ m
Moment of inertia (proof mass)	$I = 0.00022$ kg m^2
Spring constant	$k = 186.3$ N/m

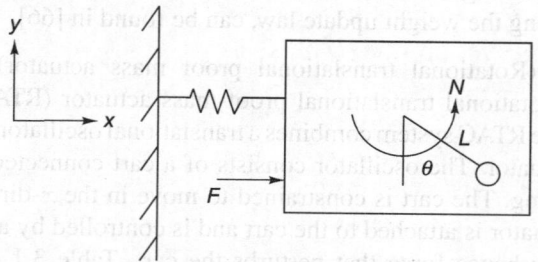

Figure 3.32 RTAC system

Considering the state variable as $x = [x_1 \ x_2 \ x_3 \ x_4]^T = [\xi \ \dot{\xi} \ \theta \ \dot{\theta}]^T$, the control variable as u, and the disturbance as w, the dynamics can be written as

$$\dot{x} = \begin{bmatrix} x_2 \\ \dfrac{-x_1 + \epsilon x_4^2 \sin x_3}{1 - \epsilon^2 \cos^2 x_3} \\ x_4 \\ \dfrac{\epsilon(x_1 - \epsilon x_4^2 \sin x_3)\cos x_3}{1 - \epsilon^2 \cos^2 x_3} \end{bmatrix} + \begin{bmatrix} 0 \\ \dfrac{-\epsilon \cos x_3}{1 - \epsilon^2 \cos^2 x_3} \\ 0 \\ \dfrac{1}{1 - \epsilon^2 \cos^2 x_3} \end{bmatrix} u + \begin{bmatrix} 0 \\ \dfrac{1}{1 - \epsilon^2 \cos^2 x_3} \\ 0 \\ \dfrac{-\epsilon \cos x_3}{1 - \epsilon^2 \cos^2 x_3} \end{bmatrix} w \qquad (3.11)$$

where $\epsilon = \dfrac{mL}{\sqrt{(I + mL^2)(M + m)}}$ is a very small number. It is required to derive a T–S fuzzy model for this system from the data generated from this model.

Solution

Considering the disturbance w as zero, Eqn (3.11) is discretized using Euler approximation where the sampling time is taken as $T = 0.01$ second. The RTAC system will be identified in terms of a discrete T–S fuzzy using an FNN. A total of 4000 data points are generated using the Euler representation of the system dynamics. The system states x_1, x_2, and x_4 are fuzzified in the range $[-1.5, 2.5]$, while the state x_3 is fuzzified in the range $[6, 30]$. Each fuzzy variable is partitioned into three fuzzy zones thus resulting in a total number of 81 local models. The local submodel parameters are obtained by updating the linear model weights using the gradient descent technique with a learning rate $\eta = 0.001$. Among various fuzzy rules, two rules are given as follows:

Rule 1: If $x_1 = -1.5$, $x_2 = -1.5$, $x_3 = 6$, and $x_4 = -1.5$, then

$$x(k+1) = \begin{bmatrix} 0.0062 & 0.0093 & 0.0043 & 0.0099 \\ 0.0037 & 0.0005 & 0.0092 & 0.0006 \\ 0.0086 & 0.0024 & 0.0081 & 0.0057 \\ 0.0024 & 0.0048 & 0.0022 & 0.0021 \end{bmatrix} x(k) + \begin{bmatrix} 0.01 \\ 0.02 \\ 0.005 \\ 0.008 \end{bmatrix} u(k)$$

Rule 2: If $x_1 = -0.5$, $x_2 = -0.5$, $x_3 = 12$ and $x_4 = -0.5$, then

$$x(k+1) = \begin{bmatrix} 1.0048 & 0.0094 & -0.0003 & -0.002 \\ 0.0025 & 0.9866 & -0.0003 & 0.0011 \\ 0.0149 & -0.0002 & 0.9727 & 0.0082 \\ -0.002 & -0.0010 & -0.0001 & 0.9661 \end{bmatrix} x(k) + \begin{bmatrix} 0.009 \\ 0.013 \\ 0.016 \\ 0.007 \end{bmatrix} u(k)$$

After identification, 4000 new data points are generated to validate the model. The result of model identification and validation is shown in Figure 3.33. The RMS error for model prediction is 0.003, whereas for model validation it is 0.007.

3.5.2 The T–S Fuzzy Model Using Linearization

The parameters A_i and B_i can be directly obtained from the system dynamics by linearizing the non-linear dynamics at different operating points. This approach is valid for both continuous and discrete time systems. The following two examples will illustrate this approach of deriving a T–S fuzzy model from the non-linear equations.

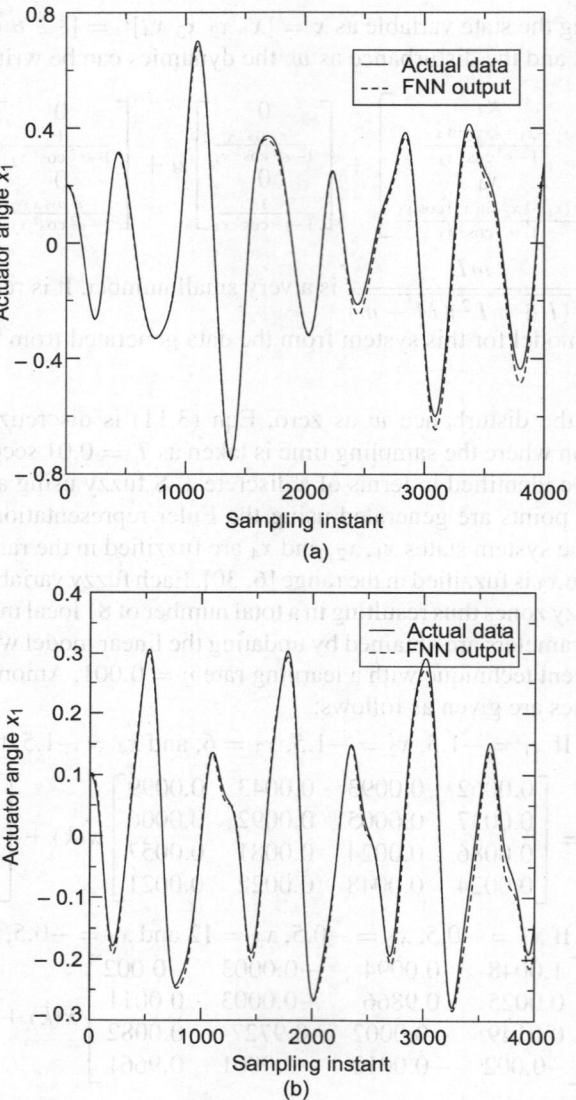

Figure 3.33 RTAC–system identification: (a) model prediction and (b) model validation

Example 3.11 (Cart–pole system) The continuous time T–S fuzzy model will be described for an inverted pendulum mounted on a cart [67, 68]. The cart–pole dynamics has been derived in Chapter 1. The dynamical equations for the pendulum subsystem can be written as

$$\dot{x}_1 = x_2$$

$$\dot{x}_2 = \frac{g \, \sin(x_1) - amlx_2^2 \, \sin(2x_1)/2 - a \, \cos(x_1)u}{4l/3 - aml \, \cos^2(x_1)} \tag{3.12}$$

where x_1 is the angle of the pendulum from vertical, x_2 is the angular velocity of the pendulum, u is the control input applied to the cart, m is mass of the pendulum, M is mass of the cart, $2l$ is length of the pendulum, g is the acceleration due to gravity, and $a = 1/(m + M)$. The system parameters are taken as $m = 2.0$ kg, $M = 8.0$ kg, $2l = 1.0$ m and $g = 9.8$ m/s^2. Derive a T–S fuzzy model for this system.

Solution

Pendulum angle x_1 is fuzzified in the operating region $[-\pi/2, \pi/2]$ and x_2 is always considered as 0. The system is approximated with four fuzzy regions in positive half with centres 0, $\dfrac{\pi}{6}$, $\dfrac{\pi}{3}$, and $\dfrac{88\pi}{180}$. The Gaussian function is chosen as fuzzy membership function and spread of the Gaussian function is chosen in such a way that at each operating point only two fuzzy subsystems will be fired. Around the equilibrium point $(0, 0)$, the system is linearized using the standard Taylor series expansion. At other operating points, linear models are obtained using the technique as described in Chapter 1. Similar approximation is done in negative half of the operating region. Controllers are designed assuming that the approximation is fair enough. Thus, T–S fuzzy model of system (8.64) is described by the following four rules.

Rule 1: If $x(t)$ is around $[0\ 0]^T$

Then $\dot{x} = \begin{bmatrix} 0 & 1 \\ 17.2941 & 0 \end{bmatrix} x + \begin{bmatrix} 0 \\ -0.1765 \end{bmatrix} u$

Rule 2: If x is around $[\pm\frac{\pi}{6}\ 0]^T$

Then $\dot{x} = \begin{bmatrix} 0 & 1 \\ 15.8169 & 0 \end{bmatrix} x + \begin{bmatrix} 0 \\ -0.1464 \end{bmatrix} u$

Rule 3: If x is around $[\pm\frac{\pi}{3}\ 0]^T$

Then $\dot{x} = \begin{bmatrix} 0 & 1 \\ 12.6304 & 0 \end{bmatrix} x + \begin{bmatrix} 0 \\ -0.0779 \end{bmatrix} u$

Rule 4: If x is around $[\pm\frac{88\pi}{180}\ 0]^T$

Then $\dot{x} = \begin{bmatrix} 0 & 1 \\ 9.6193 & 0 \end{bmatrix} x + \begin{bmatrix} 0 \\ -0.0065 \end{bmatrix} u$

where $x = [x_1\ x_2]^T$. Thus, the T-S fuzzy model of system (8.64) will be a fuzzy-weighted average of the above four rules.

MATLAB code

```
CartPole.m
function out = CartPole(x);
m=2.0; M=8.0; l=0.5; g=9.8;
a=1/(m+M);
tmp1=g*sin(x(1))-a*m*l*x(2)*x(2)*sin(2*x(1))/2;
tmp2=4*l/3-a*m*l*cos(x(1))*cos(x(1));
f = [x(2);(tmp1-a*cos(x(1))*u)/tmp2];
```

```
gradf = [0   ((g*cos(x(1))-a*m*l*x(2)*x(2)*cos(2*x(1))))
*tmp2-tmp1*a*m*l*sin(2*x(1)))/(tmp2*tmp2);
1   a*m*l*x(2)*sin(2*x(1))/tmp2];
for i=1:2
A(i,:)=(gradf(:,i) + ((f(i)-
x'*gradf(:,i))/(norm(x)*norm(x)))*x)';
end
A
B=-a*cos(x(1))/tmp2;
```

The MATLAB command to execute the above code for operating point $[0\ 0]^T$ is 'CartPole([0;0])', which gives A and B matrices as $A = \begin{bmatrix} 0 & 1 \\ 17.2941 & 0 \end{bmatrix}$ and $B = \begin{bmatrix} 0 \\ -0.1765 \end{bmatrix}$.

Example 3.12 [Series direct current (DC) motor]   The discrete time T–S fuzzy model of a series DC motor [69] will be derived using the dynamics of the system. The field circuit of a series DC motor is connected in series with its armature circuit. The dynamical equations of a series DC motor with $i_a - i_f = i$ are described by

$$J\frac{d\omega}{dt} = -D\omega + K_m L_f i^2 - \tau_L'$$

$$L\frac{di}{dt} = -K_m L_f \omega i - Ri + V \tag{3.13}$$

where ω is the speed to be controlled, V is the input voltage, τ_L is the load torque, and $J = 0.000704$ kg $-$ m^2 is the moment of inertia of the motor, $D = 0.0004$ N $-$ m/rad/s is the viscous friction coefficient, K_m is the motor torque constant, L_f is the field inductance $L = L_a + L_f = 0.0917$ H, and $R = R_a + R_f = 7.2$ Ω, where L_a and R_a are the armature inductance and resistance, respectively. $K_m L_f = 0.1236$ N-m/Wb-A. Derive a T–S fuzzy model.

Solution

Considering the states as $x_1 = \omega$ and $x_2 = i$ and input as $u = V$, the state–space model of the system can be written as

$$\dot{x}_1 = -\frac{D}{J}x_1 + \frac{K_m L_f}{J}x_2^2 - \tau_L$$

$$\dot{x}_2 = -\frac{K_m L_f}{L}x_1 x_2 - \frac{R}{L}x_2 + \frac{1}{L}u \tag{3.14}$$

The system dynamics is modelled as T–S fuzzy system by fuzzifying the states x_1 and x_2. Each state is fuzzified in five equally spaced regions in the range [0 20] m/s and [0 1] A, respectively. Thus, the discrete T–S fuzzy model will have 25 fuzzy rules in total. Around the operating point $(0, 0)$, we have linearized the continuous time system model using the standard Taylor series expansion. At other operating points, linear models are obtained by the technique described in section 1.5.2 of chapter 1. Then the subsystem models are discretized for the sampling interval of 0.01 s using MATLAB. For clear understanding, two typical subsystems using linearization of the dynamics are presented below.

Rule 1: IF x_1 is around 0 and x_2 is 0

$$\text{THEN } \dot{x} = \begin{bmatrix} -0.568 & 0 \\ 0 & -78.517 \end{bmatrix} x + \begin{bmatrix} 0 \\ 10.905 \end{bmatrix} u$$

Rule 2: IF x_1 is around 5 and x_2 is 0.25

$$\text{THEN } \dot{x} = \begin{bmatrix} -2.755 & 87.6 \\ -0.0008 & -85.239 \end{bmatrix} x + \begin{bmatrix} 0 \\ 10.905 \end{bmatrix} u$$

where $x = [x_1 \ x_2]^T$. The corresponding discrete time models are obtained as

Rule 1: IF $x_1(k)$ is around 0 and $x_2(k)$ is 0

$$\text{THEN } x(k+1) = \begin{bmatrix} 0.994 & 0 \\ 0 & 0.456 \end{bmatrix} x(k) + \begin{bmatrix} 0 & 0.076 \end{bmatrix} u(k)$$

Rule 2: IF $x_1(k)$ is around 5 and $x_2(k)$ is 0.25

$$\text{THEN } x(k+1) = \begin{bmatrix} 0.973 & 0.58 \\ 0 & 0.426 \end{bmatrix} x(k) + \begin{bmatrix} 0.036 \\ 0.073 \end{bmatrix} u(k)$$

The MATLAB code is as follows.

MATLAB code

```
LinDcmotor.m
function out=LinDcmotor(x);
R=7.2; L=0.0917; D=0.0004; KL=0.1236; J=0.0007046;
f = [-D/J*x(1)+KL/J*x(2)*x(2); -KL/L*x(1)*x(2)-
R/L*x(2)];
gradf = [-D/J -KL/L*x(2); 2*KL/J*x(2)
-KL/L*x(1)-R/L];
if(norm(x)==0)
a=gradf';
else
for i=1:2
a(i,:)=(gradf(:,i) + ((f(i)-
x'*gradf(:,i))/(norm(x)*norm(x)))*x)';
end
end
a
b=[0;1/L];
c=[1 0];
d=0;
sysc=ss(a,b,c,d);
sysd=c2d(sysc,0.01)
```

SUMMARY

This chapter has introduced basic concepts regarding fuzzy sets, fuzzy relations, inference mechanism, membership functions, and rule base, so that the readers can easily grasp various fuzzy logic control architectures existing in the literature. Two important models, namely the Mamdani model and the T–S fuzzy model are presented in detail. The conceptual ideas for the development of the Mamdani type fuzzy PI and fuzzy PD controllers have been enumerated. Nonlinear system identification using T–S fuzzy models has been illustrated using examples.

EXERCISES

1. How close is the geometry of a square to that of a rectangle of length a and width b? Define a suitable fuzzy membership function for the same.
2. Consider the following two fuzzy sets representing a car image and a truck image:

$$\text{Car} = \left\{ \frac{0.4}{\text{truck}}, \frac{0.3}{\text{scooter}}, \frac{0.2}{\text{boat}}, \frac{0.9}{\text{car}} \right\}$$

$$\text{Truck} = \left\{ \frac{1}{\text{truck}}, \frac{0.1}{\text{scooter}}, \frac{0.3}{\text{boat}}, \frac{0.5}{\text{car}} \right\}$$

Find the following:
   (a) Car $\cup$ truck   (b) Car $\cap$ truck   (c) $\overline{\text{Car}}$
   (d) Truck $\cup \overline{\text{truck}}$ (e) $\overline{\text{Car} \cap \text{truck}}$ (f) Car $\cap \overline{\text{car}}$

3. Consider the rule base of a simple fuzzy logic controller using an error signal e and change in error Δe as inputs:
   IF $e = $ P AND $\Delta e = $P THEN $c = $N
   IF $e = $ P AND $\Delta e = $N THEN $c = $Z
   IF $e = $ N AND $\Delta e = $P THEN $c = $Z
   IF $e = $ N AND $\Delta e = $N THEN $c = $P
   where c is the output of the controller and P, Z, and N denote positive, negative, and zero, respectively. Assume that the input variables have the following membership values in the input fuzzy sets:
   $\mu_N(e) = 0.4; \quad \mu_P(e) = 0.6$
   $\mu_N(\Delta e) = 0.3; \quad \mu_P(\Delta e) = 0.7$
   (a) Use the Mamdani inference to show the overall implied fuzzy output set graphically.
   (b) Defuzzify using (i) maximum value and (ii) centroid.
   (c) Find the individual implied output fuzzy sets N, Z, P. Use the centre of gravity defuzzification.

4. Given a fuzzy set $A = \left\{ \frac{0.2}{1}, \frac{0.3}{-1}, \frac{0.1}{2}, \frac{0.5}{3}, \frac{0.4}{-3} \right\}$ and a function $f(x) = 1 + x^2$ determine the fuzzy set $B = f(A)$ by using the extension principle.

5. Find the X- and Y- projections of the following relational matrix:

$$R = X \downarrow$$

	5	7	9
1	0.2	0.8	0.6
2	0.5	0.5	0.4
3	0.6	0.7	0.3

$Y \rightarrow$

6. Given the following rule:
    IF heatflow is excessive,
    THEN temperature is high
   and the relational matrix

$$R = \text{heatflow} \downarrow$$

	$100°C$	$200°C$	$300°C$
50 cal	0.8	0.7	0.9
100 cal	0.6	0.5	0.4
150 cal	0.3	0.5	0.2

temperature →

determine the membership distribution of very high temperature if the membership distribution of very excessive heat flow is given by

$$\mu_{\text{very excessive}} (\text{heatflow}) = \left\{ \frac{0.7}{50 \text{ cal}}, \frac{0.8}{100 \text{ cal}}, \frac{0.9}{200 \text{ cal}} \right\}$$

7. Following is the dynamical equation to control the velocity of a single-stage rocket:

$$\frac{dv(t)}{dt} = c(t) \left(\frac{m}{M - mt} \right) - g \left(\frac{R}{R + y(t)} \right) - 0.5v^2(t) \left(\frac{\rho_a A C_d}{M - mt} \right)$$
(3.15)

   where $v(t)$ is the rocket velocity at time t (output of the plant), $y(t)$ is the rocket altitude, and $c(t)$ is the velocity of the exhaust gases. In general, the exhaust gas velocity is proportional to the cross-sectional area of the nozzle, so we take it as the input. The system parameters are tabulated below.

   M, the initial mass of the rocket and fuel = 15000 kg
   m, the exhaust gases mass flow rate = 100 kg/s
   A, the maximum cross sectional area of the rocket = 1 m^2
   g, the acceleration due to gravity = 9.81 m/s^2
   R, the radius of earth = 6.37 × 10^6 m
   ρ_a, the air density = 1.21 kg/m^3
   C_d, the drag coefficient of the rocket = 0.3.

   (a) Using the Euler method of discretization, write down the discrete state–space model of the system.
   (b) Generate 1000 input-output data points using the derived discrete model and identify the T–S fuzzy model of the system using a fuzzy neural network.

8. The model of a magnetic ball suspension system is given by

$$M \frac{d^2 y(t)}{dt^2} = Mg - \frac{i^2(t)}{y(t)}$$

$$v(t) = Ri(t) + L \frac{di(t)}{dt}$$

   where $y(t)$ is the ball position in metres, $M = 0.1$ kg is the mass of the ball, $g = 9.81$ m/s^2 is the gravitational acceleration, $R = 50 \ \Omega$ is the resistance of the winding, $L = 0.5$ H is the winding inductance, $v(t)$ is the input voltage, and $i(t)$ is the winding current. The position of the ball is measured by a position sensor. We assume that the ball will stay between the magnetic coil and the ground level. The state space model of the system can be written as

$$\frac{dx_1(t)}{dt} = x_2(t)$$

$$\frac{dx_2(t)}{dt} = g - \frac{x_3^2(t)}{Mx_1(t)}$$

$$\frac{dx_3(t)}{dt} = -\frac{R}{L}x_3(t) + \frac{1}{L}v(t)$$

where the states are $x_1(t) = y(t)$, $x_2(t) = \dot{y}(t)$, and $x_3(t) = i(t)$.

(a) Linearize the system for the operating points $x_1 \in [0, 1, 2]$, $x_2 \in [0, 2, 4]$, and $x_3 \in [1, 5, 10]$ (the total number of rules is 27). Find the continuous time T–S fuzzy model of the system.

(b) See if the model and the original system generate the same output for an input given by $u(t) = 5 + 2\sin(t)$. If the model does not predict actual system response properly, then revisit step 1 with more number of rules.

4

Indirect Adaptive Control Using Neural Networks

The neural control schemes can be broadly classified as direct adaptive and indirect adaptive schemes following the similar notion within the classical adaptive control framework. Historically, neural control schemes were developed using indirect adaptive frameworks as methodologies for direct adaptive control were developed later. In the direct adaptive schemes, the controller is parametrized in terms of a neural network (NN) and the weight update law is derived such that the combination of the NN controller and the plant model in the feedback configuration is stable. In contrast, the indirect adaptive schemes use the explicit neural model of the plant to tune controller parameters. The conceptual framework for the indirect adaptive control for the continuous time system is given in Figure 4.1. The plant model identifier is captured using a neural network. Assuming that the plant model identifier represents the actual plant properly, the controller is designed such that the combination stabilizes the plant model identifier although the state feedback is taken from the actual plant.

Figure 4.1 Schematic structure of indirect adaptive control.

Thus, the indirect adaptive control is implemented in following two steps:

1. System identification: The neural model of the plant is explicitly learnt

4

Indirect Adaptive Control Using Neural Networks

The neural control schemes can be broadly classified as direct adaptive and indirect adaptive schemes following the similar notions within the classical adaptive control framework. Historically, neural control schemes were developed using indirect adaptive framework, as methodologies for direct adaptive control were developed later. In the direct adaptive schemes, the controller is parametrized in terms of a neural network (NN) and the weight update law is derived such that the combination of the NN controller and the plant model in the feedback configuration is stable. In contrast, the indirect adaptive schemes use the explicit neural model of the plant to tune controller parameters. The conceptual framework for the indirect adaptive control for the continuous time system is given in Figure 4.1. The plant model identifier is captured using a neural network. Assuming that the plant model identifier represents the actual plant properly, the controller is designed such that the controller stabilizes the plant model identifier although the state feedback is taken from the actual plant.

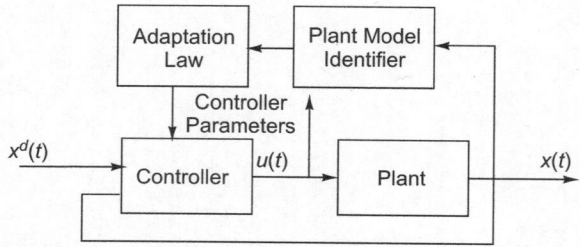

Figure 4.1 Schematic structure of indirect adaptive control

Thus, the indirect adaptive control is implemented in following two stages.

1. System identification: The neural model of the plant is explicitly learnt.

2. The control: The controller is designed so that the explicit model identified in step 1 is stable in the closed loop.

The scheme for discrete time system is also similar except that the variables $x^d(t)$, $u(t)$, and $x(t)$ are replaced by discrete time variables $x^d(k+1)$, $u(k)$, and $x(k+1)$, respectively. The indirect control concepts introduced in this chapter can be found in earlier works [70, 71, 72, 73, 74, 75].

Example 4.1 (Indirect adaptive control of a first-order linear system) The first-order plant is represented as

$$x(k+1) = ax(k) + bu(k) \tag{4.1}$$

It is assumed that the plant parameters a and b are unknown. Design an indirect adaptive control scheme for the above system.

Solution

The parameter vector is represented as $\hat{\theta} = [\hat{a}\ \hat{b}]^T$ that consists of unknown parameters. The regression vector is represented as $\phi(k) = [x(k)\ u(k)]^T$ which consists of the previous plant state and previous control input. Using the recursive least square (RLS) algorithm, the plant parameters can be estimated as

$$\hat{\theta}(k+1) = \hat{\theta}(k) + P(k)\phi(k)[1 + \phi^T(k)P(k)\phi(k)]^{-1}$$

$$[y(k+1) - \phi^T(k)\hat{\theta}(k)] \tag{4.2}$$

$$P(k+1) = P(k) - P(k)\phi(k)[1 + \phi^T(k)P(k)\phi(k)]^{-1}\phi^T(k)P(k) \tag{4.3}$$

where $P(k)$ is the covariance matrix. The derivation of the RLS algorithm is given in the appendix. Let us assume that the plant follows a desired trajectory $x^d(k+1)$, the one-step ahead controller is designed as

$$u(k) = \frac{1}{\hat{b}}\left(x^d(k+1) - \hat{a}x(k)\right) \tag{4.4}$$

Every sampling instant, the parameter vector is updated using (4.2), and is substituted in the control law (4.4). For clarity, the parameter vector is updated using the regression vector $\hat{\theta} = [\hat{a}\ \hat{b}]^T$ and the control input $u(k)$ is computed using (4.4). The simulation results for actual parameter vectors $a = 0.6$ and $b = 1$ are shown in Figure 4.2, where (a) shows how the parameters are learnt through the time instant k, (b) shows how $x(k)$ tracks the desired trajectory $x(k)$, and (c) shows the corresponding control input. The MATLAB code is given as follows:

```
clear all
a=0.6;b=1;
thh(1,1)=0.1;    thh(2,1)=0.1;
P=10*eye(2,2);    x(1)=0.5;
for k=1:50
xd(k+1)=sin(0.5*(k+1));
ah=thh(1,k);    bh=thh(2,k);
u(k)=(1/bh)*(xd(k+1)-ah*x(k));
x(k+1)=a*x(k)+b*u(k);
phi(:,k)=[x(k) u(k)]';
L(:,k)=P*phi(:,k)*inv(1+(phi(:,k))'*P*phi(:,k));
thh(:,k+1)=thh(:,k)+L(:,k)*(x(k+1)-(phi(:,k))'*thh(:,k));
P=P-L(:,k)*(phi(:,k))'*P;
end
thh(:,50)
```

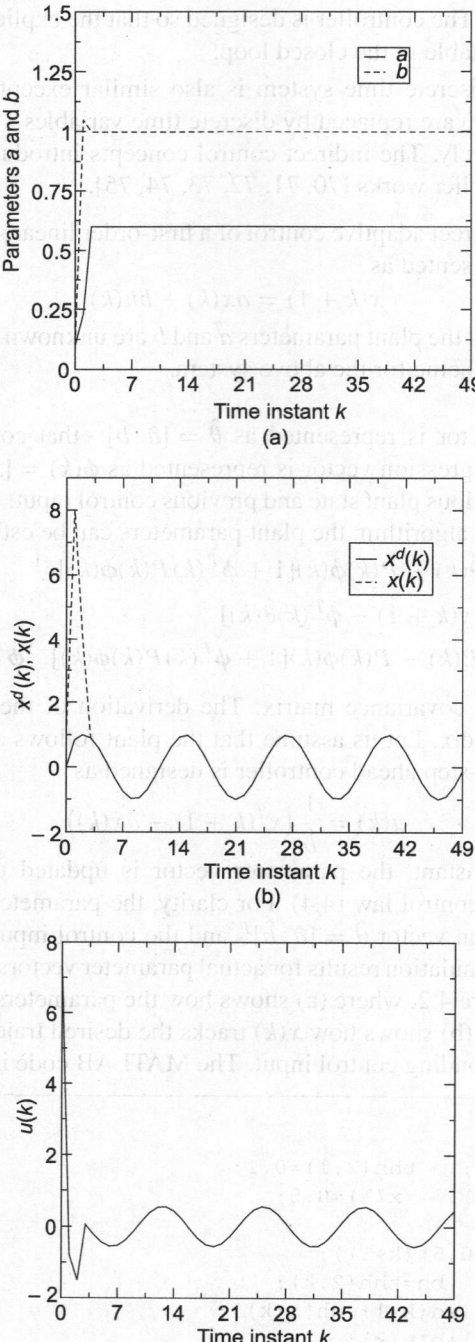

Figure 4.2 Example 4.1: (a) the learning of parameters a and b, (b) the trajectory tracking result, and (c) the control input $u(k)$

4.1 CONTINUOUS TIME AFFINE SYSTEMS

Consider the following non-linear affine continuous time system:

$$\dot{x} = f(x) + g(x)u \tag{4.5}$$

where $x \in R^n$ is the system state vector and $u \in R^m$ is the input vector. $f(x)$ is an $n \times 1$ vector while $g(x)$ is an $n \times m$ matrix. It is assumed that the non-linear functions $f(x)$ and $g(x)$ are unknown. Thus the first step is to perform explicit system identification of (4.5).

4.1.1 Model Identification

Before the parametric structure for the system identification can be selected, model (4.5) is expressed as

$$\dot{x} = -x + \overline{f}(x) + g(x)u \tag{4.6}$$

where $\overline{f}(x) = f(x) + x$. Let two RBF networks represent functions $\overline{f}$ and g, respectively. Then the neural network-based model of system (4.5) is given by

$$\dot{\hat{x}} = -\hat{x} + \hat{W}_1^T \phi_1(x) + \hat{W}_2^T \phi_2(x)u \tag{4.7}$$

where $\hat{W}_1 \in R^{L_1 \times n}$ and $\hat{W}_2 \in R^{L_2 \times n}$ are the weight matrices and $\phi_1 \in R^{L_1 \times 1}$ and $\phi_2 \in R^{L_2 \times m}$ are the basis functions of the networks, respectively. L_1 and L_2 are the number of hidden neurons in the networks.

The NN-based model (4.7) fairly approximates (4.5) if we can derive the appropriate update laws for $\hat{W}_1$ and $\hat{W}_2$ such that the system response of (4.7) follows that of actual plant (4.5).

Let us assume that there exists two ideal weight matrices W_1 and W_2 such that $\overline{f}(x)$ and $g(x)$ can be written as

$$\overline{f}(x) = W_1^T \phi_1(x), \quad g(x) = W_2^T \phi_2(x)$$

Readers should note that it may not be possible to find a NN that can exactly approximate a non-linear function. However, the approximation errors have not been considered here to present the concept in a simplified manner. Thus the actual plant dynamics (4.6) in terms of ideal NN weights are given by

$$\dot{x} = -x + W_1^T \phi_1(x) + W_2^T \phi_2(x)u \tag{4.8}$$

The error vector between the system states and model states is defined as $e = x - \hat{x}$. Thus the error dynamics between the actual plant and the identified model can be derived by subtracting (4.7) from (4.8):

$$\dot{e} = \dot{x} - \dot{\hat{x}}$$

$$= -x + W_1^T \phi_1(x) + W_2^T \phi_2(x)u + \hat{x} - \hat{W}_1^T \phi_1(x) - \hat{W}_2^T \phi_2(x)u$$

$$= -e + \tilde{W}_1^T \phi_1(x) + \tilde{W}_2^T \phi_2(x)u$$

To derive the weight update laws, let us consider a Lyapunov function candidate. V as

$$V = \frac{1}{2} e^T e + \text{tr} \left\{ \frac{1}{\alpha_1} \tilde{W}_1^T \tilde{W}_1 + \frac{1}{\alpha_2} \tilde{W}_2^T \tilde{W}_2 \right\}$$

where tr represents trace of a matrix. Taking the derivative

$$\dot{V} = e^T \dot{e} + \text{tr} \left\{ \frac{1}{\alpha_1} \tilde{W}_1^T \dot{\tilde{W}}_1 + \frac{1}{\alpha_2} \tilde{W}_2^T \dot{\tilde{W}}_2 \right\}$$

Since W_1 and W_2 are constant matrices, we can write

$$\dot{V} = e^T \dot{e} + \text{tr} \left\{ -\frac{1}{\alpha_1} \tilde{W}_1^T \dot{\hat{W}}_1 - \frac{1}{\alpha_2} \tilde{W}_2^T \dot{\hat{W}}_2 \right\}$$

$$= e^T \left(-e + \tilde{W}_1^T \phi_1(x) + \tilde{W}_2^T \phi_2(x) u \right) +$$

$$\text{tr} \left\{ -\frac{1}{\alpha_1} \tilde{W}_1^T \dot{\hat{W}}_1 - \frac{1}{\alpha_2} \tilde{W}_2^T \dot{\hat{W}}_2 \right\}$$

Using the properties of trace, we can write

$$\dot{V} = - e^T e +$$

$$\text{tr} \left\{ \tilde{W}_1^T \phi_1(x) e^T + \tilde{W}_2^T \phi_2(x) u e^T - \frac{1}{\alpha_1} \tilde{W}_1^T \dot{\hat{W}}_1 - \frac{1}{\alpha_2} \tilde{W}_2^T \dot{\hat{W}}_2 \right\}$$

The following update laws

$$\dot{\hat{W}}_1 = \alpha_1 \phi_1(x) e^T \qquad (4.9)$$

$$\dot{\hat{W}}_2 = \alpha_2 \phi_2(x) u e^T \qquad (4.10)$$

will make $\dot{V} = -e^T e$ that is negative definite. Since V is taken to be positive definite and $\dot{V}$ turns out to be negative definite, e will converge to 0. Thus the model states will match the system states as time progresses which serves the purpose of system identification.

4.1.2 Controller Design

Once the system is identified using the above-described method, feedback linearization can be applied to design a control law based on the identified model provided g is invertible. Suppose that g is a square matrix and is invertible. Also suppose that the system should follow a desired trajectory x^d. In feedback linearization technique, as described in Chapter 1, the control law is designed such that the error dynamics become linear as well as stable. Here the objective is to design a control law for the identified model such that the model state vector tracks x^d. If the model is an accurate representation of the system, the system states will also follow the same desired trajectory. Let us consider the following control law:

$$u = \left(\hat{W}_2^T \phi_2(x) \right)^{-1} \left[\dot{\hat{x}} - \hat{W}_1^T \phi_1(x) + \dot{x}^d + K_v \hat{e} \right] \qquad (4.11)$$

where x^d is the desired state vector and $\hat{e} = x^d - \hat{x}$, K_v is a design parameter. Putting the above control law in Eqn (4.7), we get

$$\dot{\hat{x}} = -\hat{x} + \hat{W_1}^T \phi_1(x) + \hat{W_2}^T \phi_2(x) \left(\hat{W_2}^T \phi_2(x) \right)^{-1}$$
$$\left[\hat{x} - \hat{W_1}^T \phi_1(x) + \dot{x}^d + K_v \hat{e} \right]$$
$$\Rightarrow \dot{\hat{e}} = -K_v \hat{e} \tag{4.12}$$

If we consider a Lyapunov function candidate as $V = \frac{1}{2}\hat{e}^T \hat{e}$, then $\dot{V}$ can be written as

$$\dot{V} = \hat{e}^T \dot{\hat{e}} = -\hat{e}^T K_v \hat{e}$$

that is negative definite. Thus, $\hat{e}$ will go to 0 with time, which serves the purpose of tracking.

When the matrix $g(x)$ is a rectangular matrix, the negative definiteness of $\dot{V}$ cannot be ensured because the use of pseudo-inverse of $g(x)$ will not exactly cancel the nonlinearity $f(x)$. For those cases, if the system is a single input single output system, then mostly it can be modelled in a strict feedback form like

$$\dot{\hat{x}}_1 = \hat{x}_2$$
$$\dot{\hat{x}}_2 = \hat{x}_3$$
$$\vdots$$
$$\dot{\hat{x}}_n = \hat{f}(x) + \hat{g}(x)u \tag{4.13}$$
$$\hat{y} = \hat{x}_1 \tag{4.14}$$

where $\hat{f}(x)$ and $\hat{g}(x)$ are scalar functions. Let us define the output tracking error as $\hat{e} = y^d - \hat{y} = x_1^d - \hat{x}_1$ and another variable $r = \hat{e}^{(n-1)} + \lambda_1 \hat{e}^{(n-2)} + \cdots + \lambda_{n-1}\hat{e}$ (power denotes respective derivatives). r is often known as filtered tracking error in the literature. If the control law is chosen as

$u = \frac{1}{\hat{g}(x)}[-\hat{f}(x) + k_v r + \lambda_1 \hat{e}^{(n-1)} + \cdots + \lambda_{n-1}\hat{e}^{(1)} + \dot{x}_{nd}]$, we can write

$$\dot{r} = -k_v r$$

which is a linear and stable dynamics. Considering a Lyapunov function $V = \frac{1}{2}r^2$ we can show that r goes to 0 with time. $\lambda_1 \cdots \lambda_{n-1}$ are the design parameters which can be chosen such that the dynamics $r = \hat{e}^{(n-1)} + \lambda_1 \hat{e}^{(n-2)} + \cdots + \lambda_{n-1}\hat{e}$ is stable.

Example 4.2 Let us consider the example of a liquid level control system for the surge tank shown in Figure 4.3.

The dynamics of surge tank is represented by the following differential equation:

$$\frac{dh(t)}{dt} = \frac{-\sqrt{2gh(t)}}{A(h(t))} + \frac{1}{A(h(t))}u(t) \tag{4.15}$$

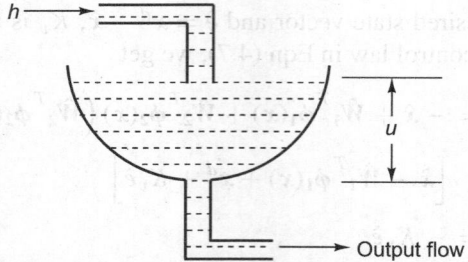

Figure 4.3   Schematic diagram of a surge tank system

where $u(t)$ is the input flow, $h(t)$ is the liquid level of the tank (output of the system), and $A(h(t)) = \sqrt{ah^2(t) + b}$ is the cross-sectional area of the tank. The input $u(t)$ can be both positive or negative, i.e., it can pull liquid out of the tank as well as put it in the tank. The parameters a and b are given as $a = 1$ and $b = 2$. The objective is to learn the dynamics of the surge tank using the affine model (4.7) and to find a suitable controller using feedback linearization as in (4.11) for set point tracking.

Solution

Defining the system output as $h(t)$ and model output as $h^m(t)$, we can write

$$\dot{h}^m(t) = -h^m(t) + \hat{W}_1^T \phi_1(h) + \hat{W}_2^T \phi_2(h)u \qquad (4.16)$$

where $\hat{W}_1$ and $\hat{W}_2$ are two weight vectors of dimensions 30 and 20, respectively. ϕ_1 and ϕ_2 are taken as Gaussian functions. The learning parameters are taken as $\alpha_1 = 0.8$ and $\alpha_2 = 0.5$. For a sinusoidal input u with the random magnitude between 5 and 10, the system and model dynamics are evolved while updating the weights using the update laws (4.9) and (4.10), where $e = h - h^m$. The identification result is shown in the Figure 4.4(a), where the solid line represents the system output and the dashed line represents the model output. Based on the model dynamics,

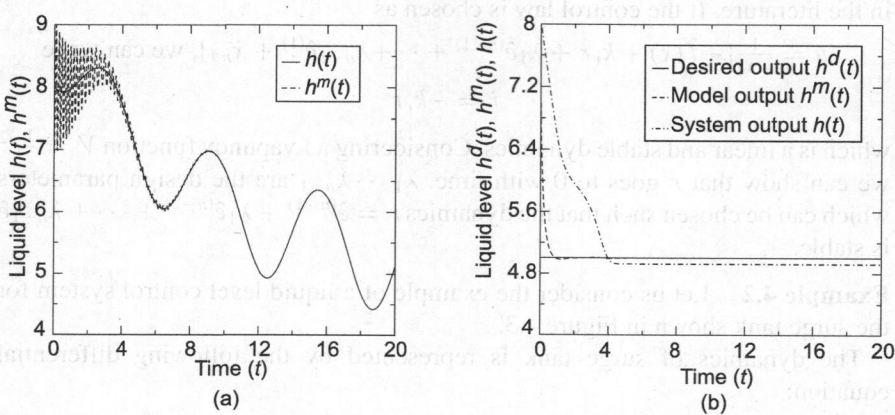

Figure 4.4   (a) The identification of surge tank system using an affine neural model and (b) the set point tracking result using a feedback linearization-based control law based on the identified model

a control input is generated using Eqn (4.11) to achieve a desired liquid level of 5 m. The set point tracking result is shown in Figure 4.4(b). The corresponding control input is shown in Figure 4.5.

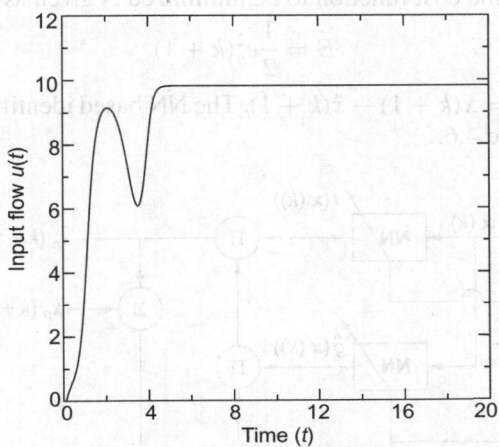

Figure 4.5 Control input for tracking a set point of 5 m

4.2 DISCRETE TIME AFFINE SYSTEMS

A large class of single input single output discrete time affine non-linear systems can be represented in the strict feedback form as follows:

$$x_1(k + 1) = x_2(k)$$

$$x_2(k + 1) = x_3(k)$$

$$\vdots$$

$$x_n(k + 1) = f(x(k)) + g(x(k))u(k) \qquad (4.17)$$

where $x(k) = [x_1(k)\, x_2(k) \cdots x_n(k)]^T \in R^n$, $y(k) \in R$, and $u(k) \in R$. $f(x(k))$ and $g(x(k))$ are two smooth functions and it is assumed that $g(x(k))$ does not take 0 value, i.e., the system is feedback linearizable.

4.2.1 Model Identification

The system (4.17) can be learnt using a neural network model as

$$\hat{x}_1(k + 1) = x_2(k)$$

$$\hat{x}_2(k + 1) = x_3(k)$$

$$\vdots$$

$$\hat{x}_n(k + 1) = \hat{f}(x(k)) + \hat{g}(x(k))u(k) \qquad (4.18)$$

where $\hat{f}(x(k))$ and $\hat{g}(x(k))$ are the outputs of two NNs. Since the model is so chosen that the first $n-1$ states of the model follow the first $n-1$ states of the system, the weights of the NNs should be updated such that $\hat{x}_n(k+1)$ follows $x_n(k+1)$. Thus, the cost function to be minimized is given as follows:

$$E = \frac{1}{2}e_n^2(k+1)$$

where $e_n(k+1) = x(k+1) - \hat{x}(k+1)$. The NN based identification framework is shown in Figure 4.6.

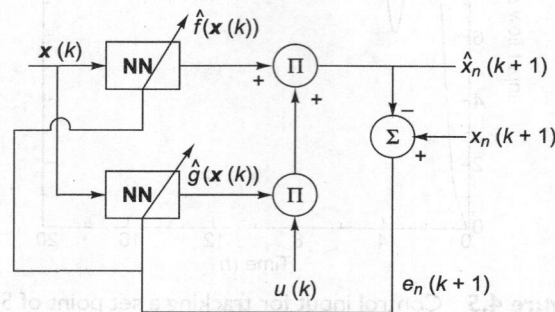

Figure 4.6 Identification of affine systems using feed-forward networks

If the weights of the neural networks are denoted by W_1 and W_2, then we can write

$$\hat{x}_n(k+1) = \hat{f}(W_1, x(k)) + \hat{g}(W_2, x(k))u(k)$$

The weights W_1 and W_2 can be updated either by using an RLS algorithm as described in the appendix or by using a gradient descent algorithm, i.e.,

$$W_1(k+1) = W_1(k) - \eta_1 \frac{\partial E}{\partial W_1} \tag{4.19}$$

$$W_2(k+1) = W_2(k) - \eta_2 \frac{\partial E}{\partial W_2} \tag{4.20}$$

4.2.2 Controller Design

Suppose that a system follows a desired state trajectory $x_d(k)$. Since the model is identified such that the model states follow the system states, the controller can be designed such that the model state vector $\hat{x}(k)$ follows the desired state vector $x_d(k)$. Using the concept of feedback linearization, the control law can be obtained as follows:

$$u(k) = \frac{1}{\hat{g}(x(k))}[-\hat{f}(x(k)) + k_v r(k) +$$

$$\lambda_1 \hat{e}_n(k) + \cdots + \lambda_{n-1}\hat{e}_2(k) + x_{nd}(k+1)] \tag{4.21}$$

where $\hat{e}_i(k) = x_{id}(k) - \hat{x}_i(k)$ and $r(k) = \hat{e}_n(k) + \lambda_1 \hat{e}_{n-1}(k) + \cdots + \lambda_{n-1}\hat{e}_1(k)$. The detail of feedback linearization design technique for discrete time system can be found in Chapter 5. At each sampling instant the weight vectors of the neural networks are updated using (4.19) and (4.20), and is substituted in the control law (4.21). The parameters k_v and λs are design parameters which are chosen heuristically. We should note that if $\hat{g}(x(k))$ is 0 then $u(k)$ is unbounded. This can be avoided by adding a small constant in the denominator. Stability analysis for such a control law can be found in Chapter 5.

Example 4.3   Consider the dynamics of a single link manipulator:

$$\dot{x}_1(t) = x_2(t)$$

$$\dot{x}_2(t) = -10\sin(x_1) + u(t)$$

Represent the system in a discrete strict feedback form. Identify the discrete dynamics using a NN model as given in Eqn (4.18). Design an adaptive control law for the system using the control law (4.21) such that $x_1(k)$ follows a desired trajectory of $\sin(0.1k)$.

Solution
The discretized manipulator dynamics is as follows:

$$x_1(k+1) = x_1(k) + \text{DT}\, x_2(k)$$

$$x_2(k+1) = x_2(k) + \text{DT}\,(-10\sin(x_1(k)) + u(k))$$

where DT is the sampling interval. Considering $z_1(k) = x_1(k)$ and $z_2(k) = x_1(k) + \text{DT}\, x_2(k)$ we can write

$$z_1(k+1) = z_2(k)$$

$$z_2(k+1) = 2z_2(k) - z_1(k) + \text{DT}^2(-10\sin(z_1(k)) + u(k))$$

Note that g is constant in this case. Thus we only approximate f for the above system. Let us assume that there exist an RBF network weight vector W_1 such that the system can be approximated using the following model:

$$\hat{z}_1(k+1) = z_2(k)$$

$$\hat{z}_2(k+1) = W_1^T \phi_1(z(k)) + \text{DT}^2 u(k)$$

where $z(k) = [z_1(k)\ z_2(k)]^T$. The cost function to be minimized is

$$E = \frac{1}{2}(z_2(k+1) - \hat{z}_2(k+1))^2$$

The weight update law using the gradient descent algorithm can be computed as follows:

$$W_1(k+1) = W_1(k) - \eta_1 \frac{\partial E}{\partial W_1}$$

$$= W_1(k) + \eta_1(z_2(k+1) - \hat{z}_2(k+1))\frac{\partial \hat{z}_2(k+1)}{\partial W_1}$$

$$= W_1(k) + \eta_1(z_2(k+1) - \hat{z}_2(k+1))\phi_1$$

The system dynamics is learnt using the above weight update law where the control input $u(k)$ is generated from the output of a PD controller. The number of RBF neurons is taken as 50. The results of system identification are shown in Figure 4.7, where the solid line represents the system state z_2 and the dashed line represents the model state $\hat{z}_2$.

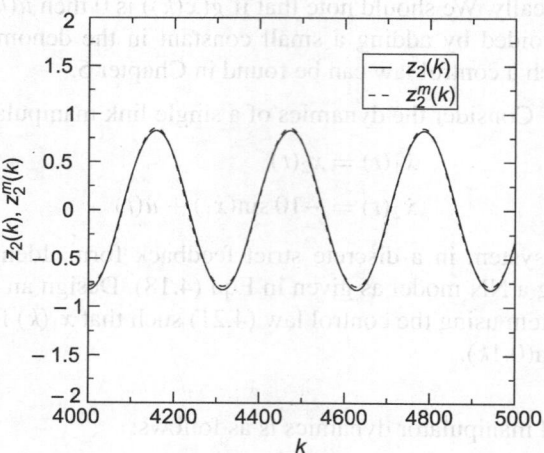

Figure 4.7 Identification result

Once the dynamics is learnt, the control law to track a desired trajectory of $z_{1d}(k) = \sin(0.1\,k)$ is given by

$$u(k) = \frac{1}{DT^2}[-W_1^T \boldsymbol{\phi}_1(z(k)) + k_v r(k) + \lambda_1 \hat{e}_2(k) + z_{2d}(k+1)]$$

where $r(k) = \hat{e}_2(k) + \lambda_1 \hat{e}_1(k)$, $e_1(k) = z_{1d}(k) - z_1(k)$, $\hat{e}_2(k) = z_{2d}(k) - z_2(k)$. k_v and λ_1 are taken as 0.5 and 0.1, respectively.

The tracking result and corresponding control input are shown in Figure 4.8. The RMS tracking error is found to be 0.023.

4.3 DISCRETE TIME NONAFFINE SYSTEM

Consider the following nonaffine discrete time system:

$$x(k+1) = f(x(k), u(k)) \tag{4.22}$$

where $x(k) \in R^n$ and $u(k) \in R^p$ represent respectively the state and input vectors of the system at the kth sampling instant.

4.3.1 Model Identification

The states of the systems are assumed to be accessible, and non-linear function $f(\cdot)$ is assumed to be unknown. The unknown mapping $f(\cdot)$ can be modelled using an $n + p$ input and n output RBF network as shown in Figure 4.9. The ith output of such a network can be expressed as

$$\hat{x}_i = \hat{f}_i(v) = \sum_{j=1}^{L} W_{ij}\phi_j(\| v - c_j \|) \tag{4.23}$$

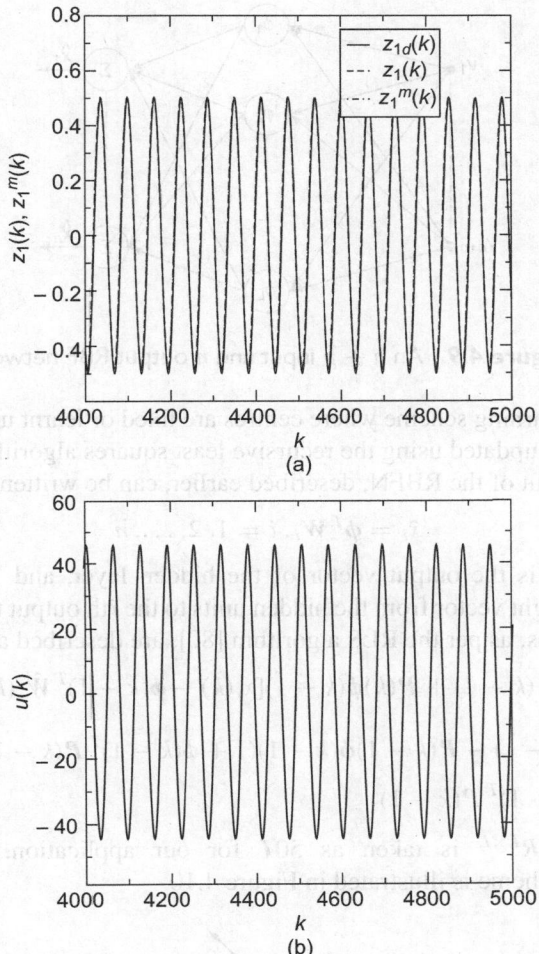

Figure 4.8   Tracking result: (a) the desired output, system output, and model output and (b) the corresponding control input

where $v \in R^m$ is the network input vector, $\| \cdot \|$ denote the Euclidean norm, $c_j \in R^{n+p}$, $1 \leq j \leq L$, are RBF centres, $\phi(\cdot)$ is the jth activation function of the hidden layer, W_{ij}, $1 \leq j \leq L$, $1 \leq i \leq n$ are the connection weights from the hidden layer to the output layer, and L is the number of hidden units in the first layer.

The centres and weights of an RBF network can be tuned using ideas from a non-linear system identification theory such as the parallel recursive prediction error (PRPE) algorithm [76] or the extended Kalman filtering (EKF) algorithm [72, 77]. The simplest approach is to update the centres using the gradient descent algorithm, and the weights can be updated using the simple LMS [78] algorithm. Although computational requirement increases by adjusting centres, the number of radial centres can be substantially reduced by this approach [79]. The generalization performance of such a network is much better as compared

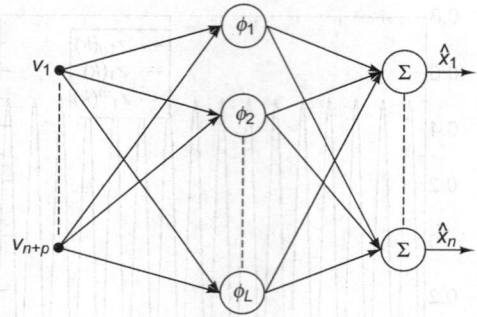

Figure 4.9 An $n + p$ input and n output RBF network

to the hybrid learning scheme where centres are fixed or learnt unsupervised and the weights are updated using the recursive least squares algorithm.

The ith output of the RBFN, described earlier, can be written as

$$\hat{x}_i = \boldsymbol{\phi}^T \boldsymbol{W}_i, \, i = 1, 2, \ldots, n \tag{4.24}$$

where $\boldsymbol{\phi} \in R^L$ is the output vector of the hidden layer, and $\boldsymbol{W}_i \in R^L$ is the connection weight vector from the hidden units to the ith output unit. The weight update equations, as per the RLS algorithm [80], are described as

$$\hat{\boldsymbol{W}}_i(k) = \hat{\boldsymbol{W}}_i(k - 1) + \boldsymbol{P}(k)\boldsymbol{\phi}(k - 1)[x_i(k) - \boldsymbol{\phi}(k - 1)^T \hat{\boldsymbol{W}}_i(k - 1)] \tag{4.25}$$

$$\boldsymbol{P}(k) = \boldsymbol{P}(k - 1) - \boldsymbol{P}(k - 1)\boldsymbol{\phi}(k - 1)(1 + \boldsymbol{\phi}(k - 1)^T \boldsymbol{P}(k - 1)\boldsymbol{\phi}(k - 1))^{-1}$$
$$\boldsymbol{\phi}(k - 1)^T \boldsymbol{P}(k - 1) \tag{4.26}$$

where $\boldsymbol{P}(k) \in R^{L \times L}$ is taken as $50\boldsymbol{I}$ for our application. The complete identification scheme is illustrated in Figure 4.10.

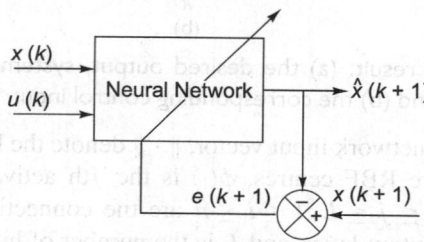

Figure 4.10 System identification using a feed-forward network

4.3.2 Controller Design: Traditional NN Approach

The NN-based indirect adaptive control follows the scheme as shown in Figure 4.11. The control law $u(k)$ is parametrized using an RBFN as

$$u(k) = \boldsymbol{W}_C^T \boldsymbol{\psi} \tag{4.27}$$

where $\boldsymbol{W}_C \in R^{l_2 \times p}$ is the weight matrix of the network, l_2 is the number of hidden layer neurons, and $\boldsymbol{\psi} \in R^{p \times 1}$ is the basis function. The input to this network is the

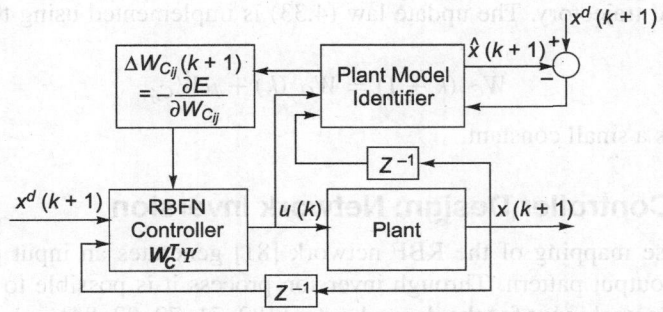

Figure 4.11 NN-based indirect adaptive control

desired state trajectory $x^d(k + 1)$ and the actual plant feedback $x(k)$. The weight vector W_C is updated using a simple gradient descent algorithm as

$$W_{C_{ij}}(k + 1) = W_{C_{ij}}(k) - \eta \frac{\partial E}{\partial W_{C_{ij}}} \quad \text{for } i = 1, \ldots, p \text{ and } j = 1, \ldots, l_2 \quad (4.28)$$

where the error function E is given as

$$E = \frac{1}{2}(\tilde{x}^T \tilde{x}) \quad \text{where} \quad \tilde{x} = x^d - \hat{x} \quad (4.29)$$

Readers should note that the term $\frac{\partial E}{\partial W_{C_{ij}}}$ can be computed using the system model $\hat{W}^T \phi$ as follows:

$$\frac{\partial E}{\partial W_{C_{ij}}} = \frac{\partial E}{\partial u_i} \frac{\partial u_i}{\partial W_{C_{ij}}} \quad (4.30)$$

Another approach to find the update law for W_C is using a Lyapunov function approach as discussed in chapter 2. The Lyapunov function candidate V is selected as the quadratic error function in desired trajectories:

$$V = \frac{1}{2}(\tilde{x}^T \tilde{x}) \quad \text{where} \quad \tilde{x} = x^d - \hat{x} \quad (4.31)$$

where x^d is the desired output activation and $\hat{x}$ is the actual output activation of the model. The time derivative of the Lyapunov function V is given by

$$\dot{V} = -\tilde{x}^T \sum \frac{\partial \hat{x}}{\partial u_i} \frac{\partial u_i}{\partial W_{C_{ij}}} \dot{W}_{C_{ij}} \quad (4.32)$$

Let the term $\frac{\partial \hat{x}}{\partial u_i} \frac{\partial u_i}{\partial W_{C_{ij}}}$ be denoted by J_u. If the weight update law is given by the expression

$$\dot{W}_{C_{ij}} = \frac{\| \tilde{x} \|^2}{\| J_u^T \tilde{x} \|^2} J_u^T \tilde{x} \quad \text{for } i = 1, \ldots, p \text{ and } j = 1, \ldots, l_2 \quad (4.33)$$

then $\dot{V} = - \| \tilde{x} \|$. Thus, the update law is stable and the controller will drive the model to follow the desired trajectory. With the assumption that the model approximates the actual plant properly, it can be said that the plant will also follow the desired trajectory. The update law (4.33) is implemented using the discrete version as

$$W_{C_{ij}}(k + 1) = W_{C_{ij}}(k) + \mu \dot{W}_{C_{ij}} \tag{4.34}$$

where μ is a small constant.

4.3.3 Controller Design: Network Inversion

The inverse mapping of the RBF network [81] generates an input pattern for a desired output pattern. Through inversion process it is possible to obtain the requisite control input for the desired output [82, 71, 72, 83, 84].

The RBFN model (as given by 4.23) represents a non-linear mapping from an $(n + p)$-dimensional input space to an n-dimensional output space. The objective of the inverse operation on this model is to predict only p inputs out of $n + p$ number of total inputs. The remaining n inputs are known *a priori* (present system states). The predicted input can be mathematically expressed as

$$\hat{u}(k) = g\left(\hat{x}(k), x^d(k + 1), c, W\right) \tag{4.35}$$

The inversion algorithm predicts these p inputs by updating input activation $\hat{u}(k)$ iteratively till the desired output activation is achieved or the number of iterations reaches a maximum t_{max}. This upper bound is decided by sampling the interval and computation time required per iteration. The initial guess of the input activation $\hat{u}(k)$ during each sampling interval is taken as the input activation $u(k - 1)$ predicted in the previous sampling instant. For the case of first sampling interval, the initial guess is selected arbitrarily from the input space. The controller structure is given in Figure 4.12.

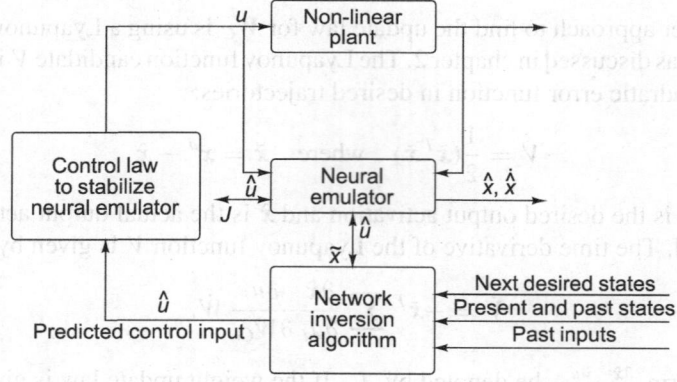

Figure 4.12 Indirect adaptive controller using network inversion

Three different inversion algorithms that will be enumerated in this chapter are as follows:

1. Gradient search (GS) in the input space
2. Lyapunov function (LF) approach
3. Extended Kalman filter (EKF) based approach

Gradient Search in Input Space

The iterative inversion of the RBF network can be carried out using the gradient descent algorithm as proposed by Linden and Kindermann [17]. The iterative rule is as follows:

$$\hat{u}_i^{t+1}(k) = \hat{u}_i^t(k) - \eta \frac{\partial E}{\partial \hat{u}_i^t(k)} + \alpha[\hat{u}_i^t(k) - \hat{u}_i^{t-1}(k)]$$

where t = iterative step, η = learning rate, and α = momentum rate. The error function is given by

$$E = \frac{1}{2} \sum_{i=1}^{n} (x_i^d - \hat{x}_i)^2$$

Then the partial derivative of error function is

$$\frac{\partial E}{\partial u_i} = - \sum_{j=1}^{n} (x_j^d - \hat{x}_j) \frac{\partial \hat{x}_j}{\partial u_i}$$

$$\frac{\partial \hat{x}_j}{\partial u_i} = \sum_{k=1}^{l} \frac{\partial \hat{x}_j}{\partial \phi_k} \frac{\partial \phi_k}{\partial u_i} \qquad (4.36)$$

Lyapunov Function Approach

The inverse mapping based on the Lyapunov function as a general means of achieving a recall process for selective attention as applicable to a pattern recognition process has been presented by Lee. We adapt the same concept for control applications in the following way. Network inversion can be achieved using the Lyapunov function approach very efficiently. The advantage of this approach is that convergence is guaranteed since the algorithm is derived using the Lyapunov stability concept.

If we choose a Lyapunov function candidate $V(x(t), t)$ such that

1. $V(x(t), t)$ is positive definite
2. $\dot{V}(x(t), t)$ is negative definite

then the system is *asymptotically stable.*

The Lyapunov function candidate V is chosen to be a quadratic error function in desired trajectories:

$$V = \frac{1}{2}(\tilde{x}^T \tilde{x}) \quad \text{where} \quad \tilde{x} = x^d - \hat{x} \qquad (4.37)$$

where x^d is the desired output activation and $\hat{x}$ is the actual output activation of the RBFN model. The time derivative of the Lyapunov function V is given by

$$\dot{V} = -\tilde{x}^T \frac{\partial \hat{x}}{\partial u} \dot{u} = -\tilde{x}^T J \dot{u} \qquad (4.38)$$

where

$$J = \frac{\partial \hat{x}}{\partial u} \quad J \in R^{n \times p} \tag{4.39}$$

Theorem 4.1   If an arbitrary initial input activation $u(0)$ is updated by

$$u(t') = u(0) + \int_0^{t'} \dot{u} \, dt \tag{4.40}$$

where

$$\dot{u} = \frac{\| \tilde{x} \|^2}{\| J^T \tilde{x} \|^2} J^T \tilde{x} \tag{4.41}$$

then $\tilde{x}$ converges to zero under the condition that $\dot{u}$ exists along the convergence trajectory.

Proof: Substitution for u in $\dot{V}$ yields

$$\dot{V} = - \| \tilde{x} \|^2 \leq 0$$

where $\dot{V} < 0$ for all $\tilde{x} \neq 0$ and $\dot{V} = 0$ if and only if $\tilde{x} = 0$. Thus the update law is stable and $\tilde{x}$ converges to 0 in time. The iterative input activation update rule can be given as

$$u(t') = u(t' - 1) + \mu \dot{u}(t' - 1) \tag{4.42}$$

where μ is a small constant representing the update rate and t' represents the iterative index. The possible numerical instability associated with the weight update law can be avoided by adding a small positive constant ϵ in the denominator. In that case $\dot{u}$ modifies to

$$\dot{u} = \frac{\| \tilde{x} \|^2}{\| J^T \tilde{x} \|^2 + \epsilon} J^T \tilde{x} \tag{4.43}$$

Using the above equation, $\dot{V}$ becomes

$$\dot{V} = -\| \tilde{x} \|^2 \frac{\| J^T \tilde{x} \|^2}{\| J^T \tilde{x} \|^2 + \epsilon} = -\alpha \| \tilde{x} \|^2$$

where $0 < \alpha < 1$. Since α is positive, $\dot{V}$ is negative semi-definite. Thus V will decrease with the update of u, so as the tracking error $\tilde{x}$. Once the update of u is over, input $u(k)$ at kth instant is assigned to the updated value $u(i')$ and applied to the actual system.

Extended Kalman Filtering Based Inversion

Extended Kalman Filtering (EKF) is a method of estimating the state vector. We assign the unknown input vector $u(k)$ as the state vector to be estimated. Assuming update of individual input u_i to be independent of other input updates, the RBFN model as a function of only the ith input to be estimated is given by

$$\hat{x}(k + 1) = \hat{f}(x(k), u(k), c, W) = h(u_i(k)) \tag{4.44}$$

It should be noted that during the network inversion the estimation of control input $u(k)$ is completely independent of the feedback loop of the control system. Thus argument k is redundant. Now we introduce a new argument t, which refers to the iterative step in carrying out the desired inversion. Thus the RBFN model can be described by following non-linear equations where the vector $u(t)$ refers to the states of the neural network:

$$u_i(t+1) = u_i(t) \tag{4.45}$$

$$x^d = h(u_i(t)) + \xi(t) = \hat{x}(t) + \xi(t) \tag{4.46}$$

Here $\xi(t)$ is assumed to be the white noise vector with $n \times n$ covariance matrix $R(t)$. The application of EFK to Eqns (4.45) and (4.46) gives the following inversion algorithm using the matrix inversion lemma:

$$\hat{u}_i(t) = \hat{u}_i(t-1) + K_i(t)(x^d - \hat{x}(t)) \tag{4.47}$$

$$K_i(t) = P(t-1)H_i(t)^T \left(H_i(t)P_i(t-1)H_i(t)^T + R_i(t) \right)^{-1} \tag{4.48}$$

$$P_i(t) = P_i(t-1) - P_i(t-1)K_i(t)H_i(t) \tag{4.49}$$

where $K_i(t)$, the $(1 \times n)$ matrix, is called the Kalman gain and $H_i(t)$, the $(n \times 1)$ matrix, is defined as

$$H_i(t) = \left[\frac{\partial h}{\partial u_i} \right]_{u = \hat{u}(t-1)} \tag{4.50}$$

The jth element of $H_i(t)$ can be expressed for the case of thin plate spline activation as follows:

$$H_{ij} = \frac{\partial h_j}{\partial u_i} = \frac{\partial \hat{x}_j}{\partial u_i} = \sum_{k=1}^{L} W_{jk}(1 + 2\log(d(k)))(u_i - c_{k,i}) \tag{4.51}$$

where $d(k) = \| v - c_k \|$.

The covariance matrix $R(t)$ is assumed to be diagonal. This assumption avoids the matrix inversion involved in Eqn (4.48). Applying the matrix inversion lemma, we have

$$\left(H_i(t)P_i(t-1)H(t)^T + R_i(t) \right)^{-1} = \frac{1}{\lambda} \left(I - \frac{P_i(t-1)H_i(t)H_i(t)^T}{\lambda + P_i(t-1)H_i(t)^T H_i(t)} \right) \tag{4.52}$$

The inversion algorithm is thus simplified as

$$\hat{u}_i(t) = \hat{u}_i(t-1) + K_i(t)(x^d - \hat{x}(t)) \tag{4.53}$$

$$K_i(t) = \frac{1}{\lambda(t)} P_i(t-1)H_i(t)^T \left(I - \frac{P_i(t-1)H_i(t)H_i(t)^T}{\lambda(t) + P_i(t-1)H_i(t)^T H_i(t)} \right) \tag{4.54}$$

$$P_i(t) = P_i(t-1) - K_i(t)H_i(t)P_i(t-1) \tag{4.55}$$

As covariance matrix $R(t)$ is unknown *a priori*, λ is estimated on-line using the following recursion:

$$\hat{\lambda}(t) = \hat{\lambda}(t-1) + \nu(t) \left(\frac{(x^d - \hat{x}(t))^T (x^d - \hat{x}(t))}{n} - \hat{\lambda}(t-1) \right) \qquad (4.56)$$

where $\nu(t) = \frac{1}{t}$

Example 4.4 (Surge tank) Let us consider the previous example of a surge tank system as described in Eqn (4.57). The objective is to learn the dynamics of the surge tank using an RBFN model and to find a suitable controller both for the set point and trajectory tracking using the network inversion algorithm.

System identification: Since the identification of the system is done in discrete time, the Euler representation of the system with a sampling time $T = 0.1$ s is used for generating the data. The input flow is assumed to vary between -50 and $+50$. A total of 3000 training data $u(k), h(k), h(k+1)$ have been generated using the following discrete model:

$$h(k+1) = h(k) + T \left[\frac{-\sqrt{2gh(k)}}{\sqrt{(h^2(k)+1)}} + \frac{1}{\sqrt{(h^2(k)+1)}} u(k) \right] \qquad (4.57)$$

where $u(k)$ and $h(k)$ are the inputs and $h(k+1)$ is the desired output for the RBFN model. The number of hidden layer neurons for the RBFN is taken as 50. The basis functions are assumed to be Gaussian for which the centres are fixed randomly within its input range. The input–output data are normalized. The recursive least square technique is applied to train the network. Once the identification is over, the model is validated for a sinusoidal input $10 + 5 \sin(Tk)$. The identification result on the training data set is shown in Figure 4.13(a) and for the test input is shown in Figure 4.13(b), where $h(k)$ is the system output and $h^m(k)$ is the RBFN output. The RMS error for training is found to be 2.87%, while for testing it is 4.56%.

Controller Design

Using the identified model, a suitable controller has to be designed both for tracking a set point and a trajectory. The network inversion technique would be used to design the controller where the update rule is based on the Lyapunov approach. The desired trajectory for the level $h(t)$ is given as

$$h^d(t) = 6 + 2 \sin t$$

The Lyapunov function to find the controller update law is considered as

$$V = \frac{1}{2}\tilde{h}^2 = \frac{1}{2}(h^d - h^m(k))^2$$

where h^d is the desired output of the system which remains unchanged during the input update. $h_m(k)$ is the output of the RBFN model expressed as

$$h^m(k) = \sum_{j=1}^{50} w_j \phi_j,$$

where $\qquad \phi_j = \exp(-[(h_N(k) - c_{1j})^2 + (u_N(k) - c_{2j})^2])$

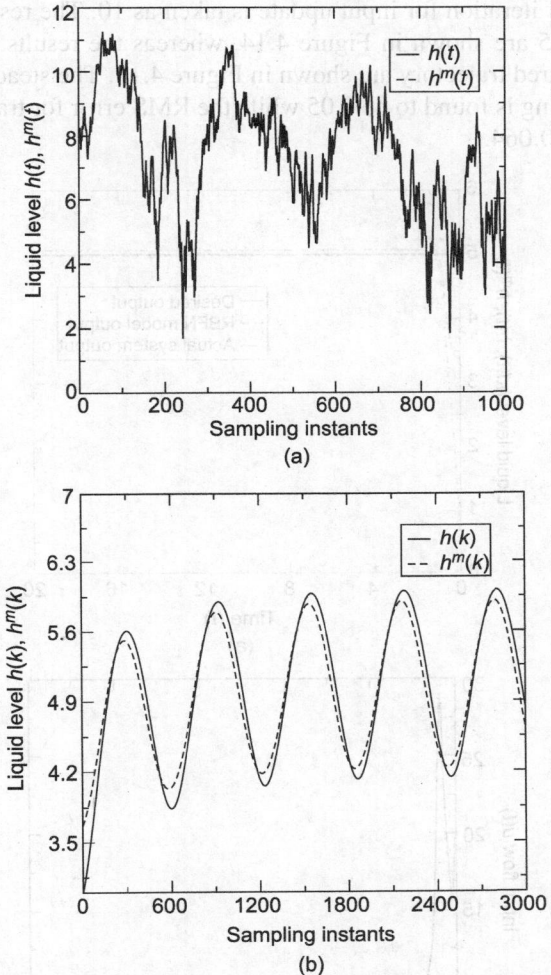

Figure 4.13 Surge tank–identification result: (a) how the model output matches with the actual output for the training data set and (b) the same for test data

where u_N and h_N are the normalized input and output of the actual system, which are inputs to the RBFN model. Using the above equation and Eqns (4.39), (4.41), and (4.42), the update law for u is derived as

$$u(t') = u(t' - 1) + d_N \mu \frac{\tilde{h} J}{J^2 + \alpha}$$

where μ and α are taken as 0.8 and 0.1, respectively, d_N is the denormalizing factor and J is computed as

$$J = -2 \sum_{j=1}^{50} \phi_j (u_N(t' - 1) - c_{2j}) w_j$$

The number of iteration for input update is taken as 10. The results for tracking a set point of 5 are shown in Figure 4.14, whereas the results for tracking the sinusoidal desired trajectory are shown in Figure 4.15. The steady state error for set point tracking is found to be 0.05 while the RMS error for trajectory tracking is found to be 0.064.

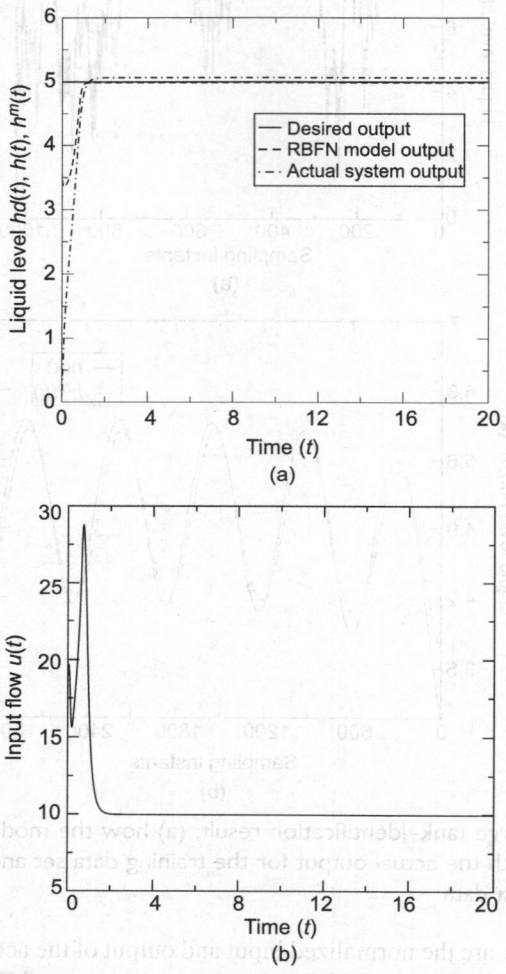

Figure 4.14 Surge tank–set point tracking: (a) the desired and actual outputs of both the system and model and (b) the corresponding control input

Let us now consider that the input $u(k)$ is predicted using another RBFN as

$$u(k) = \sum_{j=1}^{L} w_{2j} \psi_j \tag{4.58}$$

where

$$\phi_j = \exp(-[\|\text{in}(k) - c_{1j}\|^2)$$

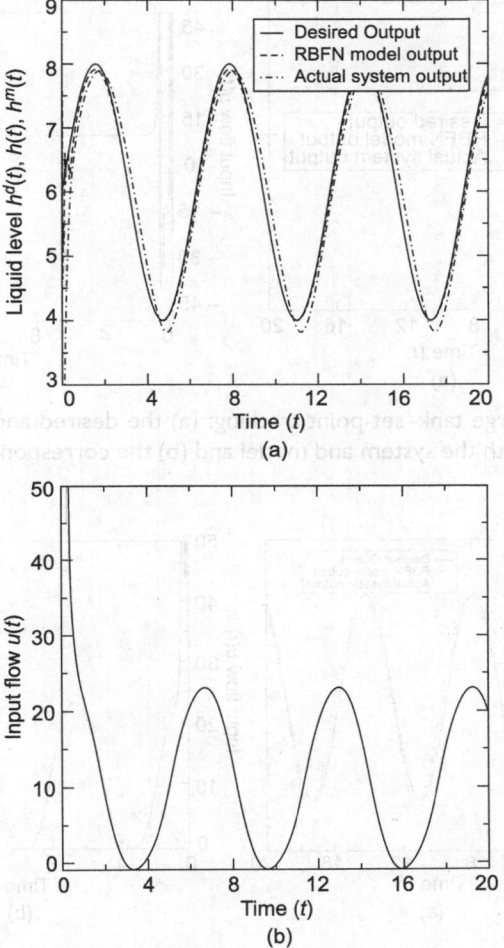

Figure 4.15 Surge tank–trajectory tracking: (a) the desired and actual outputs of both the system and model and (b) the corresponding control input

and in(k) is the input to this RBFN which consists of the system output and error. L is taken as 30. The weights w_{2j}s are updated using the update law (4.33) which for this case turns out to be

$$w_{2j}(t') = w(t' - 1) + d_N \mu_u \frac{\tilde{h} J}{J^2 + \alpha} \psi_j$$

where J is same as that in the previous case. The number of iteration to update the weights is taken as 50 with a learning rate of 0.02. The tracking results to track the same set point and trajectory are given in Figures 4.16 and 4.17. The steady state error for set point tracking is found to be 0.045, while the RMS error for trajectory tracking is found to be 0.057. However, the control input required for set point tracking is more in this case.

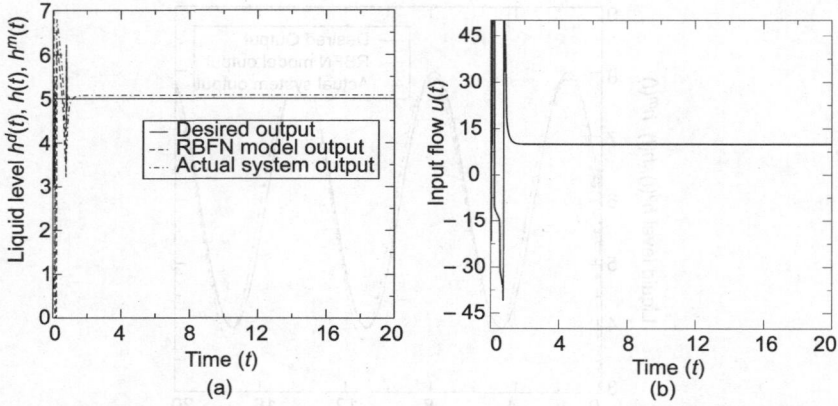

Figure 4.16 Surge tank–set point tracking: (a) the desired and actual outputs of both the system and model and (b) the corresponding control input

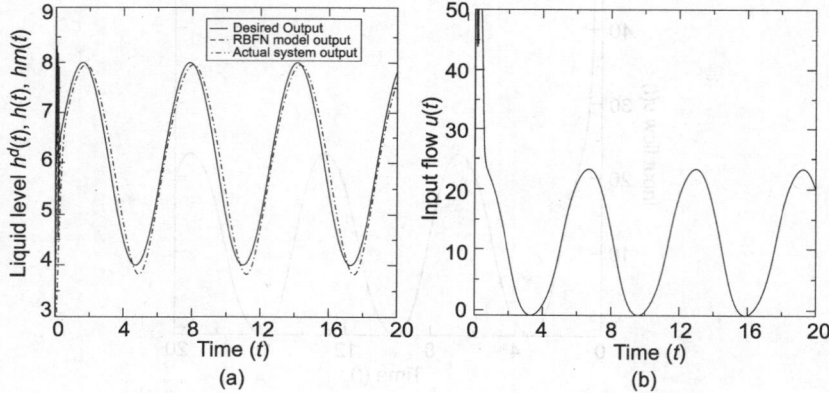

Figure 4.17 Surge tank–trajectory tracking for parametrized control: (a) the desired and actual outputs of both the system and model and (b) the corresponding control input

Example 4.5 (Single link manipulator)   Let us consider the following model of a single link manipulator:

$$\dot{x}_1 = x_2 \tag{4.59}$$

$$\dot{x}_2 = -g\sin(x_1) + u \tag{4.60}$$

where x_1 is the manipulator angle and u is the control torque and g is the acceleration due to gravity, which is taken as 9.81 m/s^2. The objective is to learn the dynamics of the manipulator system using an RBFN network and to find a suitable controller using the network inversion algorithm for regulating the manipulator angle.

System identification: The Euler representation of the system with a sampling time $T = 0.01$ s, is used for generating the data. Since the system is open loop unstable, a PD controller is used for data generation. Random sinusoidal trajectories are taken as the desired trajectories. Various dither signals like noise, impulse, and step are added to the PD controller output to make the identification proper. The number of hidden layer neurons for the RBFN is taken as 50. The basis functions are assumed to be Gaussian for which the centres are fixed randomly within its input range. The input–output data are normalized. The recursive least square technique is applied to train the network. Once the identification is over, the model is validated for a same control input to the system and model. The identification result on the training data set is shown in Figure 4.18, where (a) shows the manipulator angle x_1 and (b) shows the angular velocity x_2. The input training data are shown in Figure 4.19.

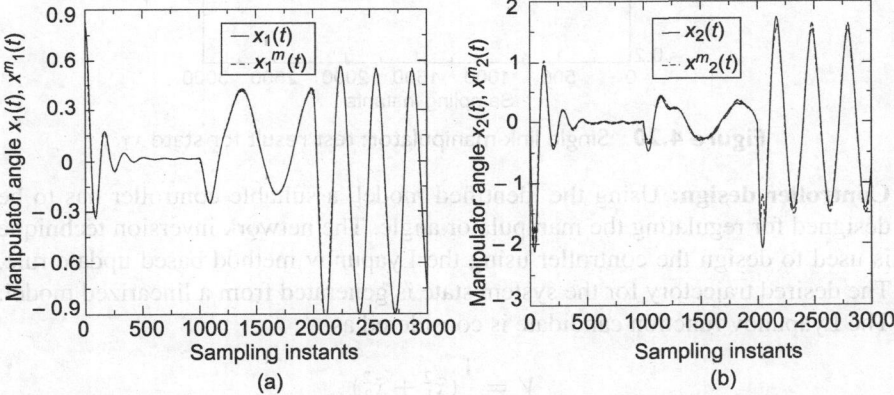

Figure 4.18 Single link manipulator: identification result: (a) the state x_1 and (b) the state x_2

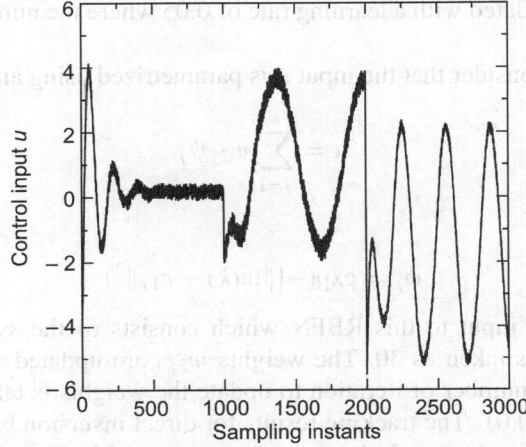

Figure 4.19 Single link manipulator: training data for input u

The system and model output for the test input are shown in Figure 4.20 where x_1 is the system output and x_1^m is the RBFN output. The RMS errors for training are found to be 1.36% and 4.77%, while for testing they are 2.81% and 6.24%.

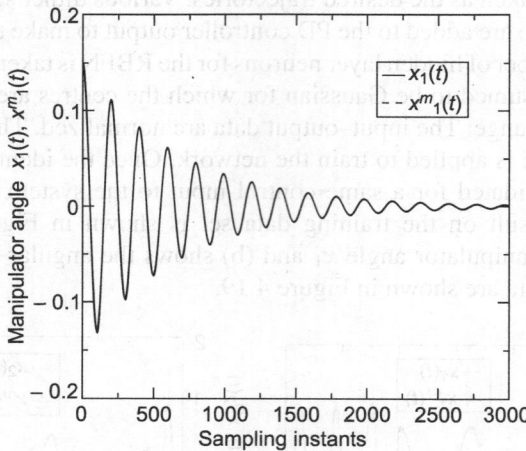

Figure 4.20   Single link manipulator: test result for state x_1

Controller design: Using the identified model, a suitable controller has to be designed for regulating the manipulator angle. The network inversion technique is used to design the controller using the Lyapunov method based update rule. The desired trajectory for the system state is generated from a linearized model. The Lyapunov function candidate is considered as

$$V = \frac{1}{2}(\tilde{x}_1^2 + \tilde{x}_2^2)$$

where $\tilde{x}_1 = x_1^d - x_1^m$ and $\tilde{x}_2 = x_2^d - x_2^m$, d represents the desired output of the system, and m represents the model. The desired states remain unchanged during the input update. Using the RBFN model and using Eqns (4.39), (4.41), and (4.42), the input u is updated with a learning rate of 0.05 where the number of iterations is taken as 30.

Let us now consider that the input u is parametrized using another RBFN as

$$u = \sum_{j=1}^{L} w_{2j} \psi_j \qquad (4.61)$$

where

$$\phi_j = \exp(-[\|\text{in}(k) - c_{1j}\|^2)$$

and in(k) is the input to this RBFN which consists of the system states and error vector. L is taken as 30. The weights w_{2j}s are updated using the update law (4.33). The number of iteration to update the weights is taken as 50 with a learning rate of 0.01. The tracking results for direct inversion based control and parametrized control to track the trajectory generated by a reference model are shown in Figure 4.21.

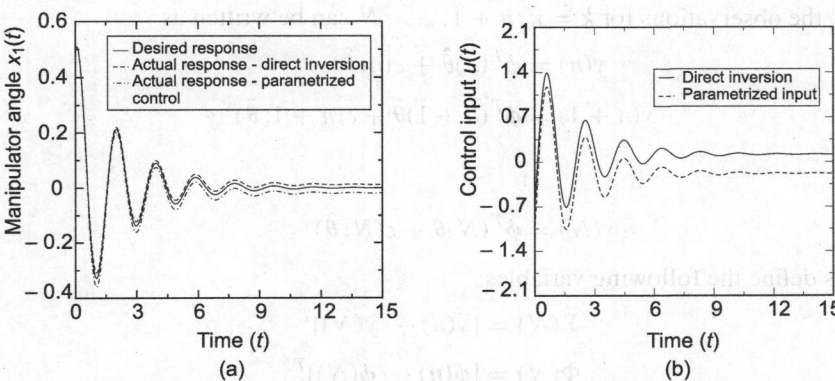

Figure 4.21   Single link manipulator–regulation: (a) the state x_1 converges to 0 and (b) the corresponding control input

SUMMARY

The indirect adaptive control schemes use explicit model of the plant. This chapter presents different indirect adaptive control schemes both for continuous time systems and discrete time systems. For continuous time systems, the plants are assumed to be in input–affine forms. The use of network inversion is also discussed in designing an indirect adaptive controller. Here are some interesting works [85, 86, 87] that readers may find interesting in terms of both techniques and applications.

APPENDIX

Recursive Least Squares

Let us consider the following discrete time system:

$$y(k) + a_1 y(k-1) + a_2 y(k-2) + \cdots + a_n y(k-n) = b_1 u(k-1)$$
$$+ \cdots b_n u(k-n) \qquad (4.62)$$

where the parameters as and bs are unknown. We assume that up to Nth observations of input and output are available to us. The objective is to estimate the parameters such that they best fit to the observed data set. Let us define two vectors $\hat{\theta}$ and $\phi(k)$ such that $\hat{\theta} = [\hat{a}_1 \cdots \hat{a}_n \ \hat{b}_1 \cdots \hat{b}_n]^T$ where $\hat{a}$s and $\hat{b}$s are estimates of as and bs and $\phi(k) = [-y(k-1) - \cdots y(k-n) \ u(k-1) \cdots u(k-n)]^T$. Denoting the estimation error at kth instant by $e(k; \hat{\theta})$, we can write

$$y(k) + \hat{a}_1 y(k-1) + \hat{a}_2 y(k-2) + \cdots + \hat{a}_n y(k-n)$$
$$= \hat{b}_1 u(k-1) + \cdots \hat{b}_n u(k-n) + e(k; \hat{\theta}) \qquad (4.63)$$

Thus, the observations for $k = n, n+1, \ldots, N$ can be written as

$$y(n) = \boldsymbol{\phi}^T(n)\hat{\boldsymbol{\theta}} + e(n; \hat{\boldsymbol{\theta}})$$

$$y(n+1) = \boldsymbol{\phi}^T(n+1)\hat{\boldsymbol{\theta}} + e(n+1; \hat{\boldsymbol{\theta}})$$

$$\vdots$$

$$y(N) = \boldsymbol{\phi}^T(N)\hat{\boldsymbol{\theta}} + e(N; \hat{\boldsymbol{\theta}})$$

Let us define the following variables:

$$\boldsymbol{Y}(N) = [y(n) \cdots y(N)]^T$$

$$\Phi(N) = [\boldsymbol{\phi}(n) \cdots \boldsymbol{\phi}(N)]^T$$

$$\boldsymbol{\epsilon}(N; \hat{\boldsymbol{\theta}}) = [e(n) \cdots e(N)]^T$$

In terms of the above variables, the error equations can be combined as

$$Y = \Phi\hat{\boldsymbol{\theta}} + \boldsymbol{\epsilon}(N; \hat{\boldsymbol{\theta}}) \tag{4.64}$$

Let us define a cost function $J(\hat{\boldsymbol{\theta}}) = \boldsymbol{\epsilon}^T \boldsymbol{\epsilon}$ such that the estimated parameter vector $\hat{\boldsymbol{\theta}}$ minimizes this cost. One can write

$$J = \boldsymbol{\epsilon}^T \boldsymbol{\epsilon}$$

$$= (Y - \Phi\hat{\boldsymbol{\theta}})^T (Y - \Phi\hat{\boldsymbol{\theta}})$$

$$= Y^T Y - \hat{\boldsymbol{\theta}}^T \Phi^T Y - Y^T \Phi\hat{\boldsymbol{\theta}} - \hat{\boldsymbol{\theta}}^T \Phi^T \Phi\hat{\boldsymbol{\theta}}$$

To minimize J with respect to $\hat{\boldsymbol{\theta}}$ we need to take the derivative of J with respect to $\hat{\boldsymbol{\theta}}$ and equate it to 0 which gives the following expression:

$$\Phi^T \Phi\hat{\boldsymbol{\theta}} = \Phi^T Y$$

The solution to the above equation is

$$\hat{\boldsymbol{\theta}} = (\Phi^T \Phi)^{-1} \Phi^T Y \tag{4.65}$$

which is the least squares estimate of $\hat{\boldsymbol{\theta}}$. In recursive least squares (RLS), instead of estimating the parameters from an observed data set off-line, a recursive relation for updating the parameters is derived. From the definition of Φ, we can write for data up to $N + 1$,

$$\Phi^T(N+1)\Phi(N+1) = \Phi^T(N)\Phi(N) + \boldsymbol{\phi}(N+1)\boldsymbol{\phi}^T(N+1) \tag{4.66}$$

It is clear from Eqn (4.65) that we need to find out the inverse of $\Phi^T\Phi$. Let us define a matrix P as

$$P(N+1) = [\Phi^T(N+1)\Phi(N+1)]^{-1} \tag{4.67}$$

From Eqns (4.66) and (4.67) we can write

$$P(N+1) = [P^{-1}(N) + \boldsymbol{\phi}(N+1)\boldsymbol{\phi}^T(N+1)]^{-1} \tag{4.68}$$

We know from the matrix inversion lemma that

$$(A + BCD)^{-1} = A^{-1} - A^{-1}B(C^{-1} + DA^{-1}B)^{-1}DA^{-1}$$

In this case, $A = P^{-1}(N)$, $B = \phi(N+1)$, $C = 1$, and $D = \phi^T(N+1)$. Thus, we can write

$$
P(N+1) = P(N) - P(N)\phi(N+1)[1 + \phi^T(N+1)P(N)\phi(N+1)]^{-1}
$$
$$
\phi^T(N+1)P(N) \tag{4.69}
$$

Again

$$
\Phi^T(N+1)Y(N+1) = \Phi^T(N)Y(N) + \phi(N+1)y(N+1) \tag{4.70}
$$

Combining (4.65), (4.69), and (4.70)

$$
\hat{\theta}(N+1) = \hat{\theta}(N) + P(N)\phi(N+1)[1 + \phi^T(N+1)P(N)\phi(N+1)]^{-1}
$$
$$
[y(N+1) - \phi^T(N+1)\hat{\theta}(N)]
$$

which is the recursive least squares parameter update law.

EXERCISES

1. Following is the dynamical equation to control the velocity of a single-stage rocket:

$$
\frac{dv(t)}{dt} = c(t)\left(\frac{m}{M-mt}\right) - g\left(\frac{R}{R+y(t)}\right) - 0.5v^2(t)\left(\frac{\rho_a A C_d}{M-mt}\right) \tag{4.71}
$$

where $v(t)$ is the rocket velocity at time t (output of the plant), $y(t)$ is the rocket altitude, and $c(t)$ is the velocity of the exhaust gases. In general, the exhaust gas velocity is proportional to the cross-sectional area of the nozzle, so we take it as the input. The system parameters are tabulated below.

M, the initial mass of the rocket and fuel	=	15000 kg
m, the exhaust gases mass flow rate	=	100 kg/s
A, the maximum cross sectional area of the rocket	=	1 m^2
g, the acceleration due to gravity	=	9.81 m/s^2
R, the radius of earth	=	6.37×10^6 m
ρ_a, the air density	=	1.21 kg/m^3
C_d, the drag coefficient of the rocket	=	0.3

   (a) Derive the state–space model of the system.
   (b) If the derived state space model is in the input affine form, identify the system dynamics using the concept as given in Section 4.1.

2. Consider the model of an induction motor given by

$$
\dot{\omega} = \frac{n_p M}{J L_r}(\psi_a i_b - \psi_b i_a) - \frac{T_L}{J}
$$

$$
\dot{\psi}_a = -\frac{R_r}{L_r}\psi_a - n_p\omega\psi_b + \frac{R_r}{L_r}M i_a
$$

$$
\dot{\psi}_b = -\frac{R_r}{L_r}\psi_b + n_p\omega\psi_a + \frac{R_r}{L_r}M i_b
$$

where ω is the rotor angular rate, ψ_a and ψ_b are rotor fluxes, and i_a and i_b are the input currents. $M = 0.001$ H is the mutual inductance and $R_r = 8\Omega$ is the rotor resistance, $L_r = 0.05$ H is the rotor inductance, and $n_p = 50$ is the number of poles.

(a) Assuming the load torque as 0, express the state–space model of the system in input affine form.

(b) Identify the system dynamics using continuous time system identification as given in Section 4.1.

3. Consider the dynamics of a series DC motor.

$$J\frac{d\omega}{dt} = -D\omega + K_m L_f i^2 - \tau_L$$

$$L\frac{di}{dt} = -K_m L_f \omega i - Ri + V$$

where ω is the motor speed to be controlled, V is the input voltage, τ_L is the load torque, $J = 0.000704$ kg m^2 is the moment of inertia of the motor, $D = 0.0004$ N $-$ m/rad/s is the viscous friction coefficient, K_m is the motor torque constant, L_f is the field inductance, and $L = L_a + L_f = 0.0917$ H, $R = R_a + R_f = 7.2$ Ω where L_a and R_a are the armature inductance and resistance, respectively. $K_m L_f = 0.1236$ Nm/WbA.

(a) Considering the load torque as 0, derive the state space model of the system in discrete time.

(b) Identify the system dynamics using a neural network.

(c) Design an adaptive controller for the system based on the identified model.

4. The model of a two-phase permanent magnet stepper motor is given by

$$\dot\theta = \omega$$

$$J\dot\omega = -K_m i_a \sin(N\theta) + K_m i_b \cos(N\theta) - B\omega - T_L$$

$$L_a \dot i_a = -R_a i_a + K_b \omega \sin(N\theta) + v_a$$

$$L_b \dot i_b = -R_b i_b - K_b \omega \cos(N\theta) + v_b$$

where θ is the angular position, ω is the angular rate and i_a, i_b are the currents for phases A and B, J is the moment of inertia, K_m is the motor torque constant, N is the number of teeth on the rotor, B is the viscous friction coefficient, L_a, L_b and R_a, R_b are the inductances and resistances of phases A and B, K_b is the back *emf* constant, and T_L is the load torque. Model the load torque for a mechanical link of length l and mass m as $mgl\sin(\theta)$ where m is the mass of the shaft.

The system parameters are tabulated below:

$J = 0.0733$ kg $-$ m^2, $B = 0.002$ Nms/rad

$m = 0.4014$ kg, $N = 50, l = 1$m

$L_a = L_b = 0.7$ H, $R_a = 0.9$ Ω, $R_b = 1.2\Omega$

$K_m = 0.25$ Nm/A, $K_b = 0.025$ Vs/rad, $g = 9.81$ m/s^2

(a) Derive the state–space model of the system.

(b) Discretize the system dynamics using the Euler method.

    (c) Identify the system dynamics using an RBFN.

    (d) Design an adaptive control scheme for voltages v_a and v_b such that θ follows a desired trajectory $r(t)$.

5. The dynamics of an automobile are given by

$$\dot{v}(t) = \frac{1}{m}(-A_\rho v^2(t) - d + f(t))$$

$$\dot{f}(t) = \frac{1}{\tau}(-f(t) + u(t))$$

where u is the control input. $u > 0$ represents a throttle input and $u < 0$ represents a brake input, m is the mass of the vehicle, A_ρ is the aerodynamic drag, d is a constant frictional force, f is the driving or braking force, and τ is the engine/brake time constant. Assume that $u \in [-1000, \ 1000]$. Take the system parameters as $m = 1300 \ kg$, $A_\rho = 0.3 \ \text{N s}^2/\text{m}^2$, $d = 100$ N, and $\tau = 0.2$ s.

    (a) Derive the state–space model of the system.

    (b) Discretize the system dynamics using the Euler method.

    (c) Identify the system dynamics using an RBFN.

    (d) Design an adaptive controller for the system using the concept of network inversion such that $v(t)$ tracks a reference of $r(t) = 50$.

5

Direct Adaptive Control Using Neural Networks

Feedback linearization, as discussed in Chapter 1, is a useful control design technique in the control system literature where a large class of non-linear systems can be made linear by non-linear state feedback. The controller can be proposed in such a way that the closed loop error dynamics becomes linear as well as stable [15]. The main problem with this control scheme is that the cancellation of the non-linear dynamics depends on the exact knowledge of the system non-linearities. When the system non-linearities are not known completely, they can be approximated either by the neural (NNs) networks or by fuzzy systems. The controller then uses these estimates to linearize the system. The parameters of the controller are updated such that the output tracking error converges to zero with time. This design technique is popularly known as direct adaptive control technique. This chapter presents a direct adaptive neural control scheme for a class of affine non-linear systems which are exactly input–output linearizable by non-linear state feedback.

5.1   DIRECT ADAPTIVE CONTROL

Let us consider the following example of a first-order system:

$$\dot{y} = -a_p y + b_p u \tag{5.1}$$

where y is the plant output, u is the plant input, and a_p and b_p are the plant parameters. Let us assume that the plant parameters are unknown. The desired performance of the system is specified by the following model:

$$\dot{y}_m = -a_m y_m + b_m r \tag{5.2}$$

where r is the reference signals and a_m, b_m are known constants. The objective of the adaptive control design is to formulate a control law and an adaptation law such that the error between the system output and the model output converges to 0. It is assumed that the sign of b_p is known to us. Let us choose the following control law:

$$u = \hat{a}_r r + \hat{a}_y y \tag{5.3}$$

where $\hat{a}_r$ and $\hat{a}_y$ are variable controller parameters for which we need to derive an update law. Combining (5.1) and (5.3), the closed loop dynamics becomes

$$\dot{y} = -(a_p - \hat{a}_y b_p)y + \hat{a}_r b_p r \tag{5.4}$$

The choice of control law (5.3) allows the possibility of perfect model matching if the plant parameters are known. In that case if we choose the following values of controller parameters

$$a_r^* = \frac{b_m}{b_p} \qquad a_y^* = \frac{a_p - a_m}{b_p} \tag{5.5}$$

the closed loop dynamics would become

$$\dot{y}_m = -a_m y_m + b_m r \tag{5.6}$$

which is identical to the reference model. But since the parameters a_p and b_p are unknown, the control input will achieve this objective adaptively based on the tracking error $e = y - y_m$. Let us define the parameter errors as $\tilde{a}_r = \hat{a}_r - a_r^*$ and $\tilde{a}_y = \hat{a}_y - a_y^*$. Combining (5.4) and (5.2), the error dynamics can be written as

$$\dot{e} = -a_m e + b_p(\tilde{a}_r r + \tilde{a}_y y) \tag{5.7}$$

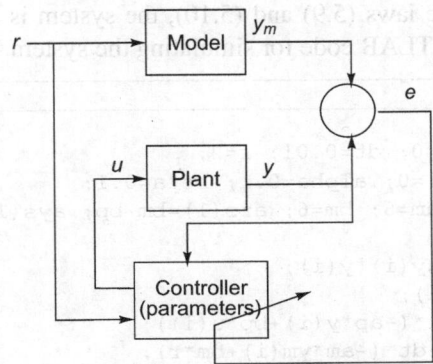

Figure 5.1 A model reference adaptive control scheme

To analyse the error convergence, consider the following Lyapunov function candidate:

$$V = \frac{1}{2}e^2 + \frac{1}{2\alpha}|b_p|\tilde{a}_r^2 + \frac{1}{2\beta}|b_p|\tilde{a}_y^2 \tag{5.8}$$

Taking derivative of the above equation, we can write

$$\dot{V} = e\dot{e} + \frac{1}{\alpha}|b_p|\tilde{a}_r\dot{\tilde{a}}_r + \frac{1}{\beta}|b_p|\tilde{a}_y\dot{\tilde{a}}_y$$

$$= e(-a_m e + b_p(\tilde{a}_r r + \tilde{a}_y y)) + \frac{1}{\alpha}|b_p|\tilde{a}_r\dot{\hat{a}}_r + \frac{1}{\beta}|b_p|\tilde{a}_y\dot{\hat{a}}_y$$

If we choose the following adaptation laws

$$\dot{\hat{a}}_r = -\alpha \, \text{sgn}(b_p) \, e \, r \tag{5.9}$$

$$\dot{\hat{a}}_y = -\beta \, \text{sgn}(b_p) \, e \, y \tag{5.10}$$

then $\dot{V} = -a_m e^2 + b_p e(\tilde{a}_r r + \tilde{a}_y y) - |b_p|\tilde{a}_r \text{sgn}(b_p)er - |b_p|\tilde{a}_y \text{sgn}(b_p)ey.$
Since $b_p = |b_p| \, \text{sgn}(b_p)$, $\dot{V}$ becomes $-a_m e^2$ which is negative definite. Thus,
the signals e, $\tilde{a}_r$, and $\tilde{a}_y$ are bounded. Furthermore, because of the boundedness
of $\dot{e}$, according to Barbalat's lemma [15], e will converge to 0. Since it is assumed
that the sign of b_p is known, the update laws (5.9) and (5.10) are implementable.
A general block diagram of the above-described adaptive control scheme is
shown in Figure 5.1.

Example 5.1   Consider the following first-order linear plant:

$$\dot{y} = 2y + 5u$$

Design an adaptive controller of the form (5.3). Simulate the response of the
closed loop system for a reference signal of $r = 2$ using the adaptation laws (5.9)
and (5.10).

Solution
The plant parameters $a_p = -2$ and $b_p = 5$ are assumed to be unknown. The
reference model parameters are chosen as $a_m = 3$ and $b_m = 3$. The initial values
of the controller parameters $\hat{a}_r$ and $\hat{a}_y$ (refer to (5.3)) are chosen as 0. The initial
conditions of both the plant and model are also taken as 0. Using the control
law (5.3) and update laws (5.9) and (5.10), the system is simulated for $r = 2$.
Following is the MATLAB code for simulating the system.

```
LinAdap.m
clear all
y(1)=0; ym(1)=0; dt=0.01; r=2;
ar(1)=0; ay(1)=0; alpha=0.1; beta=0.1;
ap=-2; bp=3; am=6; bm=6; ars(1)=bm/bp; ays(1)=(ap-am)/bp;
for i=1:500
u(i)=ar(i)*r+ay(i)*y(i);
e(i)=y(i)-ym(i);
y(i+1)=y(i)+dt*(-ap*y(i)+bp*u(i));
ym(i+1)=ym(i)+dt*(-am*ym(i)+bm*r);
ar(i+1)=ar(i)-alpha*sign(bp)*e(i)*r;
ay(i+1)=ay(i)-beta*sign(bp)*e(i)*y(i);
t(i)=i*dt;
ars(i+1)=bm/bp;
ays(i+1)=(ap-am)/bp;
end
t(i+1)=(i+1)*dt;
plot(t,y,'r',t,ym,'b'); figure;
plot(t,ay,'r',t,ar,'b'); hold on;
plot(t,ays,'y',t,ars,'g')
```

The results are shown in Figure 5.2 where (a) shows the tracking performance and
(b) shows the evolution of the adaptive parameters. We should note that though the

final parameters are not exactly the same as that of the desired ones, the tracking is achieved as time progresses.

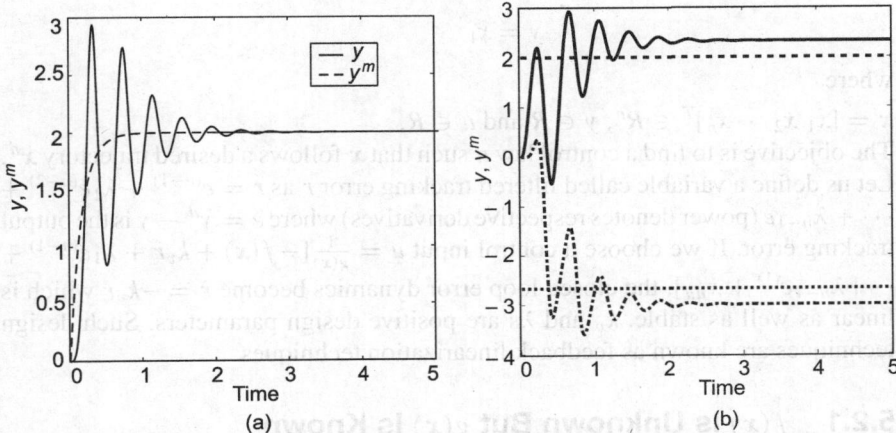

Figure 5.2 Adaptive control: (a) tracking performance and (b) parameters estimation

The above-described method of adaptive control can be extended for a first-order non-linear system of the following form:

$$\dot{y} = -a_p y - c_p f(y) + b_p u$$

where $f(y)$ is a known non-linear function. The control law in this case can be chosen as

$$u = \hat{a}_r r + \hat{a}_y y + \hat{a}_f f(y)$$

with the following adaptation laws:

$$\dot{\hat{a}}_r = -\alpha \operatorname{sgn}(b_p) e\, r$$

$$\dot{\hat{a}}_y = -\beta \operatorname{sgn}(b_p) e\, y$$

$$\dot{\hat{a}}_f = -\gamma \operatorname{sgn}(b_p) e\, f(y)$$

Till now we have considered that the structure of the plant is known, but the parameters of the plant are unknown. Following sections will discuss various adaptive control schemes for non-linear systems when the non-linearities of the plant are also unknown to us.

5.2 SINGLE INPUT SINGLE OUTPUT AFFINE SYSTEMS

A large class of single input single output affine systems is represented by the following equations:

$$\dot{x}_1 = x_2$$

$$\dot{x}_2 = x_3$$

$$\dot{x}_n = f(x) + g(x)u \qquad (5.11)$$

$$y = x_1$$

where

$x = [x_1 \ x_2 \cdots x_n]^T \in R^n$, $y \in R$ and $u \in R$.

The objective is to find a control law u such that x follows a desired trajectory x^d. Let us define a variable called filtered tracking error r as $r = e^{(n-1)} + \lambda_1 e^{(n-2)} + \cdots + \lambda_{n-1} e$ (power denotes respective derivatives) where $e = y^d - y$ is the output tracking error. If we choose a control input $u = \frac{1}{g(x)}[-f(x) + k_v r + \lambda_1 e^{(n-1)} + \cdots + \lambda_{n-1} e^{(1)} + \dot{x}_{nd}]$, the closed loop error dynamics become $\dot{r} = -k_v r$ which is linear as well as stable. k_v and λs are positive design parameters. Such design techniques are known as feedback linearization techniques.

5.2.1 $f(x)$ Is Unknown But $g(x)$ Is Known

The problem arises with feedback linearization control techniques when the nonlinear functions $f(x)$ or $g(x)$ or both are unknown. For such cases, different function approximators like the NNs or fuzzy systems can be used to estimate these non-linear functions. The controller parameters can be updated such that the closed loop error converges to zero. In this section, we have assumed that the function $f(x)$ is unknown, while the function $g(x)$ is known. $f(x)$ is approximated by a radial basis function network (RBFN). The pictorial diagram of the RBFN when approximating $f(x)$ is shown in Figure 5.3. As can be seen in the figure, the RBFN has a single output. Thus, the network weight constitutes a vector $\hat{W} \in R^{L \times 1}$ in this case.

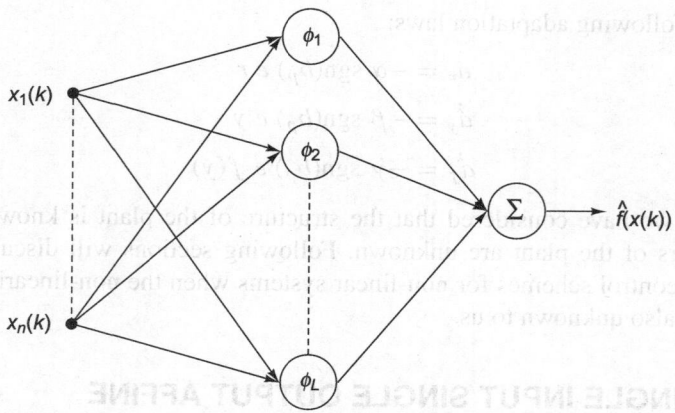

Figure 5.3 An n input and 1 output RBF network

The following theorem summarizes the control design process.

Theorem 5.1 Suppose that the non-linear function $f(x)$ of system (5.11) is unknown, while the function $g(x)$ is known. Let $f(x)$ be approximated as $\hat{f}(x) = \hat{W}^T \phi(x)$ using a radial basis function network. Then the control law $u = \frac{1}{g(x)}[-\hat{f}(x) + k_v r + \lambda_1 e^{n-1} + \cdots + \lambda_{n-1} e^{(1)} + \dot{x}_{nd}]$ will stabilize the system (5.11) in the sense of Lyapunov provided $\hat{W}$ is updated using the update law $\dot{\hat{W}} = -F \phi \, r$, where F is a positive definite matrix.

Proof: The output tracking error is defined as $e = y_d - y = x_{1d} - x_1$ where $y_d = x_{1d}$ is the desired output of the system. The filtered tracking error r is defined as

$$r = e^{(n-1)} + \lambda_1 e^{(n-2)} + \cdots + \lambda_{n-1} e \qquad (5.12)$$

where $e^{(n-1)}$ is the $n-1$th derivative of e and so on. $\lambda_1, \ldots, \lambda_{n-1}$ are chosen such that Eqn (5.12) represents a stable dynamics. The objective is to achieve a satisfactory tracking result as well as to maintain boundedness of the error and the neural network weight vector. The control law u is defined as follows:

$$u = \frac{1}{g(x)}[-\hat{f}(x) + k_v r + \lambda_1 e^{(n-1)} + \cdots + \lambda_{n-1} e^{(1)} + \dot{x}_{nd}] \qquad (5.13)$$

where $\hat{f}(x) = \hat{W}^T \phi(x)$. Let us assume that there exists an ideal weight vector W such that the original function $f(x)$ can be represented as $f(x) = W^T \phi(x)$. According to the universal approximation property of the neural network, for any smooth function $f(\cdot)$, there exist weights such that $f(x) = W^T \phi(x) + \epsilon$, where ϵ is the estimation error. It is assumed that the estimation error ϵ is zero. This assumption will hold for comparatively less complex non-linear functions if we take a sufficiently large number of adjustable weights. Putting the control law u (Eqn (5.13)) into the system Eqn (5.11), we get

$$\dot{x}_n = f(x) + g(x) \cdot \frac{1}{g(x)}[-\hat{f}(x) + k_v r + \lambda_1 e^{(n-1)} + \cdots + \lambda_{n-1} e^{(1)} + \dot{x}_{nd}]$$

$$= W^T \phi - \hat{W}^T \phi + k_v r + \lambda_1 e^{(n-1)} + \cdots + \lambda_{n-1} e^{(1)} + \dot{x}_{nd} \qquad (5.14)$$

Defining $\tilde{W}^T = W^T - \hat{W}^T$ we can write

$$\dot{x}_n = \tilde{W}^T \phi + k_v r + \lambda_1 e^{(n-1)} + \cdots + \lambda_{n-1} e^{(1)} + \dot{x}_{nd}$$

$$\text{or } \dot{x}_{nd} - \dot{x}_n = e^{(n)} = -\tilde{W}^T \phi - k_v r - \lambda_1 e^{(n-1)} - \cdots - \lambda_{n-1} e^{(1)} \qquad (5.15)$$

Differentiation of Eqn (5.12) yields

$$\dot{r} = e^n + \lambda_1 e^{(n-1)} + \cdots + \lambda_{n-1} e^{(1)} \qquad (5.16)$$

Substituting e^n from (5.16) into (5.15)

$$\dot{r} - \lambda_1 e^{(n-1)} - \cdots - \lambda_{n-1} e^{(1)} = -\tilde{W}^T \phi - k_v \dot{r} - \lambda_1 e^{(n-1)} - \cdots - \lambda_{n-1} e^{(1)}$$

$$\text{or} \qquad \dot{r} = -k_v r - \tilde{W}^T \phi \qquad (5.17)$$

Consider a Lyapunov function

$$V = \frac{1}{2}r^2 + \frac{1}{2}\tilde{W}^T F^{-1} \tilde{W} \tag{5.18}$$

where F is a positive definite matrix. Differentiating (5.18),

$$\dot{V} = r\dot{r} + \tilde{W}^T F^{-1} \dot{\tilde{W}} \tag{5.19}$$

Substituting $\dot{r}$ from (5.17) into (5.19),

$$\dot{V} = r(-k_v r - \tilde{W}^T \phi) + \tilde{W}^T F^{-1} \dot{\tilde{W}} \tag{5.20}$$

Since W is constant, we can write $\dot{\tilde{W}} = \dot{W} - \dot{\hat{W}} = -\dot{\hat{W}}$. Thus,

$$\dot{V} = -k_v r^2 - \tilde{W}^T \phi r - \tilde{W}^T F^{-1} \dot{\hat{W}}$$

$$= -k_v r^2 - \tilde{W}^T (\phi r + F^{-1} \dot{\hat{W}}) \tag{5.21}$$

Equating the second term of (5.21) to zero, we get

$$\phi r + F^{-1} \dot{\hat{W}} = 0$$

$$\text{or } \dot{\hat{W}} = -F\phi r \tag{5.22}$$

Using the update law (5.22), (5.21) becomes

$$\dot{V} = -k_v r^2 \tag{5.23}$$

Since $V > 0$ and $\dot{V} \le 0$, this shows stability in the sense of Lyapunov so that r and $\tilde{W}$ (hence $\hat{W}$) are bounded. Hence the proof. Moreover $\int_0^\infty -\dot{V} dt < \infty$, $\ddot{V} = -2r k_v \dot{r}$ and all signals on the right hand side of (5.17) verify the boundedness of $\dot{r}$ and hence of $\ddot{V}$. Therefore $\dot{V}$ is uniformly continuous. Thus, according to Barbalat's lemma [15], $\dot{V} \to 0$, as $t \to \infty$. Since $\dot{V} = -k_v r^2$, $\dot{V}$ will be zero only when r is zero. Thus, r converges to zero with time. Since (5.12) represents a stable dynamics, the output tracking error $e(t)$ also vanishes with time.

Inclusion of non-zero RBFN approximation error: Let us now assume a non-zero RBFN approximation error ϵ which is bounded by a small positive number ϵ_N, i.e., the original function $f(x)$ can be written as

$$f(x) = W^T \phi(x) + \epsilon$$

where $\|\epsilon\| < \epsilon_N$. With this modification the time derivative of the Lyapunov function becomes

$$\dot{V} = r(-k_v r - \tilde{W}^T \phi - \epsilon) - \tilde{W}^T F^{-1} \dot{\hat{W}} \tag{5.24}$$

Considering the same update law $\dot{\hat{W}} = -F \phi r$, we get

$$\dot{V} = -k_v r^2 - \epsilon r$$

$\dot{V}$ can be further expressed as

$$\dot{V} \leq -k_v r^2 + \|\epsilon \, r\|$$

$$\text{or,} \quad \dot{V} \leq -k_v r^2 + \epsilon_N \|r\| = -\|r\|(k_v \|r\| - \epsilon_N)$$

The above expression will be negative if $k_v \|r\| - \epsilon_N$ is positive. In other words,

$$k_v \|r\| - \epsilon_N > 0 \quad \text{or,} \quad \|r\| > \frac{\epsilon_N}{k_v}$$

This implies that $\dot{V}$ is negative definite as long as $\|r\| > \dfrac{\epsilon_N}{k_v}$. r will be inside a ball of radius $\frac{\epsilon_N}{k_v}$. By choosing a large k_v, the filtered tracking error r can be made arbitrarily small which serves the purpose of tracking.

5.2.2 $f(x)$ And $g(x)$ Both Are Unknown

It is difficult to achieve closed loop stability when both $f(x)$ and $g(x)$ are unknown. For a system to be feedback linearizable, $|g(x)|$ should be greater than zero, $\forall x$, i.e., $g(x)$ should not take the value 0 at any operating point. However, since the approximation of $g(x)$ can take any value, the control law may become unbounded. To avoid this problem, a projection algorithm is used to keep $\hat{g}(x)$ away from zero. The effect of this projection algorithm is compensated by adding a sliding mode term in the control input. It is assumed that the bounds on $g(x)$ are known to us, i.e., $g_l < g(x) < g_u$. The following theorem then summarizes the design process when the bounds are positive.

Theorem 5.2   Given that both non-linear functions $f(x)$ and $g(x)$ of system (5.11) are unknown, let $f(x)$ be approximated as $\hat{f}(x) = \hat{W}^T \phi(x)$ and $g(x)$ be approximated as $\hat{g}(x) = \hat{P}^T \psi(x)$ using two radial basis function networks. Let $u_1 = \frac{1}{\hat{g}(x)}[-\hat{f}(x) + k_v r + \lambda_1 e^{(n-1)} + \cdots + \lambda_{n-1} e^{(1)} + \dot{x}_{nd}]$ and $u_2 = \frac{|\hat{g}|}{g_l}|u_1|\text{sgn}\,(r)$. Then the control law $u = u_1 + u_2$ will stabilize system (5.11) in the sense of Lyapunov provided $\hat{W}$ and $\hat{P}$ are updated using the update laws $\dot{\hat{W}} = -F\phi\,r$ and

$$\dot{\hat{P}} = 0 \text{ when } \hat{g} - g_l < 0 \text{ and } \psi\,u_1\,r > 0,$$

$$= -G\psi u_1 r \text{ otherwise}$$

Proof: Assume that there exist two ideal weight vectors W and P such that the original functions $f(x)$ and $g(x)$ can be represented as

$$f(x) = W^T \phi(x) \qquad g(x) = P^T \psi(x) \tag{5.25}$$

The control law is defined as

$$u = u_1 + u_2 \tag{5.26}$$

where

$$u_1 = \frac{1}{\hat{g}(x)}[-\hat{f}(x) + k_v r + \lambda_1 e^{(n-1)} + \cdots + \lambda_{n-1} e^{(1)} + \dot{x}_{nd}] \qquad (5.27)$$

where $\hat{f}(x) = \hat{W}^T \phi(x)$ and $\hat{g}(x) = \hat{P}^T \psi(x)$. Also $\dot{x}_{nd}$ can be written in terms of u_1 as

$$\dot{x}_{nd} = \hat{g}(x)u_1 + \hat{f}(x) - k_v r - \lambda_1 e^{(n-1)} - \cdots - \lambda_{n-1} e^{(1)} \qquad (5.28)$$

Using (5.25) and (5.26), system (5.11) can be rewritten as

$$\dot{x}_n = W^T \phi + P^T \psi u = W^T \phi + P^T \psi u_1 + P^T \psi u_2 \qquad (5.29)$$

Since $\dot{x}_n = \dot{x}_{nd} - e^{(n)}$, substituting $\dot{x}_{nd}$ and $e^{(n)}$ from (5.28) and (5.16) into (5.29), we get

$$\hat{g}(x)u_1 + \hat{f}(x) - k_v - \lambda_1 e^{(n-1)} - \cdots - \lambda_{n-1} e^{(1)} - \dot{r}$$
$$+\lambda_1 e^{(n-1)} + \cdots + \lambda_{n-1} e^{(1)} = W^T \phi + P^T \psi u_1 + P^T \psi u_2$$

Rewriting the last equation,

$$\dot{r} = -k_v r - W^T \phi + \hat{W}^T \phi - P^T \psi u_1 - \hat{P}^T \psi u_1 - P^T \psi u_2$$

$$\text{or } \dot{r} = -k_v r - \tilde{W}^T \phi - \tilde{P}^T \psi u_1 - P^T \psi u_2 \qquad (5.30)$$

where $\tilde{W} = W - \hat{W}$ and $\tilde{P} = P - \hat{P}$. To prove the closed loop system stability, let us consider a Lyapunov function

$$V = \frac{1}{2}r^2 + \frac{1}{2}\tilde{W}^T F^{-1}\tilde{W} + \frac{1}{2}\tilde{P}^T G^{-1}\tilde{P} \qquad (5.31)$$

where F and G are two positive definite matrices. Differentiating (5.31),

$$\dot{V} = r\dot{r} + \tilde{W}^T F^{-1}\dot{\tilde{W}} + \tilde{P}^T G^{-1}\dot{\tilde{P}} \qquad (5.32)$$

Substituting $\dot{r}$ from (5.30) into (5.32),

$$\dot{V} = r(-k_v r - \tilde{W}^T \phi - \tilde{P}^T \psi u_1 - P^T \psi u_2) + \tilde{W}^T F^{-1}\dot{\tilde{W}} + \tilde{P}^T G^{-1}\dot{\tilde{P}}$$

Since W and P are constants, we can write $\dot{\tilde{W}} = \dot{W} - \dot{\hat{W}} = -\dot{\hat{W}}$ and $\dot{\tilde{P}} = \dot{P} - \dot{\hat{P}} = -\dot{\hat{P}}$, thus

$$\dot{V} = -k_v r^2 - \tilde{W}^T \phi r - \tilde{W}^T F^{-1}\dot{\hat{W}} - \tilde{P}^T \psi u_1 r - P^T \psi u_2 r - \tilde{P}^T G^{-1}\dot{\hat{P}}$$

$$= -k_v r^2 - \tilde{W}^T(\phi r + F^{-1}\dot{\hat{W}}) - \tilde{P}^T(\psi u_1 r + G^{-1}\dot{\hat{P}}) - P^T \psi u_2 r \quad (5.33)$$

Equating the second and third terms of (5.33) to zero, we get

$$\phi r + F^{-1}\dot{\hat{W}} = 0$$

$$\text{or } \dot{\hat{W}} = -F\phi r \qquad (5.34)$$

$$\psi u_1 r + G^{-1}\dot{\hat{P}} = 0$$

$$\text{or } \dot{\hat{P}} = -G\psi u_1 r \qquad (5.35)$$

From (5.27) it is clear that u_1 becomes unbounded when $\hat{g}(x) \to 0$. Using the concept of projecting $\hat{g}$ inside a set where $\hat{g} \neq 0$, we modify the update law for $\hat{P}$ in such a way that for $g > 0$ when the estimate $\hat{g}$ is less than the lower bound g_l and at the same $\dot{\hat{P}}$ is negative, then we do not update $\hat{P}$. Thus, the modified update law becomes

$$\dot{\hat{P}} = 0 \text{ when } \hat{g} - g_l < 0 \text{ and } \psi \, u_1 \, r > 0$$
$$= -G\psi u_1 r \text{ otherwise} \qquad (5.36)$$

Using update laws (5.34) and (5.35), (5.33) becomes

$$\dot{V} = -k_v r^2 - P^T \psi u_2 r = -k_v r^2 - g u_2 r \qquad (5.37)$$

Define the sliding mode term as $u_2 = \frac{|\hat{g}|}{g_l}|u_1|\text{sgn}\,(r)$. Eqn (5.37) then becomes

$$\dot{V} = -k_v r^2 - \frac{g|\hat{g}|}{g_l}|u_1|r \text{ sgn }(r) \qquad (5.38)$$

Since we have assumed that $|g| > 0$ and the lower bound g_l is known to us, the term $\frac{g|\hat{g}|}{g_l}|u_1|$ is always positive. Moreover r sgn (r) is always positive. Thus, $\dot{V}$ is negative definite. Since $V > 0$ and $\dot{V} \leq 0$, this shows stability in the sense of Lyapunov so that r, $\tilde{W}$ (hence $\hat{W}$) and $\tilde{P}$ (hence $\hat{P}$) are bounded. Hence the proof.

We should note that to make the control input bounded, the update law for $\hat{P}$ is modified according to Eqn (5.36). It follows from Eqn (5.36) that in the absence of the sliding mode term u_2, $\dot{V}$ in Eqn (5.33) equals $-k_v r^2 - \tilde{P}^T \psi u_1 r$ when $\dot{\hat{P}} = 0$. Since $-\tilde{P}^T \psi u_1 r$ can take both positive and negative values, negative definiteness of $\dot{V}$ cannot be ensured in this case. To compensate the additional term in $\dot{V}$ the sliding mode term, u_2 is introduced in the control law (5.26). When $\dot{\hat{P}} = 0$,

$$\dot{V} = -k_v r^2 + \hat{g} u_1 r - g u_1 r - \frac{g|\hat{g}|}{g_l}|u_1|r \text{ sgn }(r)$$

Since, g and $\hat{g}$ are of the same sign, $\hat{g} u_1 r$ and $-g u_1 r$ will always be of different signs. If one contributes a positive value to $\dot{V}$ other will contribute a negative one. The term that takes a positive value will be compensated by $-\frac{g|\hat{g}|}{g_l}|u_1|r$ sgn (r), which is always negative and with a magnitude greater than $|g u_1 r|$. Thus $\dot{V}$ will be negative semi-definite.

Modification in the update law for $\hat{P}$ suggests that the initial value of $\hat{P}$ should be chosen such that $\hat{g} > g_l$. When the bounds on $g(x)$ are negative, the update law for $\hat{P}$ is modified as

$$\dot{\hat{P}} = 0 \text{ when } \hat{g} - g_u > 0 \text{ and } \psi \, u_1 \, r < 0,$$
$$= -G\psi u_1 r \text{ otherwise}$$

The sliding mode term will also be modified as $u_2 = \frac{|\hat{g}|}{g_u}|u_1|\text{sgn}\,(r)$.

Example 5.2 The schematic of a ball and beam system is shown in Figure 5.4 where a servo motor is coupled to the ball and beam module. The motor drives a lever arm which is coupled to a track upon which a rolling ball rests. The purpose of design is to control the position of the ball along the track by manipulating the angular position of the servo.

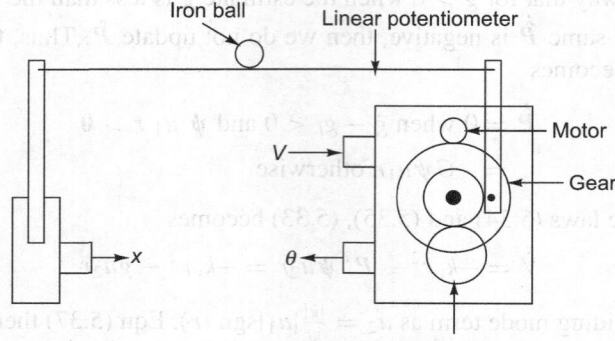

Figure 5.4   Schematic diagram of a ball–beam system

The simplified ball dynamics and motor dynamics are given by the following equations:

$$\ddot{x} = -\frac{5}{7}\overline{g}\sin(\frac{r}{L}\theta)$$

$$\ddot{\theta} = \frac{K_g K_m}{J_{eq} R_m}\left(V_i - K_b K_g \dot{\theta}\right)$$

where x is the ball position, θ is the motor angular displacement, and $\overline{g}$ is the acceleration due to gravity. Details of the parameters are given in Table 5.1. The beam angle α is related to θ by the following relation:

$$\alpha = \frac{r}{L}\theta$$

Design a direct adaptive controller using the design concepts given in this section (Theorem 5.2).

Table 5.1   Ball-beam system parameters

Motor inertia J_m	3.87×10^{-7} Kgm^2
Load inertia J_l	2.42×10^{-5} Kgm^2
Equivalent inertia J_{eq}	0.0029 Kgm^2
Gear ratio K_g	70 : 1
Motor resistance R_m	2.6 Ω
Back EMF constant K_b	0.00767 V/ (rad/s)
Motor torque constant K_m	0.00767 Nm/A
Beam length L	43.18 cm
Lever radius r	2.54 cm

Solution
Considering x, $\dot{x}$, θ, $\dot{\theta}$ as the system states and V_i as the input, the state–space model of the system can be written as

$$\dot{x}_1 = x_2$$

$$\dot{x}_2 = a_1 \sin(a_2 x_3)$$
$$\dot{x}_3 = x_4$$
$$\dot{x}_4 = a_3 x_4 + a_4 u$$

where

$a_1 = -\frac{5}{7}\overline{g} = -7$, $a_2 = \frac{r}{L} = 0.06$, $a_3 = -\frac{K_b K_m K_g^2}{J_{eq} R_m} = -38.23$, $a_4 = \frac{K_g K_m}{J_{eq} R_m} = 71.21$. Since the above model is not in the prescribed format applicable for the adaptive control scheme, we can use some state transformation before designing the controller. Let the state variables be z_1, z_2, z_3, and z_4 in the transformed domain. Let $z_1 = x_1$ and $z_2 = x_2$. Then we can write

$$\dot{z}_1 = z_2$$
$$\dot{z}_2 = a_1 \sin(a_2 x_3) = z_3 \ (say)$$
$$\dot{z}_3 = a_1 a_2 x_4 \cos(a_2 x_3) = z_4 \ (say)$$
$$\dot{z}_4 = -a_2^2 z_2 x_4^2 + a_3 z_4 + \frac{a_4 z_4}{x_4} u$$
$$y = z_1$$

where x_4 can be expressed as a function of $z = [z_1 \ z_2 \ z_3 \ z_4]^T$. The above representation is in the affine strict feedback form and the controllers (5.13) and (5.26) can be applied to the system. As mentioned earlier, the objective is to maintain the ball at a desired position or to track an admissible trajectory. In this example, we have taken the desired trajectory as

$$y_d = z_{1d} = 0.2 \sin(t)$$

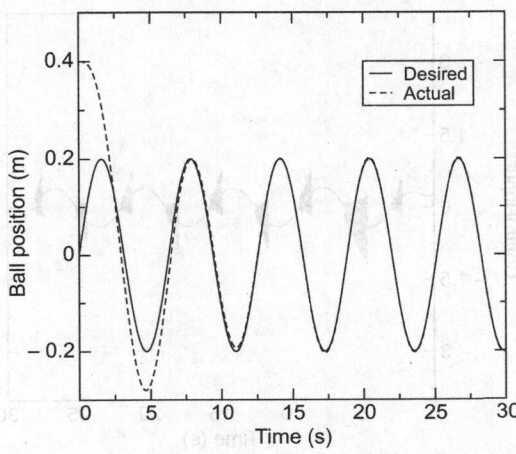

Figure 5.5 Tracking results for a ball–beam system

The controller (5.26) is successfully implemented for the system for which the parameters are taken as $k_v = 30$, $\lambda_1 = 50$, $\lambda_2 = 40$, and $\lambda_3 = 30$. The number of neurons for both RBF networks is taken as 60. The centres of the RBF network are chosen randomly between 0 and 1 and weights are initialized to very small values. The parameter matrices F_1 and F_2 are taken as the diagonal matrices of an appropriate dimension with the diagonal elements 10 and 15, respectively. The range of θ for which the dynamics is valid is given as $-45^o \leq \theta \leq 45^o$. Since $g(z) = g'(x) = a_1 a_2 a_4 \cos(a_2 x_3)$, $g'(x)$ is always negative. The bounds on g' is approximately found as $-29 < g' < -30$. Even if the exact bounds are not known, we can safely approximate some bound on g' such that the closed loop system is Lyapunov stable. Figure 5.5 shows how the ball tracks the specified desired trajectory. The corresponding beam angle and control voltage are shown in Figure 5.6.

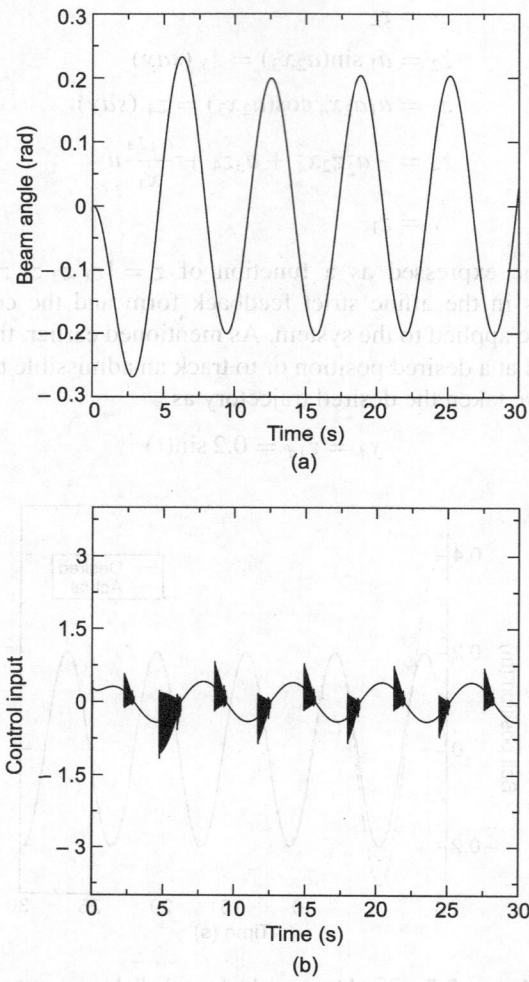

Figure 5.6 Ball–beam system: (a) beam angle and (b) control input

5.3 MULTI-INPUT MULTI-OUTPUT SYSTEMS

The control scheme, as described in the previous section, is extended for multi-input multi-output (MIMO) systems. A general MIMO system dynamics is given by

$$\dot{x} = f(x) + G(x)u \qquad (5.39)$$

$$y = Cx$$

where $x \in R^n$, $f(x) \in R^n$, $u \in R^m$, $G(x) \in R^{m \times n}$, $y \in R^p$, $C \in R^{p \times n}$. A wide class of MIMO systems can also be written in the following form:

$$\dot{x}_1 = x_2$$

$$\dot{x}_2 = f_1(x) + g_{11}(x)u_1 + \cdots + g_{1m}(x)u_m$$

$$\dot{x}_3 = x_4$$

$$\dot{x}_4 = f_2(x) + g_{21}(x)u_1 + \cdots + g_{2m}(x)u_m$$

$$\vdots$$

$$\dot{x}_{2n-1} = x_{2n}$$

$$\dot{x}_{2n} = f_n(x) + g_{n1}(x)u_1 + \cdots + g_{nm}(x)u_m$$

$$y = [x_1 \ x_3 \ \cdots \ x_{2n-1}]$$

For such cases, the system equations can be rewritten as

$$\dot{z}_1 = z_2$$

$$\dot{z}_2 = f(z) + G(z)u \qquad (5.40)$$

$$y = z_1$$

where $z_1 = [x_1 \ x_3 \ \cdots \ x_{2n-1}]$, $z_2 = [x_2 \ x_4 \ \cdots \ x_{2n}]$, $f(z) = [f_1 \ \cdots \ f_n]^T \in R^n$,

$$G(z) = \begin{bmatrix} g_{11} \cdots g_{1m} \\ \vdots \\ g_{n1} \cdots g_{nm} \end{bmatrix} \in R^{n \times m}, \ z = [z_1 \ z_2]^T \in R^{2n}.$$ The output error can be defined as

$$e = y_d - y = z_{1d} - z_1$$

$y_d = z_{1d}$ is the desired output vector. Let us define a variable r as $r = \dot{e} + \Lambda e$, where Λ is a diagonal matrix with positive diagonal elements. In the case of MIMO systems, the function $f(x)$ or $f(z)$ is a vector-valued function. Thus, the RBF network has multiple outputs as shown in Figure 5.7. The network weights constitute a matrix $\hat{W} \in R^{L \times n}$ in this case.

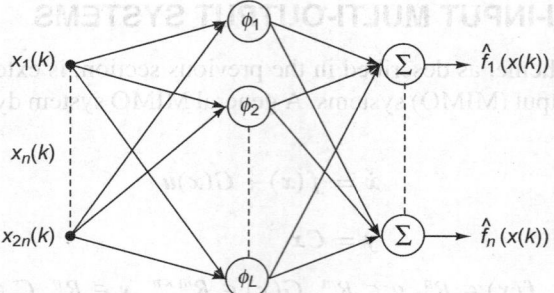

Figure 5.7 An $2n$ input and n output RBF network

Theorem 5.3 Suppose that the non-linear function $f(z)$ of system (5.40) is unknown, while the function $G(z)$ is known. Suppose also that $f(z)$ can be approximated as $\hat{f}(z) = \hat{W}^T \phi(z)$ using a radial basis function network. Then the control law $u = G^T(z)(GG^T)^{-1}[-\hat{f}(z) + K_v r + \Lambda_1 e + \dot{z}_{2d}]$ will stabilize the system (5.40) in the sense of Lyapunov provided $\hat{W}$ is updated using the update law $\dot{\hat{W}} = -F \phi\, r^T$ where F is a positive definite matrix.

Proof: The control law u is defined as follows:

$$u = G^T(z)(GG^T)^{-1}[-\hat{f}(z) + K_v r + \Lambda_1 e + \dot{z}_{2d}] \tag{5.41}$$

K_v is a positive definite diagonal matrix. Here $\hat{W}$ is the weight matrix of an appropriate dimension. Let us assume that there exists an ideal weight matrix W such that the original vector $f(z)$ can be represented as $f(z) = W^T \phi(z)$. The RBFN approximation error is assumed to be zero. Putting the control law u (5.41) in the system Eqn (5.40) and after simplification, we get

$$\dot{z}_2 = W^T \phi - \hat{W}^T \phi + K_v r + \Lambda_1 e + \dot{z}_{2d} \tag{5.42}$$

Defining $\tilde{W}^T = W^T - \hat{W}^T$, we can write

$$\dot{z}_2 = \tilde{W}^T \phi + K_v r + \Lambda_1 e + \dot{z}_{2d} \tag{5.43}$$

$$\text{Again}\quad \dot{r} = \ddot{e} + \Lambda_1 \dot{e} = \dot{z}_{2d} - \dot{z}_2 + \Lambda_1 \dot{e} \tag{5.44}$$

Combining Eqns (5.43) and (5.44),

$$\dot{r} = -K_v r - \tilde{W}^T \phi \tag{5.45}$$

Consider a Lyapunov function candidate

$$V = \frac{1}{2} r^T r + \text{trace}\left[\frac{1}{2} \tilde{W}^T F^{-1} \tilde{W} \right] \tag{5.46}$$

where F is a positive definite matrix. Since the Lyapunov function should be a scalar function, we have taken trace of $\frac{1}{2}\tilde{W}^T F^{-1}\tilde{W}$. Differentiating (5.46),

$$\dot{V} = r^T \dot{r} + \text{trace}\left[\tilde{W}^T F^{-1}\dot{\tilde{W}}\right] \tag{5.47}$$

Substituting $\dot{r}$ from (5.45) into (5.47),

$$\dot{V} = r(-K_v r - \tilde{W}^T \phi) + \text{trace}\left[\tilde{W}^T F^{-1}\dot{\tilde{W}}\right] \tag{5.48}$$

Since W is a constant matrix, we can write $\dot{\tilde{W}} = \dot{W} - \dot{\hat{W}} = -\dot{\hat{W}}$. Thus,

$$\dot{V} = -r^T K_v r - r^T \tilde{W}^T \phi + \text{trace}\left[-\tilde{W}^T F^{-1}\dot{\hat{W}}\right] \tag{5.49}$$

Using the properties of trace, $r^T \tilde{W}^T \phi = \text{trace}[\tilde{W}^T \phi r^T]$, we can further write

$$\dot{V} = -r^T K_v r + \text{trace}\left[-\tilde{W}^T \phi r^T - \tilde{W}^T F^{-1}\dot{\hat{W}}\right] \tag{5.50}$$

Equating the second term of (5.50) to zero, we get

$$\phi r^T + F^{-1}\dot{\hat{W}} = 0$$

$$\text{or } \dot{\hat{W}} = -F\phi r^T \tag{5.51}$$

Using the above update law, Eqn (5.50) becomes

$$\dot{V} = -r^T K_v r \tag{5.52}$$

Since $V > 0$ and $\dot{V} \le 0$, this shows stability in the sense of Lyapunov, so that r and $\tilde{W}$ (hence $\hat{W}$) are bounded. Hence the proof. Furthermore,

$$\int_0^\infty -\dot{V}\,dt < \infty \tag{5.53}$$

$\ddot{V} = -2r^T K_v \dot{r}$ and all signals on the right-hand side of (5.45) verify the boundedness of $\dot{r}$ and hence of $\ddot{V}$. Therefore, $\dot{V}$ is uniformly continuous. Thus, according to Barbalat's Lemma, $\dot{V} \to 0$, as $t \to \infty$ and hence r vanishes. We should note that for existence of the term $(GG^T)^{-1}$ in the control law, G should be a square matrix.

Inclusion of Non-Zero RBFN Approximation Error

As in the SISO case, if we consider a non-zero approximation error ϵ, the derivative of the Lyapunov function becomes

$$\dot{V} = -r^T K_v r + \epsilon\, r$$

To proceed further we will use the following properties of matrices:
For a positive definite matrix K_v,

$$\lambda_{\min}(K_v)\|r\|^2 \le r^T K_v r \le \lambda_{\max}(K_v)\|r\|^2$$

where $\lambda_{max}(\cdot)$ and $\lambda_{min}(\cdot)$ denote the maximum and minimum eigenvalues of a matrix. Thus, one can write

$$\dot{V} \le -\lambda_{min}(K_v)\|r\|^2 + \|\epsilon\,r\|$$
$$\le -\lambda_{min}(K_v)\|r\|^2 + \epsilon_N\,\|r\|$$

where $\|\epsilon\| < \epsilon_N$. The negative definiteness of $\dot{V}$ can be ensured as long as $\|r\| > \frac{\epsilon_N}{\lambda_{min}(K_v)}$. To improve the tracking accuracy, K_v should be chosen such that $\lambda_{min}(K_v)$ is a large number.

Example 5.3   Consider the following dynamics of a two degree of freedom horizontal manipulator [88]:

$$a_1\ddot{\theta}_1 + (a_3 C_{21} + a_4 S_{21})\ddot{\theta}_2 - a_3 S_{21}\dot{\theta}_2^{\,2} + a_4 C_{21}\dot{\theta}_2^{\,2} = \tau_1 \qquad (5.54)$$

$$(a_3 C_{21} + a_4 S_{21})\ddot{\theta}_1 + a_2\ddot{\theta}_2 + a_3 S_{21}\dot{\theta}_1^{\,2} - a_4 C_{21}\dot{\theta}_1^{\,2} = \tau_2 \qquad (5.55)$$

where θ_1, θ_2 are the link angles and τ_1, τ_2 are the joint torques. $C_{21} = \cos(\theta_2 - \theta_1)$, $S_{21} = \sin(\theta_2 - \theta_1)$. The four parameters a_1, a_2, a_3 and a_4 are taken as 0.15, 0.04, 0.03, and 0.025 kg m^2, respectively. Design a direct adaptive controller for the system using Theorem 5.3 . Take the desired trajectories as

$$\theta_1^d(t) = 1.5(1 - \cos(3t))$$
$$\theta_2^d(t) = (1 - \cos(5t)) \qquad (5.56)$$

Solution
Considering the state variables as $x_1 = \theta_1$, $x_2 = \dot{\theta}_1$, $x_3 = \theta_2$, $x_4 = \dot{\theta}_2$, and one can write

$$\dot{x}_1 = x_2$$

$$\dot{x}_2 = \frac{1}{D}[m_{22}(\tau_1 - v_1) - m_{12}(\tau_2 - v_2)]$$

$$\dot{x}_3 = x_4$$

$$\dot{x}_4 = \frac{1}{D}[-m_{21}(\tau_1 - v_1) + m_{11}(\tau_2 - v_2)]$$

where

$$m_{11} = a_1, \quad m_{22} = a_2$$
$$m_{12} = m_{21} = a_3 C_{21} + a_4 S_{21}$$
$$D = m_{11}m_{22} - m_{12}m_{21}$$
$$v_1 = -a_3 S_{21}\dot{\theta}_2^{\,2} + a_4 C_{21}\dot{\theta}_2^{\,2}$$
$$v_2 = a_3 S_{21}\dot{\theta}_1^{\,2} - a_4 C_{21}\dot{\theta}_1^{\,2}$$

Let us define $z_1 = \begin{bmatrix} x_1 \\ x_3 \end{bmatrix}$ and $z_2 = \begin{bmatrix} x_2 \\ x_4 \end{bmatrix}$. The system equations can be written in terms of these variables as

$$\dot{z}_1 = z_2$$
$$\dot{z}_2 = f(z) + G(z)\tau$$

where $\quad f(z) = \begin{bmatrix} f_1 \\ f_2 \end{bmatrix} = \frac{1}{D} \begin{bmatrix} -m_{22}v_1 + m_{12}v_2 \\ m_{21}v_1 - m_{11}v_2 \end{bmatrix} \quad$ and $\quad G(z) =$
$\begin{bmatrix} g_{11} & g_{12} \\ g_{21} & g_{22} \end{bmatrix} = \frac{1}{D} \begin{bmatrix} m_{22} & -m_{12} \\ -m_{21} & m_{11} \end{bmatrix}$. The output vector is $y = z_1$. The

desired output vector is $z_{1d} = \begin{bmatrix} x_{1d} \\ x_{3d} \end{bmatrix} == \begin{bmatrix} \theta_1^d \\ \theta_2^d \end{bmatrix}$. The control input is generated using (5.41). The controller and the neural network parameter vectors are taken as $K_v = \begin{bmatrix} 10 & 0 \\ 0 & 10 \end{bmatrix}$, $\Lambda = \begin{bmatrix} 5 & 0 \\ 0 & 5 \end{bmatrix}$. The parameter matrix F is taken as the diagonal matrix with diagonal elements 0.5. The number of neurons for the network is taken as 20. The centers of the RBF network are chosen randomly between 0 and 1 and weights are initialized to small values. The tracking results for the link angles are shown in Figure 5.8. The corresponding input torques are shown in Figure 5.9. The RMS error for the link θ_1 is found to be 0.0052, whereas for θ_2 it is 0.0048.

5.4 SINGLE-INPUT SINGLE-OUTPUT DISCRETE TIME AFFINE SYSTEMS

Discrete time SISO affine non-linear systems in the strict feedback form can be represented by the following model:

$$x_1(k + 1) = x_2(k)$$
$$x_2(k + 1) = x_3(k)$$
$$\vdots$$
$$x_n(k + 1) = f(x(k)) + g(x(k))u(k) \tag{5.57}$$
$$y(k) = x_1(k)$$

where $x(k) = [x_1(k)\, x_2(k) \cdots x_n(k)]^T \in R^n$, $y(k) \in R$ and $u(k) \in R$.

5.4.1 $f(x)$ Is Unknown, But $g(x)$ Is Known

In this section, we assume that the function $f(x)$ is unknown, while the function $g(x)$ is known. $f(x)$ is approximated by a radial basis function network (RBFN). The pictorial diagram of the RBFN when approximating $f(x)$ is shown in Figure 5.3. Discrete time counterpart of Theorem 5.1 is given as follows.

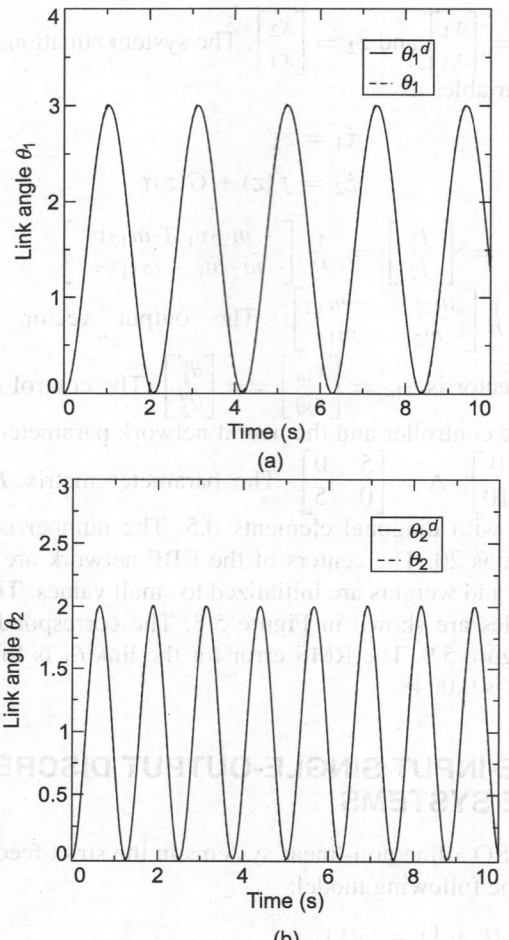

Figure 5.8   Tracking results for two link manipulator: (a) link angle θ_1 and (b) link angle θ_2

Theorem 5.4   Suppose that the non-linear function $f(x)$ of system (5.57) is unknown, while the function $g(x)$ is known. Let $f(x)$ be approximated as $\hat{f}(x) = \hat{W}^T \phi(x)$ using a radial basis function network. Then the control law $u(k) = \frac{1}{g(x)}[-\hat{f}(x) + k_v r(k) + \lambda_1 e_n(k) + \cdots + \lambda_{n-1} e_2(k) + x_{nd}(k+1)]$ will stabilize the system (5.57) in the sense of Lyapunov provided $\hat{W}$ is updated using the update law $\hat{W}(k+1) = \hat{W}(k) - \alpha \phi\, r(k+1)$ where α is a positive constant.

Proof: The output tracking error is defined as $e(k) = y_d(k) - y(k) = x_{1d}(k) - x_1(k)$, where $y_d(k) = x_{1d}(k)$ is the desired output of the system. The filtered

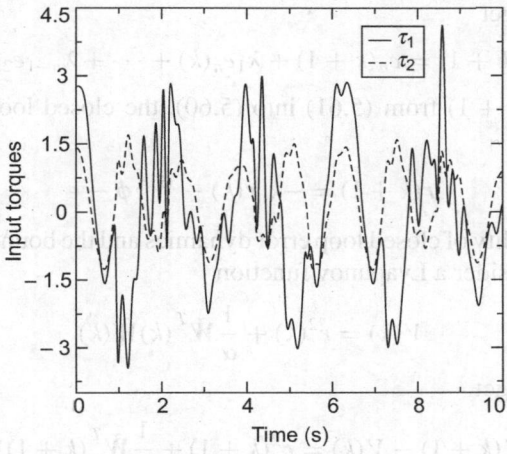

Figure 5.9   Control torques τ_1 and τ_2

tracking error is defined as

$$r(k) = e_n(k) + \lambda_1 e_{n-1}(k) + \cdots + \lambda_{n-1} e_1(k) \qquad (5.58)$$

where $e_j(k) = x_{jd}(k) - x_j(k), \quad j = 1, 2, \ldots, n. \lambda_1, \ldots, \lambda_{n-1}$ are chosen such that the above system is stable. The objective is to achieve a satisfactory tracking result as well as to maintain boundedness of the error and neural network weight vector. The control law $u(k)$ is chosen as

$$u(k) = \frac{1}{g(x)}[-\hat{f}(x) + k_v r(k) + \lambda_1 e_n(k) + \cdots + \lambda_{n-1} e_2(k) + x_{nd}(k+1)]$$

$$(5.59)$$

Putting the control law $u(k)$ [Eqn (5.59)] in system (5.57), we get

$$x_n(k+1) = f(x) + g(x)\frac{1}{g(x)}[-\hat{f}(x) + k_v r(k) + \lambda_1 e_n(k) + \cdots +$$

$$\lambda_{n-1} e_2(k) + x_{nd}(k+1)]$$

Let us assume that there exists a weight vector W such that the original function $f(x)$ can be represented as $f(x) = W^T \phi(x) + \epsilon$ where ϵ is the estimation error. Thus,

$$x_n(k+1) = W^T \phi + \epsilon - \hat{W}^T \phi + k_v r(k) + \lambda_1 e_n(k) + \cdots +$$

$$\lambda_{n-1} e_2(k) + x_{nd}(k+1)$$

Defining $\tilde{W}^T = W^T - \hat{W}^T$, we can write

$$x_n(k+1) = \tilde{W}^T \phi + k_v r(k) + \lambda_1 e_n(k) + \cdots +$$

$$\lambda_{n-1} e_2(k) + x_{nd}(k+1)$$

$$x_{nd}(k+1) - x_n(k+1) = -\tilde{W}^T \phi - k_v r(k) - \epsilon -$$

$$\lambda_1 e_n(k) - \cdots - \lambda_{n-1} e_2(k) \qquad (5.60)$$

From (5.58) we get

$$r(k+1) = e_n(k+1) + \lambda_1 e_n(k) + \cdots + \lambda_{n-1} e_2(k) \tag{5.61}$$

Substituting $e_n(k+1)$ from (5.61) into (5.60), the closed loop error dynamics becomes,

$$r(k+1) = -k_v r(k) - \tilde{W}^T \phi - \epsilon \tag{5.62}$$

To show the stability of closed loop error dynamics and the boundedness of weight vector, let us consider a Lyapunov function

$$V(k) = r^2(k) + \frac{1}{\alpha} \tilde{W}^T(k)\tilde{W}(k) \tag{5.63}$$

From (5.63), we get

$$\Delta V = V(k+1) - V(k) = r^2(k+1) + \frac{1}{\alpha} \tilde{W}^T(k+1)\tilde{W}(k+1)$$

$$- r^2(k) - \frac{1}{\alpha} \tilde{W}^T(k)\tilde{W}(k) \tag{5.64}$$

Substituting $r(k+1)$ from (5.62) into (5.64),

$$\Delta V = (-k_v r(k) - \tilde{W}^T \phi - \epsilon)^2 - r^2(k)$$

$$+ \frac{1}{\alpha}[\tilde{W}^T(k+1)\tilde{W}(k+1) - \tilde{W}^T(k)\tilde{W}(k)]$$

$$= k_v^2 r^2(k) + 2k_v r(k)\tilde{W}^T \phi + \phi^T \tilde{W}\tilde{W}^T \phi + 2k_v r(k)\epsilon + 2\tilde{W}^T \phi\epsilon$$

$$+ \epsilon^2 + \frac{1}{\alpha}[W^T(k+1)W(k+1) - 2\hat{W}^T(k+1)W(k+1)$$

$$+ \hat{W}^T(k+1)\hat{W}(k+1) - W^T(k)W(k) + 2\hat{W}^T(k)W(k)$$

$$- \hat{W}^T(k)\hat{W}(k)] \tag{5.65}$$

Since W is constant, we can write $W(k+1) = W(k) = W$. Thus,

$$\Delta V = k_v^2 r^2(k) + 2k_v r(k)\tilde{W}^T \phi + \phi^T \tilde{W}\tilde{W}^T \phi + 2k_v r(k)\epsilon + 2\tilde{W}^T \phi\epsilon$$

$$+ \epsilon^2 + \frac{1}{\alpha}[-2\hat{W}^T(k+1)W + \hat{W}^T(k+1)\hat{W}(k+1) +$$

$$2\hat{W}^T(k)W - \hat{W}^T(k)\hat{W}(k)] \tag{5.66}$$

Using the update law $\hat{W}(k+1) = \hat{W}(k) - \alpha\phi r(k+1)$, we get

$$\hat{W}^T(k+1)W(k) = \hat{W}^T(k)W(k) - r(k+1)\phi^T\hat{W}(k)$$

$$\hat{W}^T(k+1)\hat{W}(k+1) = \hat{W}^T(k)\hat{W}(k) - 2\alpha r(k+1)\hat{W}^T(k)\phi +$$

$$\alpha^2 r^2(k+1)\phi^T \phi$$

Substituting the above equations in (5.66),

$$\Delta V = -r^2(k) + k_v^2 r^2(k) + 2k_v r(k)\tilde{W}^T \phi + \phi^T \tilde{W}\tilde{W}^T \phi + 2k_v r(k)\epsilon$$

$$+ 2\tilde{W}^T \phi \epsilon + \epsilon^2 + 2r(k+1)\tilde{W}^T \phi + \alpha r^2(k+1)\phi^T \phi \qquad (5.67)$$

Denoting $\tilde{W}^T \phi$ as $\overline{e}_1$ and substituting the expression of $r(k+1)$ in the above equation, we get

$$\Delta V = - r^2(k) + k_v^2 r^2(k) + 2k_v r(k)\overline{e}_1 + \overline{e}_1^2 + 2k_v r(k)\epsilon + 2\overline{e}_1 \epsilon + \epsilon^2$$
$$+ 2\overline{e}_1(-k_v r(k) - \overline{e}_1 - \epsilon) + \alpha\|\phi\|^2(k_v^2 r^2(k) + 2k_v r(k)\overline{e}_1 + \overline{e}_1^2$$
$$+ 2k_v r(k)\epsilon + 2\overline{e}_1 \epsilon + \epsilon^2) \qquad (5.68)$$

After simplification Eqn (5.68) becomes

$$\Delta V = - r^2(k) + k_v^2 r^2(k)(1 + \alpha\|\phi\|^2) - \overline{e}_1^2(1 - \alpha\|\phi\|^2) + \epsilon^2(1 + \alpha\|\phi\|^2)$$
$$+ 2k_v r(k)\epsilon(1 + \alpha\|\phi\|^2) + 2\overline{e}_1 \epsilon \alpha\|\phi\|^2 + 2k_v r(k)\overline{e}_1 \alpha\|\phi\|^2 \qquad (5.69)$$

Let us denote $1 - \alpha\|\phi\|^2$ by γ. Using this notation and after further simplification Eqn (5.69) can be written as

$$\Delta V = - r^2 + \frac{1}{\gamma}k_v^2 r^2 + \frac{1}{\gamma}\epsilon^2 + 2\frac{1}{\gamma}k_v r\epsilon - \gamma(\overline{e}_1 - \frac{1}{\gamma - 1}(\epsilon + k_v r))^2$$
$$\leq -|r|^2 + \frac{1}{\gamma}k_v^2|r|^2 + \frac{1}{\gamma}|\epsilon|^2 + 2\frac{1}{\gamma}k_v|r||\epsilon| - \gamma(\overline{e}_1 - \frac{1}{\gamma - 1}(\epsilon + k_v r))^2$$
$$= - (1 - \frac{1}{\gamma}k_v^2)(|r|^2 - \frac{2k_v|\epsilon||r|}{\gamma - k_v^2} - \frac{1}{\gamma - 1}|\epsilon|^2) - \gamma(\overline{e}_1 - \frac{1}{\gamma - 1}(\epsilon + k_v r))^2$$

ΔV will be negative if the following conditions are satisfied:

$\gamma > 0$
$1 - \frac{1}{\gamma}k_v^2 > 0$
$|r|^2 - \frac{2k_v|\epsilon||r|}{\gamma - k_v^2} - \frac{1}{\gamma - 1}|\epsilon|^2 > 0$

The above conditions simplify to

1. $\alpha\|\phi\|^2 < 1$
2. $k_v < \sqrt{\gamma}$
3. $|r| > \frac{1}{\gamma - k_v^2}(\sqrt{\gamma} + k_v)|\epsilon|$

Since $V(k) > 0$ and $\Delta V \leq 0$, the closed loop error dynamics (5.62) is Lyapunov stable and $r(k)$, $\tilde{W}(k)$, and $\hat{W}(k)$ are thus bounded. The learning rate α and the controller parameter k_v should be chosen such that the first two conditions are satisfied. The third condition signifies that the filtered tracking error $r(k)$ is contained in a compact set defined by $|r| < B_r$ where $B_r = \frac{1}{\gamma - k_v^2}(\sqrt{\gamma} + k_v)|\epsilon|$. It follows from conditions (1) and (2) that B_r is very small which ensures that the tracking error will always be bounded by a very small constant.

5.4.2 $f(x)$ and $g(x)$ Both Are Unknown

When both non-linear functions $f(x)$ and $g(x)$ are unknown, then $f(x)$ and $g(x)$ are estimated using two radial basis function networks. If we substitute $f(x)$

and $g(x)$ with their neural approximations $\hat{f}(x)$ and $\hat{g}(x)$, the control law can be expressed as

$$u(k) = \frac{1}{\hat{g}(x)}[-\hat{f}(x) + k_v r(k) + \lambda_1 e^{(n)}(k) + \cdots +$$

$$\lambda_{n-1} e^{(2)}(k) + x_{nd}(k+1)] \tag{5.70}$$

However, this control law cannot guarantee stability for any weight update laws derived for neural estimators of f and g because the boundedness of $u(k)$ cannot be ensured when $\hat{g}(x)$ approaches 0. This can be taken care of by a projection algorithm to keep $\hat{g}(x)$ away from zero. In that scenario, a second control component is added to maintain the closed loop stability for both continuous [89, 90] and discrete time cases [91, 92]. In this section, we show that the structure of the control law (5.70) can be retained if a small constant is added to the denominator of the control input. Let us first consider the case when $g(x) > 0$. Although $g(x)$ is positive, we cannot guarantee the estimate $\hat{g}(x)$ to be positive. It can take both positive and negative values as well as pass through zero. When $-a_1 \leq \hat{g} \leq a_1$, where a_1 is a small positive constant, the controller becomes unbounded. a_1 can be found out heuristically. In such a situation, the following cases may arise:

1. *When $\hat{g} > a_1$ or $\hat{g} < -a_1$*. The problem of unboundedness of the control input would not come so the controller can take the following form:

$$u(k) = \frac{1}{\hat{g}(x)}[-\hat{f}(x) + k_v r(k) + \lambda_1 e^{(n)}(k) + \cdots$$

$$+ \lambda_{n-1} e^{(2)}(k) + x_{nd}(k+1)]$$

2. *When $0 < \hat{g} \leq a_1$*. The controller is taken as

$$u(k) = \frac{1}{\hat{g}(x) + a_2}[-\hat{f}(x) + k_v r(k) + \lambda_1 e^{(n)}(k) + \cdots$$

$$+ \lambda_{n-1} e^{(2)}(k) + x_{nd}(k+1)]$$

where a_2 is a small positive constant. If we take $a_2 > a_1$, the denominator of $u(k)$ will be greater than a_1; thus, the control input will be bounded.

3. *When $-a_1 \leq \hat{g} < 0$*. The controller is taken as

$$u(k) = \frac{1}{\hat{g}(x) + a_3}[-\hat{f}(x) + k_v r(k) + \lambda_1 e^{(n)}(k) + \cdots$$

$$+ \lambda_{n-1} e^{(2)}(k) + x_{nd}(k+1)]$$

where a_3 is another small constant. If we take $a_3 > 2a_1$, the denominator of $u(k)$ will again be greater than a_1; thus, the control input will be bounded.

When, $g(x) < 0$, the parameter a_2 should be less than $-2a_1$ and the parameter a_3 should be less than $-a_1$. The control input switches from case to case. But, since the constants a_2 and a_3 are small, the switching will be smooth. To prove the stability of the error dynamics and establish boundedness of the control input, we have to thus prove that the controller of the form

$$u(k) = \frac{1}{\hat{g}(x) + a}[-\hat{f}(x) + k_v r(k) + \lambda_1 e^{(n)}(k) + \cdots +$$

$$\lambda_{n-1} e^{(2)}(k) + x_{nd}(k+1)] \tag{5.71}$$

will stabilize system (5.57). a can take any value from the set $[0, a_2, a_3]$. The following theorem ensures the stability of the closed loop error dynamics.

Theorem 5.5   Given that both non-linear functions $f(x)$ and $g(x)$ of system (5.57) are unknown, let $f(x)$ be approximated as $\hat{f}(x) = \hat{W}^T \phi(x)$ and $g(x)$ be approximated as $\hat{g}(x) = \hat{P}^T \psi(x)$ using two radial basis function networks. Then the control law $u(k) = \frac{1}{\hat{g}(x)+a}[-\hat{f}(x) + k_v r(k) + \lambda_1 e^{(n)}(k) + \cdots + \lambda_{n-1} e^{(2)}(k) + x_{nd}(k+1)]$ will stabilize system (5.57) in the sense of Lyapunov provided $\hat{W}$ is updated using the update law $\hat{W}(k+1) = \hat{W}(k) - \alpha_1 \phi\, r(k+1)$ and $\hat{P}$ is updated using the update law $\hat{P}(k+1) = \hat{P}(k) - \alpha_2 \psi\, r(k+1)u(k)$.

Proof: According to the universal approximation property of neural networks, let us assume that there exist two weight vectors W and P such that the original functions $f(x)$ and $g(x)$ can be represented as

$$f(x) = W^T \phi(x) + \epsilon_1 \quad g(x) = P^T \psi(x) + \epsilon_2 \tag{5.72}$$

ϵ_1 and ϵ_2 are the reconstruction errors. The output tracking error and the filtered tracking errors are defined in Eqn (5.58). The objective is to achieve a satisfactory tracking result as well as to maintain boundedness of the error and the neural network weight vector. From (5.58), we get

$$r(k+1) = e^{(n)}(k+1) + \lambda_1 e^{(n)}(k) + \cdots + \lambda_{n-1} e^{(2)}(k) \tag{5.73}$$

The control law is described in Eqn (5.71) where a is a small positive number. Rearranging Eqn (5.71), we get

$$\hat{g}(x)u(k) + au(k) = -\hat{f}(x) + k_v r(k) + \lambda_1 e^{(n)}(k)$$

$$+ \cdots + \lambda_{n-1} e^{(2)}(k) + x_{nd}(k+1)$$

$$\text{or, } x_{nd}(k+1) = \hat{g}(x)u(k) + au(k) + \hat{f}(x) - k_v r(k) -$$

$$\lambda_1 e^{(n)}(k) - \cdots - \lambda_{n-1} e^{(2)}(k)$$

Combining the above equation with Eqns (5.57), (5.73), we get

$$r(k+1) = x^{nd}(k+1) - x_n(k+1) +$$

$$\lambda_1 e^{(n)}(k) + \cdots + \lambda_{n-1} e^{(2)}(k)$$

$$= \hat{g}(x)u(k) + au(k) + \hat{f}(x) - k_v r(k)$$

$$- \lambda_1 e^{(n)}(k) - \cdots \lambda_{n-1} e^{(2)}(k) - f(x)$$

$$- g(x)u + \lambda_1 e^{(n)}(k) + \cdots + \lambda_{n-1} e^{(2)}(k)$$

$$= - k_v r(k) - W^T \phi - \epsilon_1 - P^T \psi u(k) - \epsilon_2 u(k)$$

$$\hat{W}^T \phi + \hat{P}^T \psi u(k) + au(k) \tag{5.74}$$

Defining $\tilde{W}^T = W^T - \hat{W}^T$ and $\tilde{P}^T = P^T - \hat{P}^T$, Eqn (5.74) can be rewritten as

$$r(k+1) = - k_v r(k) - \tilde{W}^T \phi - \tilde{P}^T \psi u(k) - \epsilon_1 - \epsilon_3 u(k) \tag{5.75}$$

where $\epsilon_3 = \epsilon_2 - a$. Let us now consider a Lyapunov function

$$V(k) = r^2(k) + \frac{1}{\alpha_1} \tilde{W}^T(k) \tilde{W}(k) + \frac{1}{\alpha_2} \tilde{P}^T(k) \tilde{P}(k) \tag{5.76}$$

where α_1 and α_2 are two positive design parameters. From (5.76), we get

$$\Delta V = V(k+1) - V(k)$$

$$= r^2(k+1) + \frac{1}{\alpha_1} \tilde{W}^T(k+1) \tilde{W}(k+1) + \frac{1}{\alpha_2} \tilde{P}^T(k+1) \tilde{P}(k+1)$$

$$- r^2(k) - \frac{1}{\alpha_1} \tilde{W}^T(k) \tilde{W}(k) - \frac{1}{\alpha_2} \tilde{P}^T(k) \tilde{P}(k) \tag{5.77}$$

Substituting $r(k+1)$ from (5.75) into (5.77),

$$\Delta V = (-k_v r(k) - \tilde{W}^T \phi - \tilde{P}^T \psi u(k) - \epsilon_1 - \epsilon_3 u(k))^2 - r^2(k)$$

$$+ \frac{1}{\alpha_1} [\tilde{W}^T(k+1) \tilde{W}(k+1) - \tilde{W}^T(k) \tilde{W}(k)]$$

$$+ \frac{1}{\alpha_2} [\tilde{P}^T(k+1) \tilde{P}(k+1) - \tilde{P}^T(k) \tilde{P}(k)] \tag{5.78}$$

Simplifying the above equation

$$\Delta V = k_v^2 r^2(k) + 2 k_v r(k) \tilde{W}^T \phi + \phi^T \tilde{W} \tilde{W}^T \phi + u^2 \psi^T \tilde{P} \tilde{P}^T \psi$$

$$+ 2 k_v r(k) \tilde{P}^T \psi u + 2 \tilde{W}^T \phi \tilde{P}^T \psi u + \epsilon_1^2 + \epsilon_3^2 u^2 + 2\epsilon_1 \epsilon_3 u$$

$$+ 2(k_v r + \tilde{W}^T \phi + \tilde{P}^T \psi u)(\epsilon_1 + \epsilon_3 u) + \frac{1}{\alpha_1} [W^T(k+1) W(k+1) -$$

$$2\hat{W}^T(k+1) W(k+1) + \hat{W}^T(k+1) \hat{W}(k+1) - W^T(k) W(k) +$$

$$2\hat{W}^T(k) W(k) - \hat{W}^T(k) \hat{W}(k)] + \frac{1}{\alpha_2} [P^T(k+1) P(k+1)$$

$$- 2\hat{P}^T(k+1) P(k+1) + \hat{P}^T(k+1) \hat{P}(k+1)$$

$$- P^T(k) P(k) + 2\hat{P}^T(k) P(k) - \hat{P}^T(k) \hat{P}(k)] \tag{5.79}$$

Since W and P are constant vectors, we can write $W(k+1) = W(k) = W$ and $P(k+1) = P(k) = P$. Thus,

$$\Delta V = k_v^2 r^2(k) + 2k_v r(k)\tilde{W}^T\phi + \phi^T\tilde{W}\tilde{W}^T\phi + u^2\psi^T\tilde{P}\tilde{P}^T\psi$$

$$+ 2k_v r(k)\tilde{P}^T\psi u + 2\tilde{W}^T\phi\tilde{P}^T\psi u + \epsilon_1^2 + \epsilon_3^2 u^2 + 2\epsilon_1\epsilon_3 u$$

$$+ 2(k_v r + \tilde{W}^T\phi + \tilde{P}^T\psi u)(\epsilon_1 + \epsilon_3 u) + \frac{1}{\alpha_1}[-2\hat{W}^T(k+1)W$$

$$+ \hat{W}^T(k+1)\hat{W}(k+1) + 2\hat{W}^T(k)W - \hat{W}^T(k)\hat{W}(k)] +$$

$$\frac{1}{\alpha_2}[-2\hat{P}^T(k+1)P + \hat{P}^T(k+1)\hat{P}(k+1) + 2\hat{P}^T(k)P$$

$$- \hat{P}^T(k)\hat{P}(k)]$$

Using the update laws $\hat{W}(k+1) = \hat{W}(k) - \alpha_1\phi r(k+1)$ and $\hat{P}(k+1) = \hat{P}(k) - \alpha_2\phi r(k+1)u(k)$, we can write

$$\hat{W}^T(k+1)W = \hat{W}^T(k)W - \alpha_1 r(k+1)\phi^T W$$

$$\hat{W}^T(k+1)\hat{W}(k+1) = \hat{W}^T(k)\hat{W}(k)2\alpha_1 r(k+1)\hat{W}^T(k)\phi$$

$$+ \alpha_1^2 r^2(k+1)\phi^T\phi$$

$$\hat{P}^T(k+1)P = \hat{P}^T(k)P(k) - \alpha_2 r(k+1)\psi^T P$$

$$\hat{P}^T(k+1)\hat{P}(k+1) = \hat{P}^T(k)\hat{P}(k) - 2\alpha_2 r(k+1)u\hat{P}^T(k)\psi$$

$$+ \alpha_2^2 r^2(k+1)u^2\psi^T\psi$$

Substituting the above equations and expression of $r(k+1)$ in the expression of ΔV and denoting $\tilde{W}^T\phi$ as $\overline{e}_1$ and $\tilde{P}^T\psi$ as $\overline{e}_2$,

$$\Delta V = -r^2 + k_v^2 r^2(1 + \alpha_1\|\phi\|^2 + \alpha_2 u^2\|\psi\|^2) - (\overline{e}_1 + \overline{e}_2 u)^2(1 -$$

$$\alpha_1\|\phi\|^2 - \alpha_2 u^2\|\psi\|^2) + 2k_v\epsilon_1 r(1 + \alpha_1\|\phi\|^2 + \alpha_2 u^2\|\psi\|^2)$$

$$+ (\epsilon_1 + \epsilon_3 u)^2(1 + \alpha_1\|\phi\|^2 + \alpha_2 u^2\|\psi\|^2) + 2k_v r\epsilon_3 u$$

$$(1 + \alpha_1\|\phi\|^2 + \alpha_2 u^2\|\psi\|^2) + 2k_v\overline{e}_1 r(\alpha_1\|\phi\|^2 + \alpha_2 u^2\|\psi\|^2)$$

$$+ 2k_v\overline{e}_2 ur(\alpha_1\|\phi\|^2 + \alpha_2 u^2\|\psi\|^2) + 2\overline{e}_1\epsilon_1(\alpha_1\|\phi\|^2 + \alpha_2 u^2\|\psi\|^2)$$

$$+ 2\overline{e}_2 u\epsilon_1(\alpha_1\|\phi\|^2 + \alpha_2 u^2\|\psi\|^2) + 2\overline{e}_1\epsilon_3 u(\alpha_1\|\phi\|^2 + \alpha_2 u^2\|\psi\|^2)$$

$$+ 2\overline{e}_2\epsilon_3 u^2(\alpha_1\|\phi\|^2 + \alpha_2 u^2\|\psi\|^2) \tag{5.80}$$

Denoting $\alpha_1\|\phi\|^2 + \alpha_2\|\psi\|^2 u^2$ by β and simplifying Eqn (5.80), we can write

$$\Delta V = -r^2 + \frac{1}{1-\beta}[k_v^2 r^2 + 2k_v r(\epsilon_1 + \epsilon_2 u) + (\epsilon_1 + \epsilon_2 u)^2]$$

$$- (1-\beta)\left[(\overline{e}_1 + \overline{e}_2 u) - \frac{\beta}{1-\beta}(\epsilon_1 + \epsilon_3 u + k_v r)\right]^2$$

or,

$$\Delta V \leq -(1 - \frac{1}{1-\beta}k_v^2)\left(|r|^2 - \frac{2k_v\frac{1}{1-\beta}}{1-\frac{1}{1-\beta}k_v^2}|r||\epsilon_1 + \epsilon_3 u| - \right.$$

$$\left. \frac{\frac{1}{1-\beta}}{1-\frac{1}{1-\beta}k_v^2}|\epsilon_1 + \epsilon_3 u|^2 \right)$$

$$- (1 - \beta)\left[(\overline{e}_1 + \overline{e}_2 u) - \frac{\beta}{1-\beta}(\epsilon_1 + \epsilon_3 u + k_v r)\right]^2$$

ΔV will be negative if the following conditions are satisfied:

$\beta < 1$

$1 - \eta k_v^2 > 0$

$|r|^2 - \frac{2k_v\eta}{1-\eta k_v^2}|r||\epsilon_1 + \epsilon_3 u| - \frac{\eta}{1-\eta k_v^2}|\epsilon_1 + \epsilon_3 u|^2 > 0$

where $\eta = \frac{1}{1-\beta}$. The above conditions simplify to

1. $\alpha_1\|\boldsymbol{\phi}\|^2 + \alpha_2\|\boldsymbol{\psi}\|^2 u^2 < 1$
2. $k_v < \sqrt{\frac{1}{\eta}}$
3. $|r| > \frac{1}{1-\eta k_v^2}(\sqrt{\eta} + \eta k_v)|\epsilon_1 + \epsilon_3 u|$

Since $V(k) > 0$ and $\Delta V \leq 0$, the closed loop error dynamics (5.75) is Lyapunov stable and $r(k)$, $\hat{\boldsymbol{W}}(k)$, and $\hat{\boldsymbol{P}}(k)$ are bounded. The learning rates α_1, α_2 should be small enough so that condition (1) is satisfied. Also, the number of RBFN nodes should not be very large so that the above condition is maintained in the worst case scenario. Condition (2) depends on the parameters k_v and β. For a discrete time system $|k_v|$ is always less than 1. Since the bound on β is not known *a priori*, one can choose a small k_v to maintain condition (1) along with condition (2). The smaller we choose k_v, the larger we can make the learning rates α_1 and α_2. Moreover, as u is bounded, the third condition signifies that the filtered tracking error $r(k)$ is contained in the compact set $|r| < B_r$ where $B_r = \frac{1}{1-\eta k_v^2}(\sqrt{\eta} + \eta k_v)|\epsilon_1 + \epsilon_3 u|$. B_r can be made small by properly choosing the parameters, k_v, a etc. Thus, the small constant term a in the denominator of the control law maintains the boundedness of the control input as well as keeps the tracking error within a small bound.

Example 5.4 The dynamics of a non-linear SISO discrete time system is given by the following equations:

$$x_1(k + 1) = x_2(k)$$
$$x_2(k + 1) = f(x(k)) + g(x(k))u(k) \qquad (5.81)$$
$$y(k) = x_1(k)$$

where $x(k) = [x_1(k) \ x_2(k)]^T$

$$f(x(k)) = \frac{x_2(k)}{1.2 + x_1^2(k)}$$

$$g(x(k)) = 2 + \sin(x_1(k))$$

The desired output is as follows:

$$y_d(k) = \sin(0.1 \ k)$$

Design a direct adaptive controller for the above system using Theorems 5.4 and 5.5 .

Solution
The output tracking error is defined as

$$y_d(k) - y(k) = x_{1d}(k) - x_1(k) = e_1(k) \qquad (5.82)$$

When $f(x)$ is unknown and $g(x)$ is considered to be known, the control law is

$$u(k) = \frac{1}{g(x(k))}[-\hat{f}(x(k)) + k_v r(k) + x_{2d}(k+1) + \lambda_1 e_2(k)] \qquad (5.83)$$

where $r(k) = e_2(k) + \lambda_1 e_1(k)$, $e_2(k) = e_1(k+1)$. The parameters k_v and λ_1 are chosen as 0.7 and 0.6, respectively. The number of neurons for the RBF network is taken as 20. The centres of the RBF network are chosen randomly between 0 and 1 and the weights are initialized to small values. The upper bound of the learning parameter α is found to be 0.05. The output tracking result and the corresponding control input are shown in Figure 5.10.

When both $f(x)$ and $g(x)$ are unknown, the control law is defined as :

* If $\hat{g} > a_1$ or $\hat{g} < -a_1$,

$$u(k) = \frac{1}{\hat{g}(x(k))}[-\hat{f}(x(k)) + k_v r(k) + x_{2d}(k+1) + \lambda_1 e^{(2)}(k)]$$

* If $0 < \hat{g} \leq a_1$,

$$u(k) = \frac{1}{\hat{g}(x(k)) + a_2}[-\hat{f}(x(k)) + k_v r(k) + x_{2d}(k+1) + \lambda_1 e^{(2)}(k)]$$

* If $-a_1 \leq \hat{g} < 0$,

$$u(k) = \frac{1}{\hat{g}(x(k)) + a_3}[-\hat{f}(x(k)) + k_v r(k) + x_{2d}(k+1) + \lambda_1 e^{(2)}(k)]$$

where $r(k) = e^{(2)}(k) + \lambda_1 e^{(1)}(k)$, $e^{(2)}(k) = e^{(1)}(k+1)$. The parameters k_v and λ_1 are chosen as 0.5 and 0.05, respectively. a_1 and a_2 have been chosen as 0.2 and 1.5, respectively. Simulation results show that $\hat{g}$ never becomes negative; thus, the parameter a_3 does not play any role for this particular case. The number of neurons for both RBF networks equals 30. The learning rate α_1 and α_2 are taken as 0.05 and 0.05, respectively. The centres of the RBF networks are chosen randomly between 0 and 1, and the weights are initialized such that the initial value of $\hat{g}$ is greater than a_1. The output tracking result and the corresponding control input are provided in Figure 5.11, which shows that the tracking performances are satisfactory with an RMS error of 0.1. It can also be seen from Figure 5.11 that

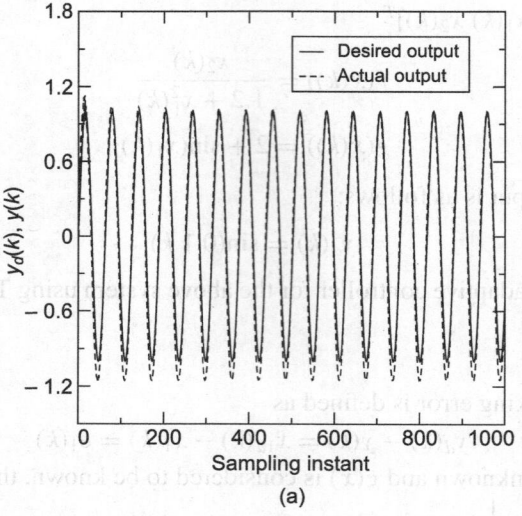

(a)

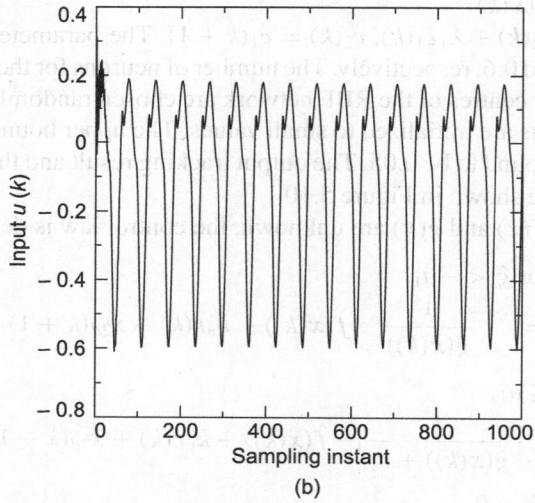

(b)

Figure 5.10 (a) Output tracking result when $f(x)$ is unknown and $g(x)$ is known and (b) the corresponding control input $u(k)$

the variation of control input is not jerky. Considering, $\epsilon_1 = 0.5$ and $\epsilon_2 = 1$, the minimum value of B_r is found as 0.2. The variations of $\beta = \alpha_1 \|\phi\|^2 + \alpha_2 \|\psi\|^2 u^2$ and $\sqrt{(\frac{1}{\eta})}$ are shown in Figure 5.12. It can be seen from the figure that β is less than 1 and $\sqrt{(\frac{1}{\eta})}$ is greater than 0.5 throughout the simulation period. Since we have taken $k_V = 0.5$, these plots show that the two conditions for stability are satisfied.

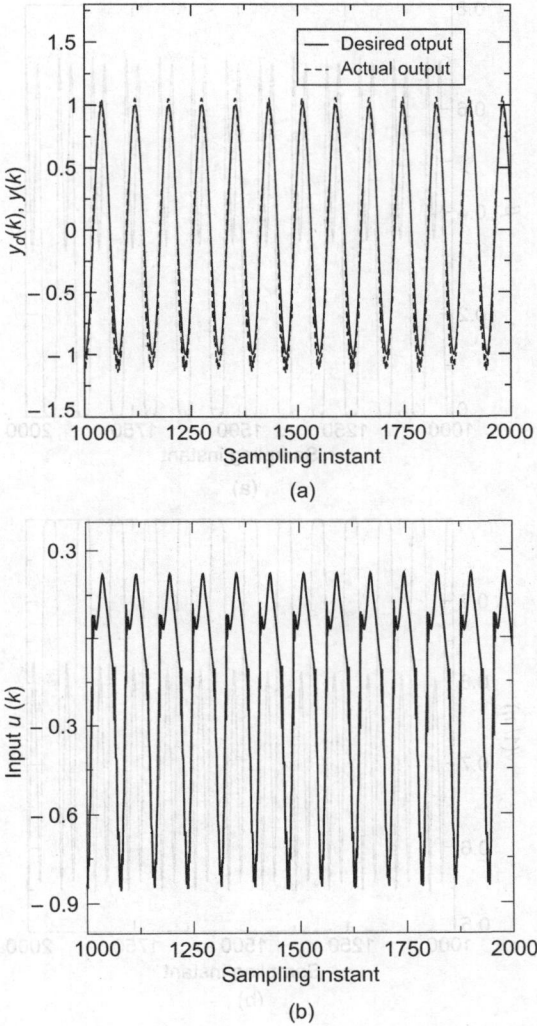

Figure 5.11 Example 5.4: (a) output tracking result of the system given in Eqn (5.81) when both $f(x)$ and $g(x)$ are unknown and (b) the corresponding control input $u(k)$

5.5 BACK-STEPPING CONTROL

Back-stepping design concepts introduced in Chapter 1 are useful for non-linear systems which can be expressed in the strict feedback form. If the non-linearity is assumed to be unknown, then the back-stepping control principles can be applied using neural networks.

5.5.1 System Description

Robust control of non-linear systems with uncertainties is of prime importance in many industrial applications. The model of many practical non-linear systems

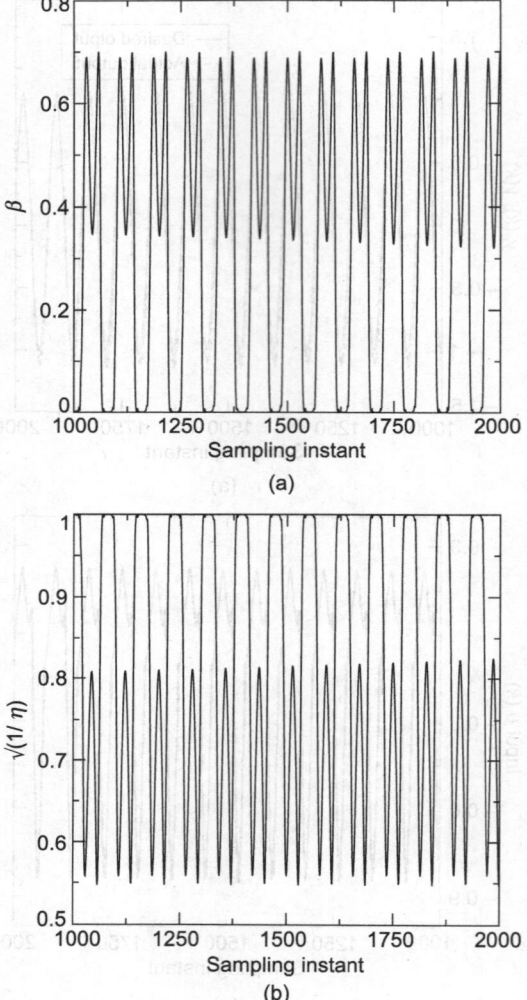

Figure 5.12 (a) Variation of β with time and (b) the variation of $\sqrt{(\frac{1}{\eta})}$ with time

can be expressed in a special state–space form

$$\dot{x}_1 = F_1(x_1) + G_1(x_1)x_2$$
$$\dot{x}_2 = F_2(x_1, x_2) + F_2(x_1, x_2)x_3$$
$$\dot{x}_3 = F_3(x_1, x_2, x_3) + F_3(x_1, x_2, x_3)x_4$$
$$\vdots = \ldots$$
$$\dot{x}_m = F_m(x_1, x_2, \ldots, x_m) + G_m(x_1, x_2, \ldots, x_m)u \qquad (5.84)$$

where $x_i \in R^n$, $i = 1, 2, \ldots, m$ denote the states of the system, $u \in R^n$ is the vector of control inputs. $F_i, G_i \in R^{n \times n}$, $i = 1, 2, \ldots, m$ are non-linear functions that contain both parametric and non-parametric uncertainties, and G_is are known and invertible. Eqn (5.84) is also in a *strict-feedback* form [15] since the non-linearities F_i, G_i depend only on $x_1, x_2, \ldots, x_i$, that is, on state variables that are *fedback*.

Definition 5.1 Stability of Systems: consider the following non-linear system:

$$\dot{x} = f(x, t) \quad y = h(x, t) \tag{5.85}$$

with state $x(t) \in R^n$. We say the solution is *uniformly ultimately bounded* (UUB) [93] if there exists a compact set $U \subset R^n$ such that for all $x(t_0) = x_0 \in U$, there exists an $\epsilon > 0$ and a number $T(\epsilon, x_0)$ such that $\|x(t)\| < \epsilon$ for all $t \geq t_0 + T$.

5.5.2 Traditional Back-stepping Design

The back-stepping design [15, 19] can be applied to the class of non-linear systems (5.84) as long as the *internal dynamics are stabilizable*. As described in Chapter 1, first select a desirable value of x_2, possibly a function of x_1, denoted by x_{2d}, such that in the ideal system $\dot{x}_1 = F_1(x_1, x_{2d})$, we have stable tracking by $x_1(t)$ of x_{1d}. Then in a second step, select x_3 to be x_{3d}, so that x_2 tracks x_{2d} and this process is repeated. Finally, select $u(t)$ such that x_m tracks x_{md}. A number of robust and adaptive procedures exist which implement the above back-stepping method. The above back-stepping procedures become more complicated when there exist parametric uncertainties in the systems. The complications are due to the following problems with the existing robust and adaptive procedures:

1. *'Regression matrices'* in each step of the back-stepping design must be determined. The computation of regression matrices is generally very tedious and time consuming.

2. One basic assumption, that the unknown system parameters must satisfy the so-called *'linearity-in-parameter'* (LIP), is quite restrictive and may not be true in many practical situations.

By using a neural network (NN), we can alleviate the disadvantages of the tedious and lengthy process of determining and computing regression matrices while retaining the merit of systematic design in the back-stepping control. As no LIP assumption is made, this design can be applied to a broader class of non-linear systems.

5.5.3 Robust Back-stepping Controller Design Using RBFN

As mentioned in previous sections, a general non-linear function $f(x) \in R^n$, $x(t) \in R^n$ can be approximated by an RBFN as

$$f(x) = W^T \phi(x) + \epsilon(x) \tag{5.86}$$

with $\epsilon(x)$ being an *RBFN reconstruction error* vector.

Controller Structure

Step 1: Design fictitious controllers for $x_2, x_3, \ldots,$ and x_m. First of all, we design the fictitious controller for x_{2d}. Recalling that

$$\dot{x}_1 = F_1(x_1) + G_1(x_1)x_2 \tag{5.87}$$

Choosing the following fictitious controller

$$x_{2d} = G_1^{-1}(-\hat{F}_1 + \dot{x}_{1d} - K_1e_1) \tag{5.88}$$

where $K_1 > 0$ is a design parameter, $\hat{F}_1$ is the estimate of F_1, and $e_1 = x_1 - x_{1d}$. Substituting (5.88) into subsystem (5.87) yields the error dynamics

$$\dot{e}_1 = F_1 - \hat{F}_1 - K_1e_1 + G_1e_2 \tag{5.89}$$

with $e_2 = x_2 - x_{2d}$. The usual adaptive back-stepping approach is to assume that the unknown parameters in F_1 are linearly parameterizable, so that the standard adaptive control can be used. Use of an NN to approximate F_1 obviates this restriction. The next step of back-stepping is to make the error $x_2 - x_{2d}$ as small as possible. Differentiating e_2 defined in (5.89) gives

$$\dot{e}_2 = \dot{x}_2 - \dot{x}_{2d} = F_2 + G_2x_3 - \dot{x}_{2d} \tag{5.90}$$

A fictitious controller for x_3 of the form

$$x_{3d} = G_2^{-1}(-\hat{F}_2 + \dot{x}_{2d} - K_2e_2 - G_1^Te_1) \tag{5.91}$$

can be chosen. Note that there is a coupling term G_1e_2 in (5.89). The purpose of the term $G_1^Te_1$ is to compensate the effect of coupling due to G_1e_2. Substituting the fictitious controller (5.91) into (5.90) gives

$$\dot{e}_2 = F_2 - \hat{F}_2 - K_2e_2 - G_1^Te_1 + G_2e_3 \tag{5.92}$$

with $e_3 = x_3 - x_{3d}$, $K_2 > 0$ a design parameter, and $\hat{F}_2$ the estimate of F_2. In a similar fashion, we can design a fictitious controller for x_m to make the error $e_{m-1} = x_{m-1} - x_{(m-1)d}$ as small as possible, i.e.,

$$x_{md} = G_{m-1}^{-1}(-\hat{F}_{m-1} + \dot{x}_{(m-1)d} - K_{m-1}e_{m-1} - G_{m-2}^Te_{m-2}) \tag{5.93}$$

The dynamics of $e_{m-1} = x_{m-1} - x_{(m-1)d}$ is then governed by

$$\dot{e}_{m-1} = F_{m-1} - \hat{F}_{m-1} - K_{m-1}e_{m-1} - G_{m-2}^Te_{m-2} + G_{m-1}e_m \tag{5.94}$$

with $e_m = x_m - x_{md}$, $K_{m-1} > 0$ a design parameter, and $\hat{F}_{m-1}$ the estimate of F_{m-1}. Here, NNs are used to approximate the complicated non-linear functions F_is, $i = 1, 2, \ldots, m$. As a result, no regression matrices are needed and the controller is reusable for different systems within the same class of non-linear systems.

Step 2: Design of actual control u. Differentiating $e_m = x_m - x_{md}$ defined in (5.94) yields

$$\dot{e}_m = \dot{x}_m - \dot{x}_{md} - F_m + G_mu - \dot{x}_{md} \tag{5.95}$$

Choosing the controller of the form

$$u = G_m^{-1}(-\hat{F}_m + \dot{x}_{md} - K_me_m - G_{m-1}^Te_{m-1}) \tag{5.96}$$

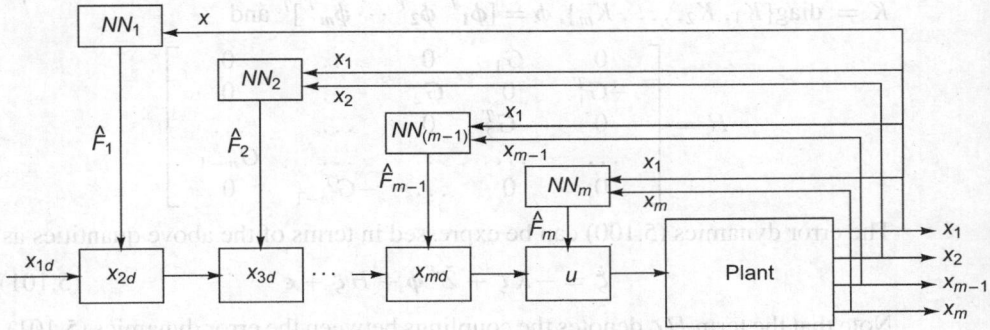

Figure 5.13 Back-stepping the NN control of non-linear systems in a 'strict-feedback' form

gives the following dynamics for error e_m:

$$\dot{e}_m = F_m - \hat{F}_m - K_m e_m - G_{m-1}^T e_{m-1} \qquad (5.97)$$

with $K_m > 0$ a design parameter and $\hat{F}_m$ the estimate of F_m. The overall control structure is shown in Figure 5.13.

Bounding Assumptions, Error Dynamics, and Weight Tuning Algorithm

Assume that the nonlinear functions F_is, $i = 1, 2, \ldots, m$ in Eqns (5.89), (5.92), (5.94), and (5.97) can be represented by m RBF networks for some constant 'ideal' weights W_i, $i = 1, 2, \ldots, m$, i.e.,

$$F_i = W_i^T \phi_i + \epsilon_i, \quad \|\epsilon_i\| < \epsilon_{iN} = constant \qquad (5.98)$$

for $i = 1, 2, \ldots, m$, where ϕ_is provide suitable basis functions for m RBFNs. The net reconstruction errors ϵ_is are bounded by known constants ϵ_{iN}, $i = 1, 2, \ldots, m$. Define the RBFN functional estimate of F_i in (5.98) by

$$\hat{F}_i = \hat{W}_i^T \phi_i \quad i = 1, 2, \ldots, m \qquad (5.99)$$

with $\hat{W}_i$ being the current RBFN weight estimates provided by the tuning algorithm. Then the error dynamics (5.89), (5.92),(5.94), and (5.97) become

$$\dot{e}_1 = \tilde{W}_1^T \phi_1 - K_1 e_1 + G_1 e_2 + \epsilon_1$$

$$\dot{e}_2 = \tilde{W}_2^T \phi_2 - K_2 e_2 - G_1^T e_1 + G_2 e_3 + \epsilon_2$$

$$\dot{e}_3 = \tilde{W}_3^T \phi_3 - K_3 e_3 - G_2^T e_2 + G_3 e_4 + \epsilon_3$$

$$\cdots = \cdots$$

$$\dot{e}_m = \tilde{W}_m^T \phi_m - K_m e_m - G_{m-1}^T e_{m-1} + \epsilon_m \qquad (5.100)$$

Define $\zeta = [e_1^T \ e_2^T \cdots e_m^T]^T$, $\epsilon = [\epsilon_1^T \ \epsilon_2^T \ \epsilon_m^T]^T$, $\tilde{Z} = \text{diag}\{\tilde{W}_1, \tilde{W}_2, \ldots, \tilde{W}_m\}$, $\Gamma = \text{diag}\{\Gamma_1, \Gamma_2, \ldots, \Gamma_m\}$,

$K = \text{diag}\{K_1, K_2, \ldots, K_m\}, \boldsymbol{\phi} = [\boldsymbol{\phi_1}^T \ \boldsymbol{\phi_2}^T \cdots \boldsymbol{\phi_m}^T]^T$ and

$$
H = \begin{bmatrix} 0 & G_1 & 0 & \cdots & 0 \\ -G_1^T & 0 & G_2 & \cdots & 0 \\ 0 & -G_2^T & 0 & \cdots & \cdots \\ \cdots & \cdots & \cdots & \cdots & G_{m-1} \\ 0 & 0 & \cdots & -G_{m-1}^T & 0 \end{bmatrix}
$$

The error dynamics (5.100) can be expressed in terms of the above quantities as

$$
\dot{\boldsymbol{\zeta}} = -K\boldsymbol{\zeta} + \tilde{Z}^T \boldsymbol{\phi} + H\boldsymbol{\zeta} + \boldsymbol{\epsilon} \tag{5.101}
$$

Note that the term $H\boldsymbol{\zeta}$ denotes the couplings between the error dynamics (5.101). The matrix H is skew-symmetric. Two assumptions, which are quite common in the neural network literature [93], are stated next.

Assumption 5.1 (Bounded reference trajectory) The desired trajectory is bounded so that

$$
\left\| \begin{matrix} q_d(t) \\ \dot{q}_d(t) \\ \ddot{q}_d(t) \end{matrix} \right\| \leq q_B \tag{5.102}
$$

with q_B being a known scalar bound.

Assumption 5.2 The ideal weights are bounded by known positive values, so that

$$
\|W_i\|_F \leq W_{iM}, \quad i = 1, 2, \ldots, m \tag{5.103}
$$

or equivalently, $\|Z\| \leq Z_m$, where Z_m is known. The symbol $\|.\|_F$ denotes the Frobenius norm.

Kwan et al. [19, 94] proposed a robust weight tuning algorithm for the two-layer NN which is given below.

Theorem 5.6 (Weight tuning algorithm) Suppose assumptions (5.1) and (5.2) are satisfied. Take the control input (5.96) with the NN weight tuning to be provided by

$$
\dot{\hat{W}}_i = \Gamma_i \boldsymbol{\phi}_i e_i^T - k_w \Gamma_i \|\boldsymbol{\zeta}\| \hat{W}_i, \quad i = 1, 2, \ldots, m \tag{5.104}
$$

with constant matrices $\Gamma_i = \Gamma_i^T > 0, i = 1, 2, \ldots, m$, and scalar positive constant k_w. Then the errors $e_i(t), i = 1, 2, \ldots, m$, and NN weights are UUB.

The errors $e_i(t), i = 1, 2, \ldots, m$, can be kept as small as possible by increasing gains K in (5.101) [19].

This tuning algorithm can be implemented online and does not require any off line training. Since no explicit system identification is done to approximate the non-linear functions, instead the error is directly minimized to adaptively tune the network parameters, NN back-stepping is another form of direct adaptive control.

Example 5.5 Consider the following mathematical model of a second order dynamical system:

$$\dot{x}_1 = x_1 \sin(x_1) + x_2$$

$$\dot{x}_2 = 0.2x_1x_2^2 + u$$

$$y = x_1 \tag{5.105}$$

Design a back-stepping control for the above system using a neural network, so that the system output y tracks the following desired trajectory:

$$y_d = x_{1d} = \sin t$$

Solution

The design example (5.105) is in strict-feedback form as described by Eqn (5.84), where $F_1(x_1) = x_1 \sin(x_1)$, $F_2(x_1, x_2) = 0.2x_1x_2^2$, and $G_1(x_1) = G_2(x_1, x_2) = 1$.

Controller Structure

First of all, we design the fictitious controller for x_{2d}. Recalling that

$$\dot{x}_1 = F_1(x_1) + G_1(x_1)x_2 \tag{5.106}$$

we choose the following fictitious controller:

$$x_{2d} = -\hat{F}_1 + \dot{x}_{1d} - k_1e_1 \tag{5.107}$$

where $k_1 > 0$ is a design parameter, $\hat{F}_1$ is the estimate of F_1, $e_1 = x_1 - x_{1d}$, and $\dot{x}_{1d} = \cos(t)$. Substituting (5.107) into subsystem (5.106) yields the error dynamics

$$\dot{e}_1 = F_1 - \hat{F}_1 - k_1e_1 + e_2 \tag{5.108}$$

with $e_2 = x_2 - x_{2d}$. To make the error e_2 as small as possible, the control input to the next stage is chosen as

$$u = -\hat{F}_2 + \dot{x}_{2d} - k_2e_2 - e_1 \tag{5.109}$$

where $k_2 > 0$ is another design parameter, $\hat{F}_2$ is the estimate of F_2. $\hat{F}_1$, and $\hat{F}_2$ are estimated using two radial basis function networks with Gaussian centres as

$$\hat{F}_1(x_1) = \hat{W}_1^T \phi(x_1)$$

$$\hat{F}_1(x_1, x_2) = \hat{W}_2^T \phi(x)$$

The network corresponding to $\hat{F}_1$ consists of five neurons and the network corresponding to $\hat{F}_2$ consists of 30 neurons. Since the inputs to the networks are the states of the system and the desired states range between -1 and 1, the centres are fixed between -1 and 1 for both networks. The update laws for $\hat{W}_1$ and $\hat{W}_2$ are given as

$$\dot{\hat{W}}_1 = \gamma_1\phi_1e_1 - k_w\gamma_1\|\zeta\|\hat{W}_1 \tag{5.110}$$

$$\dot{\hat{W}}_2 = \gamma_2\phi_2e_2 - k_w\gamma_2\|\zeta\|\hat{W}_2 \tag{5.111}$$

where ϕ_1 and ϕ_2 are the basis functions for the networks and $\|\zeta\| = \sqrt{e_1^2 + e_2^2}$. γ_1, γ_2, k_1, k_2, and k_w have been taken as 0.8, 0.6, 10, 10, and 0.2, respectively. The simulation results for this example are shown in Figures 5.14 and 5.15. Figure

5.14 shows the actual states of the system track the desired states. The RMS error for state x_1 is found to be 0.002, whereas for x_2 it is 0.007. In Figure 5.15, (a) shows the nonlinear function $F_1(x_1)$ and its neural estimate $\hat{F}_1(x_1)$, while the (b) shows the required control input. Since explicit function approximation is not done, the estimate $\hat{F}_1(x_1)$ is not very good, but it is bounded by a small number and the main objective of output tracking is achieved.

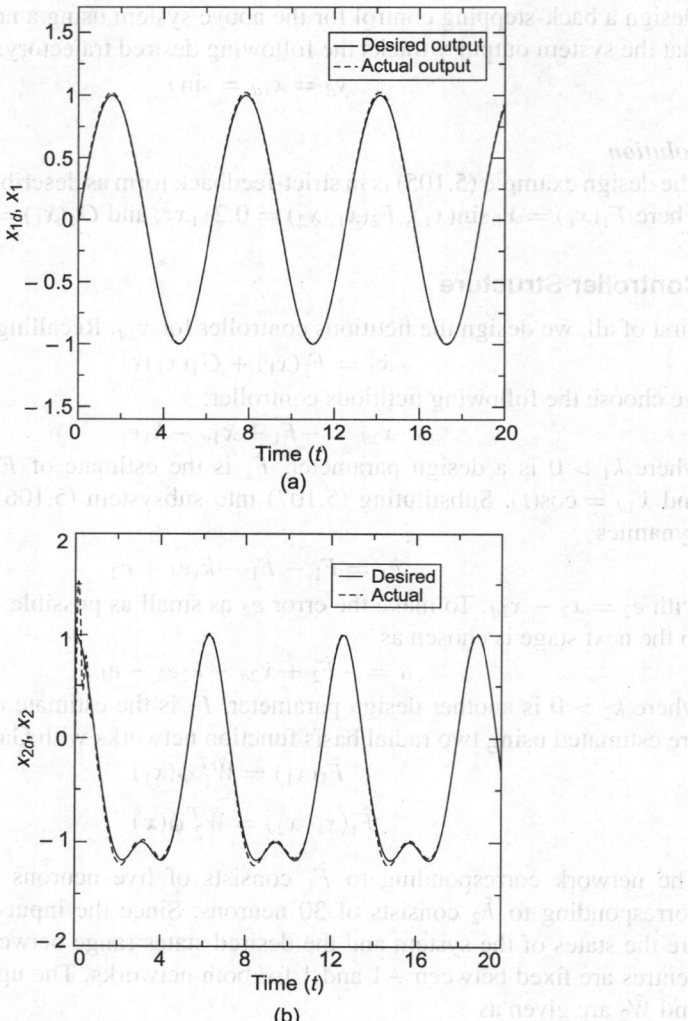

Figure 5.14 NN back-stepping: (a) state x_1 the desired trajectory x_{1d}; and (b) state x_2 the desired trajectory x_{2d}

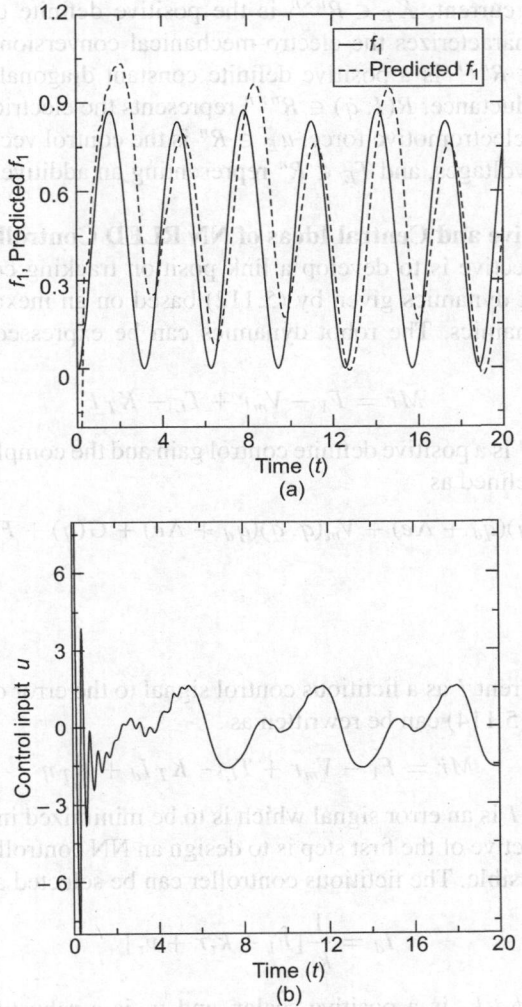

Figure 5.15 NN back-stepping: (a) approximation of the non-linear function f_1; (b) control input u

5.5.4   Back-stepping Control for a Robot Manipulator

The model for an n rigid link electrically driven robot manipulator (RLED) robot is given by

$$M(q)\ddot{q} + V_m(q,\dot{q})\dot{q} + G(q) + F(\dot{q}) + T_L = K_T I \qquad (5.112)$$

$$L\dot{I} + R(I,\dot{q}) + T_E = u_E \qquad (5.113)$$

with $q, \dot{q}, \ddot{q} \in R^n$ denoting link position, velocity, and acceleration vectors, respectively, and $M(q) \in R^{n \times n}$ is the inertia matrix, $V_m(q,\dot{q}) \in R^n$ is the centripetal–Coriolis matrix, $G(q) \in R^n$ is the gravity vector, $F(\dot{q}) \in R^n$ represents the friction terms, $T_L \in R^n$ is the additive bounded disturbance, $I \in R^n$

is the armature current, $K_T \in R^{n \times n}$ is the positive definite constant diagonal matrix which characterizes the electro-mechanical conversion between current and torque, $L \in R^{n \times n}$ is a positive definite constant diagonal matrix denoting the electrical inductance, $R(I, \dot{q}) \in R^{n \times n}$ represents the electrical resistance and the motor back-electromotive force, $u_E \in R^n$ is the control vector represents the motor terminal voltages, and $T_E \in R^n$ representing an additive bounded voltage disturbance.

Control Objective and Central Ideas of NN RLED Controller Design

The control objective is to develop a link position tracking controller [94] for the RLED robot dynamics given by (5.112) based on an inexact knowledge of manipulator dynamics. The robot dynamics can be expressed in terms of the filtered error as

$$M\dot{r} = F_1 - V_m r + T_L - K_T I \tag{5.114}$$

where $\Lambda \in R^{n \times n}$ is a positive definite control gain and the complicated non-linear function F_1 is defined as

$$F_1 = M(q)(\ddot{q}_d + \Lambda \dot{e}) + V_m(q, \dot{q})(\dot{q}_d + \Lambda e) + G(q) + F(\dot{q}) \tag{5.115}$$

and $r = \dot{e} + \Lambda e$

Design Steps

Step 1: Treat current I as a fictitious control signal to the error dynamics (5.114) (say, I_d). Then, (5.114) can be rewritten as

$$M\dot{r} = F_1 - V_m r + T_L - K_T I_d + K_T \eta \tag{5.116}$$

where $\eta = I_d - I$ is an error signal which is to be minimized in the second step. The control objective of the first step is to design an NN controller for I_d to make r as small as possible. The fictitious controller can be selected as

$$I_d = \frac{1}{k_1}[\hat{F}_1 + k_\tau r + v_\tau] \tag{5.117}$$

where $\hat{F}_1 = \hat{W}_1 \phi_1$, k_τ is a positive scalar, and v_τ is a robustifying term to be defined shortly. Substituting (5.117) into (5.116) gives

$$M\dot{r} = -V_m r + \tilde{W}_1^T \phi_1 - k_\tau \frac{K_T}{k_1} r + \varepsilon_1 + T_L$$

$$+ \left(I - \frac{K_T}{k_1}\right) \hat{W}_1^T \phi_1 - \frac{K_T}{k_1} v_\tau + K_T \eta \tag{5.118}$$

As we can see that there is an unknown term $\left(I - \frac{K_T}{k_1}\right) \hat{W}_1^T \phi_1$ in (5.118) because K_T is usually unknown. The role of the robustifying term v_τ is to suppress the effect of this signal. The form of v_τ is chosen to be

$$v_\tau = \rho_\tau \text{sgn}(\tau) \tag{5.119}$$

where

$$\rho_\tau = \|\hat{W}_1^T \phi_1\| b_k \tag{5.120}$$

and b_k in (5.120) stands for the upper bound of $\|(I - \frac{K_T}{k_1})\|$.

Step 2: The second step is to design a second NN controller for u_E such that the error signal η is as small as possible. Differentiating $\eta = I_d - I$, using (5.113), and multiplying L on both sides of the resulting expression yield

$$L\dot{\eta} = F_2 + T_E - u_E \tag{5.121}$$

where F_2 is a very complicated non-linear function of q, $\dot{q}$, r, I_d, and I. The control signal u_E can be chosen as

$$u_E = \hat{F}_2 + k_v \eta \tag{5.122}$$

where $\hat{F}_2 = W_2^T \phi_2$ and $k_v > 0$. Inserting (5.122) into (5.121) gives

$$L\dot{\eta} = \tilde{W}_2^T \phi_2 + \varepsilon_2 + T_E - k_v \eta \tag{5.123}$$

Step 3: Weight update algorithm and stability analysis

The weight tuning algorithm as stated in Theorem 5.6 is used here. It is shown that all errors such as weight updates and tracking errors are UUB. The weight tuning algorithm for two networks is restated below for convenience,

$$\dot{\hat{W}}_1 = \Gamma_1 \phi_1 r^T - k_w \Gamma_1 \|\zeta\| \hat{W}_1 \tag{5.124}$$

$$\dot{\hat{W}}_2 = \Gamma_2 \phi_2 r^T - k_w \Gamma_2 \|\zeta\| \hat{W}_2 \tag{5.125}$$

with any constant matrices $\Gamma_1 = \Gamma_1^T > 0$, $\Gamma_2 = \Gamma_2^T > 0$, $\zeta = [r^T \ \eta^T]^T$, and scalar positive constant k_w.

Example 5.6 The model for RLED is described in the form of (5.112) with

$$M(q) = \begin{bmatrix} a + b\cos(q_2) & c + \frac{b}{2}\cos(q_2) \\ c + \frac{b}{2}\cos(q_2) & c \end{bmatrix}$$

$$V_m \dot{q} = \begin{bmatrix} -b\sin(q_2)(\dot{q}_1 \dot{q}_2 + 0.5\dot{q}_2^2) \\ 0.5b\sin(q_2)\dot{q}_1^2 \end{bmatrix}$$

$$G(q) = \begin{bmatrix} d\cos(q_1) + e\cos(q_2) \\ e\cos(q_1 + q_2) \end{bmatrix}$$

$$a = l_2^2 m_2 + l_1^2(m_1 + m_2) \quad b = 2l_1 l_2 m_2,$$

$$c = l_2^2 m_2, \ d = (m_1 + m_2)l_1 g_0 \ e = m_2 l_2 g_0$$

The parameter values are $l_1 = 1m$, $l_2 = 1m$, $m_1 = 0.8$ Kg, $m_2 = 2.3$ Kg, and $g_0 = 9.8$ m/s^2. The actuator dynamics are assumed to be the permanent magnet direct-current motor (5.113). The parameters of motor are $R_j = 1\Omega$, $L_j = 0.01H$, $K_j^T = 2.0$ Nm/A, $j = 1, 2$. The desired joint trajectories are defined as $q_{d1}(t) = \sin(t)$, $q_{d2}(t) = \cos(t)$. Design a back-stepping control using neural networks as described in Section 5.5.4.

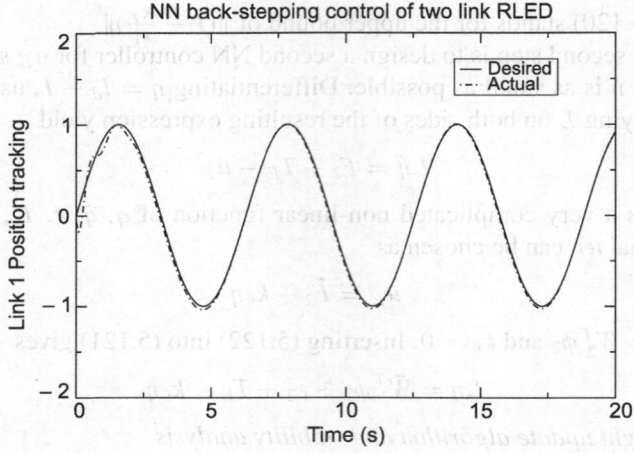

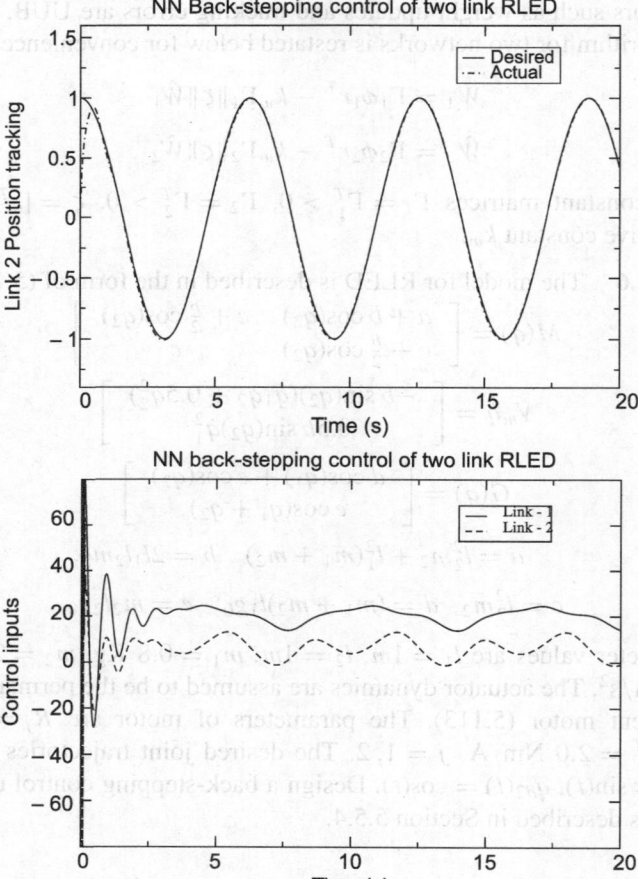

Figure 5.16 NN back-stepping control of two-link RLED

The inputs to the NNs are given by

$$x = [\zeta_1^T \ \zeta_2^T \ \cos(q)^T \ \sin(q)^T \ I^T \ 1]^T$$

where $\zeta_1 = \ddot{q}_d + \Lambda\dot{e}$ and $\zeta_2 = \dot{q}_d + \Lambda e$. Two neural networks with weight vectors W_1 and W_2 have been used in the back-stepping based control algorithm for which the weight update laws are given in Eqn (5.124). Figure 5.16 shows the simulation results for NN back-stepping control with $k_\tau = \text{diag}\{60, 60\}$ and $k_v = \text{diag}\{1.5, 1.5\}$.

SUMMARY

Neural network based direct adaptive control scheme is discussed for a class of non-linear affine systems both in continuous time and discrete time. The singularity problem of the controller is solved using a projection algorithm for continuous time case. The effect of the projection algorithm on the closed loop stability has been taken care of by introducing a sliding mode term in the controller. For discrete time case, a small constant term is added in the denominator of the usual control law and it is shown that this modification keeps the control input bounded. Neural network based back-stepping control has also been provided in this chapter. Simulation results are provided to demonstrate the design procedures. Although the chapter has presented direct adaptive control concepts mostly for input affine non-linear systems, there are many recent works [95, 96] in non-affine systems as well. In general, the literature concerning direct adaptive control using neural networks is huge. Readers are advised to refer to works [97, 98, 99, 100, 101, 102] for further study.

EXERCISES

1. Consider the following dynamics of a non-linear SISO system:

$$\dot{x}_1 = x_2$$
$$\dot{x}_2 = f(x) + g(x)u$$
$$y = x_1$$

where $x = [x_1 \ x_2]^T$ and

$$f(x) = 4\left(\frac{\sin(4\pi x_1)}{\pi x_1}\right)\left(\frac{\sin(\pi x_2)}{\pi x_2}\right)^2$$
$$g(x) = 2 + \sin(3\pi(x_1 - 0.5))$$

Design an adaptive controller using a neural network
(a) when $f(x)$ is unknown, but $g(x)$ is known
(b) when $f(x)$ and $g(x)$ both are unknown
to track a sinusoidal trajectory of unit amplitude and 1 Hz frequency.

2. The governing equations of an MEMS device [103, 104] are given as

$$\dot{Q} - \frac{1}{R}\left(V_{in} - \frac{Qg}{\epsilon A}\right) = 0$$

$$m\ddot{g} + b\dot{g} + k(g - g_0) + \frac{Q^2}{2\epsilon A} = 0$$

where Q denotes the charge, g the gap between the plate and base, and $\dot{g}$ the rate of change of the gap when the plate moves. V_{in} is the input voltage used to move the plate to a desired position. Various parameters associated with the dynamics are given in Table 5.2.

Table 5.2 Parameters used in the above equations

Parameters	Symbol	Value
Area	A	100 µm²
Permitivity	ϵ	1 C²/Nµm²
Initial gap	g_0	1 µm
Mass	m	1 mg
Damping constant	b	0.5 mg/s
Spring constant	k	1 mg/s²
Resistance	R	0.001 Ω

(a) Defining the state variables $x_1 = g$ and $x_2 = \dot{g}$, derive the state space model in strict feedback form.

(b) Design a suitable controller to maintain the gap g at a desired value of 0.5 µm.

3. The model of a magnetic ball suspension system is given by

$$M\frac{d^2 y(t)}{dt^2} = Mg - \frac{i^2(t)}{y(t)}$$

$$v(t) = Ri(t) + L\frac{di(t)}{dt}$$

where $y(t)$ is the ball position in metres, $M = 0.1$ kg is the mass of the ball, $g = 9.81$ m/s² is the gravitational acceleration, $R = 50\ \Omega$ is the resistance of the winding, $L = 0.5\ H$ is the winding inductance, $v(t)$ is the input voltage, and $i(t)$ is the winding current.

(a) Derive the state space model of the system.

(b) If the state space model is not in a strict feedback form, convert it into a strict feedback form using state transformation.

(c) Design a suitable controller such that the position of the ball tracks a desired trajectory of 5 sin t.

4. Consider the dynamics of a series DC motor:

$$J\frac{d\omega}{dt} = -D\omega + K_m L_f i^2 - \tau_L$$

$$L\frac{di}{dt} = -K_m L_f \omega i - Ri + V$$

where ω is the motor speed to be controlled, V is the input voltage, τ_L is the load torque, $J = 0.000704$ kg m^2 is the moment of inertia of the motor, $D = 0.0004$ N $-$ m/rad/s is the viscous friction coefficient, K_m is the motor torque constant, L_f is the field inductance, and $L = L_a + L_f = 0.0917$ H, $R = R_a + R_f = 7.2$ Ω where L_a and R_a are the armature inductance and resistance, respectively. $K_m L_f = 0.1236$ N m/Wb A.

(a) Considering the load torque as 0, derive the state–space model of the system in discrete time.

(b) Design a direct adaptive controller for the system using the discrete time approach, so that the motor speed achieves a desired value of 100 rad/s.

6

Approximate Dynamic Programming

Adaptive critic (AC) based control design has evolved as a powerful technique to solve optimal control problems for non-linear systems. Given a dynamical plant and the corresponding performance index, there are basically two ways of solving the associated optimal control problem; one is Pontryagin's minimum principle and the other is Bellman's dynamic programming [105]. Dynamic programming provides a computational technique to apply the principle of optimality to a sequence of decisions which define an optimal control policy. However, solving the associated Hamilton–Jacobi–Bellman (HJB) equation demands a very large computational cost in terms of time and storage requirement. The numerical complexity is overcome by using an 'approximate dynamic programming (ADP)' formulation proposed by Werbos [106]. This ADP formulation leads to the so called adaptive critic designs which utilize two parametric structures known as the action network and the critic network. The action network consists of a parametrized control law. The critic network approximates a value-related function and captures the effect that the control law will have on the future cost. At any given time, the critic provides guidance on how to improve the control law. In return, the action network can be used to update the critic. Several versions of adaptive critic have been reported in the literature [107, 108].

6.1 LINEAR QUADRATIC REGULATOR

Consider a linear time invariant system modelled by

$$\dot{x} = Ax + Bu \quad x(0) = x_0 \tag{6.1}$$

It is asked to design an optimal state feedback controller

$$u = -Kx \tag{6.2}$$

so that the performance index

$$J = \int_0^\infty \left(x^T Q x + u^T R u \right) dt \qquad (6.3)$$

is minimized. Here $Q = Q^T \geq 0$ and $R = R^T \geq 0$. This is popularly known as linear quadratic regulator (LQR) problem. The solution to the LQR problem has two parts: optimal gain K^* and optimal cost function J^*. It turns out that the optimal gain is given by [7]

$$K^* = R^{-1} B^T P \qquad (6.4)$$

where the matrix P is obtained from the algebraic Riccati equation (ARE):

$$A^T P + P A + Q - P B R^{-1} B^T P = 0 \qquad (6.5)$$

This optimal state feedback law leads to the following optimal cost function when the system is excited from the intial state x_0:

$$J^* = x_0^T P x_0 \qquad (6.6)$$

It is interesting to note that the solution to the LQR problem associated with an LTI discrete time system is also the same except that the algebraic Riccati equation (ARE) is different:

$$A^T P A - A^T P B (R + B^T P B)^{-1} + B^T P A + Q = P \qquad (6.7)$$

However, there is no such generic solution to derive an optimal cost function for any nonlinear system. Thus, this chapter is concerned with deriving an approximate optimal solution through the popular HJB formulation.

6.2 THE HJB FORMULATION

Consider the generic nonlinear system:

$$\dot{x} = f(x, u, t), \quad x(t_0) = x_0 \qquad (6.8)$$

where $x \in R^n$ and $u \in R^m$ are the state and input vector, respectively. In optimal control problem, we have to design a u such that the cost function

$$J\left(x(t_0), t_0 \right) = \phi\left(x(T), T \right) + \int_{t_0}^T L(x, u, \tau) d\tau \qquad (6.9)$$

is minimized. The instanteneous cost function $L(x, u, t)$ has a popular form $x^T Q x + u^T R u$.

Let us define

$$J\left(x(t), u(\tau), t \right) = \phi\left(x(T), T \right) + \int_t^T L(x, u, \tau) d\tau \qquad (6.10)$$

The optimal cost is

$$J^*\left(x(t), t \right) = \min_u J\left(x(t), u(\tau), t \right) \qquad (6.11)$$

If we divide the time interval $[t, \ T]$ as $[t, \ t + \Delta t] \cup [t + \Delta t, \ T]$, we can write

$$J^*\left(x(t), t \right) = \min_u \left\{ \int_t^{t+\Delta t} L \, d\tau + \int_{t+\Delta t}^T L \, d\tau + \phi\left(x(T), T \right) \right\} \qquad (6.12)$$

By the principle of optimality, the trajectory on the interval $[t + \Delta t, \ T]$ must be optimal. Thus,

$$J^*(x(t), t) = \min_u \left\{ \int_t^{t+\Delta t} L \, d\tau + J^*(x(t + \Delta t), t + \Delta t) \right\} \tag{6.13}$$

Expanding $J^*(x(t + \Delta t), t + \Delta t)$ using the Taylor series about the point $(x(t), t)$

$$J^*(x(t), t) \ = \ \min_u \left\{ \int_t^{t+\Delta t} L \, d\tau + J^* + \right.$$
$$\left. \frac{\partial J^*}{\partial t} \Delta t + \frac{\partial J^*}{\partial x}(x(t + \Delta t) - x(t)) + \text{HOT} \right\} \tag{6.14}$$

where HOT stands for higher order terms. Since $J^*(x(t), t)$ is independent of u, the terms J^* on both sides get cancelled out. Writing $x(t + \Delta t) - x(t)$ as $\dot{x}\Delta t$, we obtain

$$0 = \min_u \left\{ \int_t^{t+\Delta t} L \, d\tau + \frac{\partial J^*}{\partial t} \Delta t + \frac{\partial J^*}{\partial x} \dot{x} \Delta t + \text{HOT} \right\} \tag{6.15}$$

For small Δt, $\int_t^{t+\Delta t} L \, d\tau \approx L\Delta t$. Thus

$$0 = \min_u \left\{ L\Delta t + \frac{\partial J^*}{\partial t} \Delta t + \frac{\partial J^*}{\partial x} \dot{x} \Delta t + \text{HOT} \right\} \tag{6.16}$$

Dividing the above equation by Δt and assuming $\Delta t \to 0$, we can write

$$\frac{\partial J^*}{\partial t} + \min_u \left(L(x, u, t) + \frac{\partial J^*}{\partial x} f(x, u, t) \right) = 0 \tag{6.17}$$

subject to the boundary condition $J^*(x(T), T) = \phi(x(T), T)$. The above equation is popularly known as HJB equation and the expression $L(x, u, t) + \frac{\partial J^*}{\partial x} f(x, u, t)$ is known as the Hamiltonian. When the system dynamics is time invariant and the cost function is for a infinite time as

$$\text{Minimize } \ J(x_0) = \int_0^\infty L(x, u) d\tau \quad \text{subject to} \quad \dot{x} = f(x, u) \tag{6.18}$$

The HJB equation becomes

$$\min_u \left(L(x, u) + \frac{\partial J^*}{\partial x} f(x, u) \right) = 0 \tag{6.19}$$

6.3 HJB FOR AFFINE SYSTEMS

For an affine system $\dot{x} = f(x) + g(x)u$, the optimal control problem is to minimize

$$J(x_0) = \int_0^\infty L(x, u) d\tau \quad \text{where} \quad L = \frac{1}{2}(x^T Q x + u^T R u) \tag{6.20}$$

subject to the following constraint:

$$\dot{x} = f(x) + g(x)u \tag{6.21}$$

The HJB solution is computed as follows. Compute the optimal control law u as

$$u^* = \arg\min_u \left\{ \frac{1}{2}(x^T Q x + u^T R u) + \frac{\partial J^*}{\partial x}(f(x) + g(x)u) \right\} \quad (6.22)$$

Differentiating RHS with respect to u and setting to zero, we get

$$u^{*T} R = -\frac{\partial J^*}{\partial x} g(x) \quad (6.23)$$

$$u^* = -R^{-1} g^T(x) \left(\frac{\partial J^*}{\partial x} \right)^T \quad (6.24)$$

Note that the optimal control law contains the term J^*, the optimal cost function. The optimal cost function can be obtained from the HJB equation by setting $u = u^*$ as

$$\frac{1}{2}(x^T Q x + u^T R u) + \frac{\partial J^*}{\partial x}\left(f(x) - g(x) R^{-1} g^T(x) \left(\frac{\partial J^*}{\partial x} \right)^T \right) = 0 \quad (6.25)$$

We can note that the solution of J^* is not straightforward, and hence the application of HJB to solve optimal control law u^* has been limited. But approximate optimal solutions for the cost function are possible using the approximate dynamic programming, which will be discussed later.

Example 6.1 Find the optimal control law for the following scalar non-linear system

$$\dot{x} = -x^3 + u \quad (6.26)$$

where the cost function is

$$J = \frac{1}{2} \int_0^\infty (x^2 + u^2) dt \quad (6.27)$$

Solution
The HJB equation for the system turns out to be

$$\frac{1}{2}(x^2 + u^2) + \frac{\partial J^*}{\partial x}(-x^3 + u) = 0 \quad for \quad u^* \quad (6.28)$$

Thus optimal control law is given as

$$u^* = -\frac{\partial J^*}{\partial x} \quad (6.29)$$

The optimal cost function is obtained by replacing u^* by $-\nabla J$, where $\nabla J = \frac{\partial J^*}{\partial x}$;

$$\frac{x^2}{2} + \frac{(\nabla J)^2}{2} + \nabla J(-x^3 - \nabla J) = 0$$

$$\frac{x^2}{2} - \nabla J x^3 - \frac{(\nabla J)^2}{2} = 0$$

Since this is quadratic in ∇J, the solution is $\nabla J = -x^3 + \sqrt{x^6 + x^2}$. Note that the solution that is a positive definite function can only be accepted. Thus, the optimal control law is $u^* = x^3 - \sqrt{x^6 + x^2}$.

6.4 HDP AND DHP

Heuristic dynamic programming (HDP) and dual heuristic programming (DHP) are two popular methods of adaptive critic design. An adaptive critic design consists of three modules, namely, an actor or controller, a plant and its model, and a critic. All of them can be approximated by some parametric structures like neural network or a polynomial function. Hence, this approach might be useful where perfect information is not available for the plant. The actor computes the control input, while the critic evaluates future cost function. The parameters of these modules are updated so as to minimize the value function at every time instant.

Consider the discrete time non-linear system which is time invariant:

$$x(k + 1) = f(x(k), u(k)) \qquad (6.30)$$

where $x(k) \in R^n$ and $u(k) \in R^m$ are the state and input vector, respectively. The function $f(x(k), u(k))$ is continuously differentiable vector field defined on R^n. Design an optimal control law $\mathbf{u}^* = \alpha x$ that can minimize the folowing cost function:

$$J = \sum_0^\infty L(x(k), u(k)) \qquad (6.31)$$

where the local cost function L has the form $L = x^T(k)Qx(k) + u^T(k)Ru(k)$. Approximate dynamic programming (ADP) (or adaptive critics) is a methodology for designing an approximate optimal controller for a given plant using a learning process. Usually ADP is implemented using two neural networks, one in the role of a controller and the other in the role of a critic. The user provides design objectives through a cost function $L(k)$, the local cost.

At any time instant k, the cost-to-go function is written as a value function

$$V(k) = \sum_k^\infty L(x(k), u(k)) \qquad (6.32)$$

V_k can also be expressed as

$$V(k) = L(k) + V(k + 1) \qquad (6.33)$$

where $V(k)$ is a function of $x(k)$, while $V(k + 1)$ is a function of $x(k + 1)$.

Some of the algorithms for parameter update are

- HDP when the critic approximates $V(k)$.
- DHP when the critic approximates $\nabla V_x(k)$.

In HDP, let the actor estimates the control action using a neural network as

$$\hat{u}(k) = C(x(k), a(k)) \qquad (6.34)$$

where C represents the neural estimate of the control actions with parameter vector $a(k)$.

Similarly, the cost-to-go function is estimated by the critic as

$$\hat{V}(k) = V(x(k), w(k)) \tag{6.35}$$

where V represents the neural estimate of the cost-to-go function with parameter vector w. The block diagram of a typical adaptive critic design based control scheme is shown in Figure 6.1.

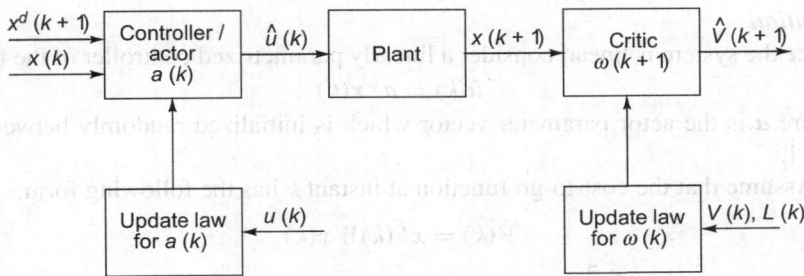

Figure 6.1 Various modules in a typical adaptive critic design

Controller parameter update: The desired control action following the optimal policy is computed by setting $\frac{\partial V(k)}{\partial u(k)}$ to 0 which leads to the following equation:

$$\frac{\partial L(x(k), u(k))}{\partial u(k)} + \left(\frac{\partial f(x(k), u(k))}{\partial u(k)}\right)^T \frac{\partial V(k+1)}{\partial x(k+1)} = 0 \tag{6.36}$$

The solution for the control action gives us the desired control action $u(k)$.

For a linear system, $x(k+1) = Ax(k) + Bu(k)$, the solution for $u(k)$ is given by

$$u(k) = -R^{-1}B^T \frac{\partial V(k+1)}{\partial x(k+1)} \tag{6.37}$$

where the last term is computed from the critic network.

Thus, the parameter vector $a(k)$ is updated as

$$a(k+1) = \arg \min_a \| u(k) - C(x(k), a(k)) \| \tag{6.38}$$

HDP critic update: The cost-to-go function as computed by the critic should follow the relation

$$V(x(k), w(k)) = L(x(k), u(k)) + V(x(k+1), w(k+1)) \tag{6.39}$$

The LHS is computed using the critic at the kth instant. L can be computed from the current values of states and input. Thus, the critic output at $k+1$ should be $V(k+1) = V(k) - L$, but it is $V(x(k+1), w(k))$. Thus, $w(k)$ must be updated such that

$$w(k+1) = \arg \min_w \| V(k+1) - V(x(k+1), w(k)) \| \tag{6.40}$$

In DHP, the actor approximates the control input, while the critic approximates the gradient of the cost function, i.e., $\nabla V_x(k)$. Since designing of the optimal controller involves computing the gradient term, it is advantageous in DHP that the neural critic directly approximates the gradient.

Example 6.2 Let us first consider a second-order linear discrte time system whose dynamics is governed by

$$x(k + 1) = Ax(k) + Bu(k) \tag{6.41}$$

where $A = \begin{bmatrix} 0.16 & 2.16 \\ -0.16 & -1.16 \end{bmatrix}$ and $B = \begin{bmatrix} -1 \\ 1 \end{bmatrix}$. Obtain a near optimal controller for this sytem using heuristic dynamic programming.

Solution

Since the system is linear, consider a linearly parametrized controller of the form

$$\hat{u}(k) = a^T x(k)$$

where a is the actor parameter vector which is initialized randomly between 0 and 1.

Assume that the cost-to-go function at instant k has the following form:

$$\hat{V}(k) = x^T(k) W x(k)$$

where $W = \begin{bmatrix} w_1 & 0 \\ 0 & w_2 \end{bmatrix}$ is the critic parameter matrix. The task of actor and critic is to estimate the parameters a and W such that the following cost function is minimized:

$$\sum_{k=0}^{N} (x^T(k) x(k) + 0.1 u^2(k))$$

The HDP algorithm is given as follows:

for $k = 1$ to N

Step 1 Find the actor output $\hat{u}(k)$ for a given state vector $x(k)$ and parameter vector $a(k)$, i.e., $\hat{u}(k) = a^T x(k)$. The initial state vector is taken as $x(k) = [0.5 \ 0.2]^T$.

Step 2 Find the utility function $L(x(k), u(k))$ for the current values of $x(k)$ and $\hat{u}(k)$,
i.e., $L(x(k), u(k)) = x^T(k) x(k) + 0.1 u^2(k)$.

Step 3 Find the current value of the cost-to-go function as $\hat{V}(k) = x^T(k) W x(k)$.

Step 4 Find the next state of the plant model $x(k + 1)$ for the current value of control input $\hat{u}(k)$ using Eqn (6.41).

Step 5 Compute the approximate cost-to-go function at instant $k + 1$ as $\hat{V}(k + 1) = x^T(k + 1) W(k) x(k + 1)$.

Step 6 The desired value of critic is given by $V^d(k) = L(x(k), u(k)) + \hat{V}(k + 1)$.

Step 7 Update critic parameters w_1 and w_2 using the BP algorithm to minimize $E_c = \frac{1}{2}(V^d(k) - \hat{V}(k))^2$. The critic parameter update law can be written as

$$w_i(k + 1) = w_i(k) + \eta_w \frac{\partial E_c}{\partial w_i} = w_i(k) + \eta_w E_c x_i^2(k), \text{ for } i = 1, 2$$

The learning rate η_w is taken as 0.7.

Step 8 Update the actor parameters a_1 and a_2 so as to minimize $E_a = \frac{1}{2}(u^d(k) - \hat{u}(k))^2$ where $u^d(k)$ is computed using Eqn (6.37). The actor parameter update law thus can be written as

$$a_i(k + 1) = a_i(k) + \eta_a \frac{\partial E_a}{\partial a_i} = a_i(k) + \eta_a E_a x_i(k), \text{ for } i = 1, 2$$

The learning rate η_a is taken as 0.7.

The results are compared with the standard LQR design for which the control law is found to be $u(k) = 0.1546 x_1(k) + 1.4537 x_2(k)$. Figure 6.2 shows the comparative results of LQR and HDP. It can be seen in Figure 6.2(e) that the actor parameters finally settle down at 0.1571 and 1.3087 which are close to the values computed using LQR.

6.5 SINGLE NETWORK ADAPTIVE CRITIC

Single network adaptive critic (SNAC) has been proposed by Padhi et al.[104]. SNAC is applicable to a large class of problems for which the optimal control equation is explicitly solvable for control in terms of state and costate variables. In such cases, it is possible to get rid of the actor network, which results in a simplified control design process. The authors also provide convergence proof for linear systems with a quadratic cost function and show that the design converges to an optimal solution.

We consider the following discrete-time nonlinear control affine system:

$$x(k + 1) = f(x(k)) + g(x(k))u(k) \tag{6.42}$$

The corresponding performance index is given by

$$J = \sum_{k=0}^{N-1} L(x(k), u(k)) \tag{6.43}$$

where $L(x(k), u(k))$ is the local cost or utility function. The objective is to design a control law $u = c(x)$ so as to minimize the cost (6.60).

At any time instant k, the cost that is going to be incurred from that moment onward can be expressed by a *value* (cost-to-go) function given by

$$V(x(k), c(x)) = \sum_{\tilde{k}=k}^{N-1} L(x(\tilde{k}), u(\tilde{k})) \tag{6.44}$$

where $c(x)$ is the control law which is going to be used for all future time instants.

One approach to solve the optimal control problem stated above consists of imbedding the minimization of the cost function (6.60) in the minimization of value function (6.44). The cost of operation from the instant k to the final time N can be written as

$$V(x(k), u(k), \ldots, u(N - 1)) = L(x(k), u(k)) + V(x(k + 1),$$
$$u(k + 1), \ldots, u(N - 1)) \tag{6.45}$$

$x(k + 1)$ depends on $x(k)$ and $u(k)$ through (6.59).

By using the *principle of optimality*, the optimal value function is given by

$$V^*(x^*(k)) = \min_{u(k)} \{L(x^*(k), u(k)) + V^*(x^*(k + 1))\} \tag{6.46}$$

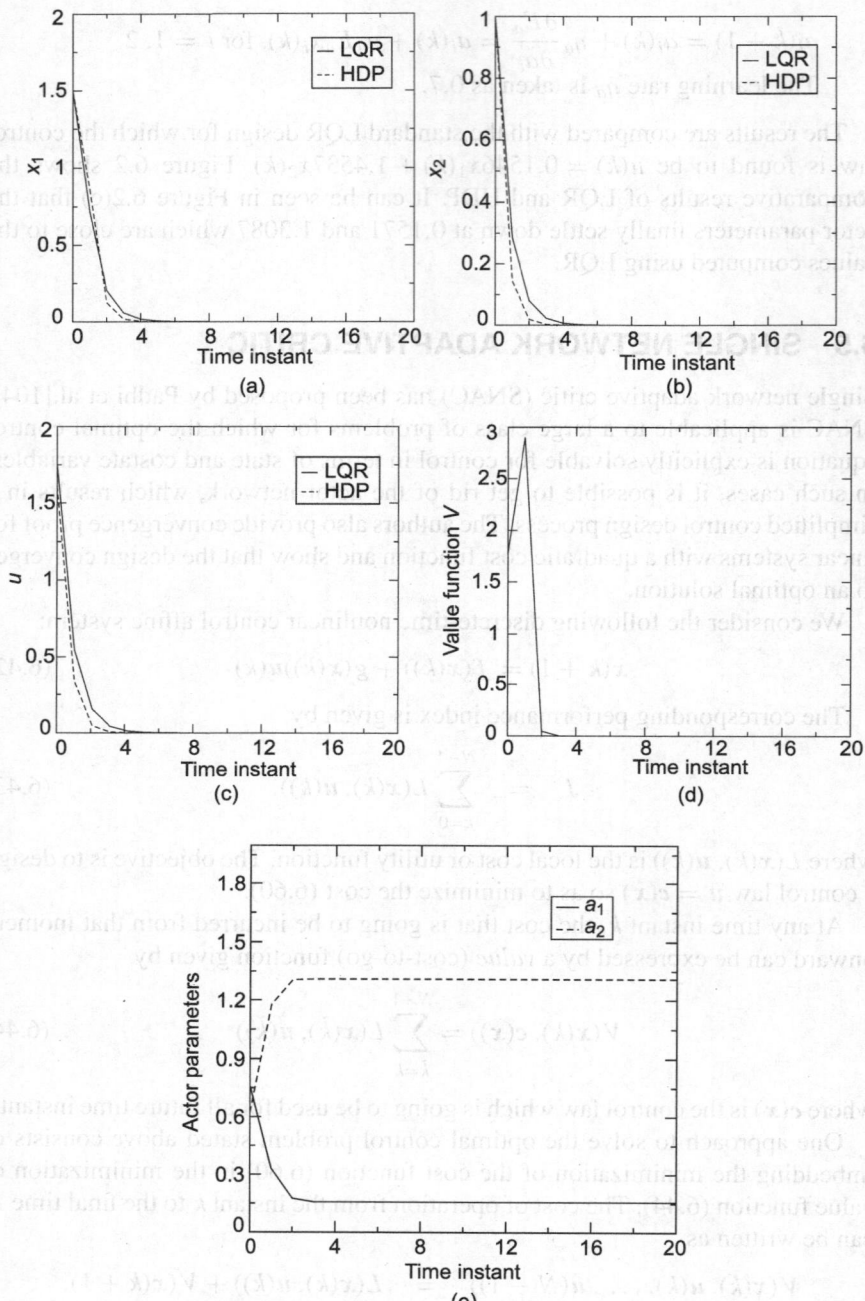

Figure 6.2 (a) and (b) show the convergence of the systems states, (c) shows the control input, (d) shows the value function for HDP algorithm, and (e) shows the update of the actor parameters for the HDP algorithm

which is known as the *recurrence relation of dynamic programming* [107]. In the adaptive critic approach, we use some approximation of this optimal cost function. Hence, we drop asterisk in the above expression and rewrite it as follows:

$$V(x(k)) = L(x(k), u(k)) + V(x(k+1)) \tag{6.47}$$

The control input which would minimize this cost-to-go function must satisfy the necessary condition of optimality given by

$$\frac{\partial V(k)}{\partial u(k)} = 0 \tag{6.48}$$

However,

$$\frac{\partial V(k)}{\partial u(k)} = \frac{\partial L}{\partial u(k)} + \frac{\partial V(k+1)}{\partial u(k)}$$

$$= \frac{\partial L}{\partial u(k)} + \left(\frac{\partial x(k+1)}{\partial u(k)}\right)^T \left(\frac{\partial V(k+1)}{\partial x(k+1)}\right)$$

$$= \frac{\partial L}{\partial u(k)} + \left(\frac{\partial x(k+1)}{\partial u(k)}\right)^T \lambda(k+1) \tag{6.49}$$

where $\lambda(k+1) = \frac{\partial V(k+1)}{\partial x(k+1)}$ is the *costate* vector. Combining Eqns (6.48) and (6.49), we get

$$\frac{\partial L}{\partial u(k)} + \left(\frac{\partial x(k+1)}{\partial u(k)}\right)^T \lambda(k+1) = 0 \tag{6.50}$$

Taking partial derivative of (6.47) with respect to $x(k)$, we get

$$\lambda(k) = \frac{\partial V(k)}{\partial x(k)} = \frac{\partial L}{\partial x(k)} + \frac{\partial V(k+1)}{\partial x(k)}$$

$$= \frac{\partial L}{\partial x(k)} + \left(\frac{\partial u(k)}{\partial x(k)}\right)^T \frac{\partial L}{\partial u(k)}$$

$$+ \left[\frac{\partial x(k+1)}{\partial x(k)} + \left(\frac{\partial x(k+1)}{\partial u(k)}\right)\left(\frac{\partial u(k)}{\partial x(k)}\right)\right]^T \frac{\partial V(k+1)}{\partial x(k+1)}$$

$$= \frac{\partial L}{\partial x(k)} + \left(\frac{\partial x(k+1)}{\partial x(k)}\right)^T \lambda(k+1)$$

$$+ \left(\frac{\partial u(k)}{\partial x(k)}\right)^T \left[\frac{\partial L}{\partial u(k)} + \left(\frac{\partial x(k+1)}{\partial u(k)}\right)^T \lambda(k+1)\right] \tag{6.51}$$

By using Eqn (6.50), the costate equation on the *optimal path* may be written as

$$\lambda(k) = \frac{\partial L}{\partial x(k)} + \left(\frac{\partial x(k+1)}{\partial x(k)}\right)^T \lambda(k+1) \tag{6.52}$$

Consider a quadratic utility function given by

$$L(x(k), u(k)) = \frac{1}{2}\left(x(k)^T Q x(k) + u(k)^T R u(k)\right) \tag{6.53}$$

For an affine system with a quadratic cost function, the costate equation may be written as

$$\lambda(k) = Qx(k) + \left(\frac{\partial x(k+1)}{\partial x(k)}\right)^T \lambda(k+1) \quad (6.54)$$

the optimal control Eqn (6.50) may be written as

$$\frac{\partial V(k)}{\partial u(k)} = Ru(k) + g^T(x(k))\lambda(k+1) = 0 \quad (6.55)$$

This gives us the following optimal control for the affine systems:

$$u(k) = -R^{-1}g^T\lambda(k+1) \quad (6.56)$$

Differentiating Eqn (6.55) once again w.r.t. $u(k)$, we get

$$\frac{\partial^2 V(k)}{\partial u^2(k)} = R > 0 \quad (6.57)$$

where $\lambda(k+1)$ is a function of $x(k)$ through (6.58). It is seen that the sufficient condition of optimality is easily satisfied for affine systems and, thus, the control law (6.56) minimizes the cost-to-go function (6.47) at every time-step k.

The difficulty in the execution of the above control law lies in the fact that the current value of control input requires the information of a future value of costate function. This can be circumvented by using an estimate of costate vector $\lambda(k+1)$ from the present states using a neural network given by

$$\lambda(k+1) = \mathcal{N}(W, x(k)) \quad (6.58)$$

where W represents the weights of the network. Since this neural network gives information about the costate vector which is related to the performance of the system, it is called a critic. The critic must be trained to learn the costate dynamics given by (6.52). The schematic of training is shown in Figure 6.3.

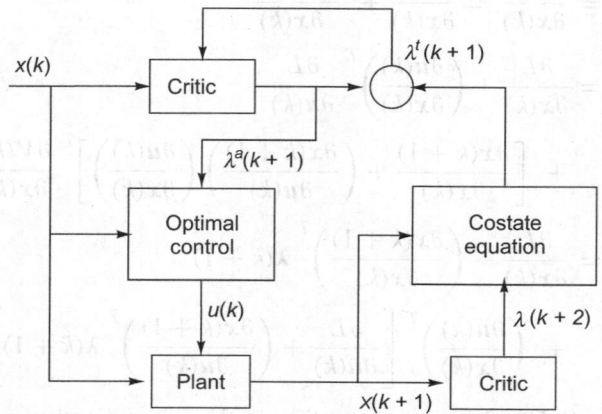

Figure 6.3   Training of a single network adaptive critic

The steps for training critic network which captures the relationship between $x(k)$ and $\lambda(k+1)$ are as follows [104]:

1. Generate the state $x(k)$ in the domain of interest.
2. For each element $x(k)$, follow the steps below:
   (a) Input $x(k)$ to critic network to obtain $\lambda(k+1) = \lambda^a(k+1)$.

    (b) Calculate the optimal control $u(k)$ using (6.56).

    (c) Get the next state $x(k + 1)$ from (6.59) using $x(k)$ and $u(k)$.

    (d) Input $x(k + 1)$ to critic network to get $\lambda(k + 2)$.

    (e) Using $x(k + 1)$ and $\lambda(k + 2)$, calculate $\lambda^t(k + 1)$ from the costate equation (6.52).

3. Train the critic network for all states $x(k)$ in the domain of operation; the output being corresponding $\lambda^t(k + 1)$.

4. Check the convergence of the critic network. If the convergence is achieved, revert to step 2 with $n = n + 1$. Otherwise repeat steps 2–3.

5. Continue steps 2–4 until the convergence is achieved.

The systems, where we can express the optimal control explicitly in terms of states and costate vector as in Eqn (6.56), we can implement optimal control policy using only a single network critic. Control affine systems with a quadratic cost function fall into this category. Since only one network (critic) is needed for implementing this control strategy, it is named as SNAC. An interested reader can find more details about this recently developed architecture in [104], including initialization of network, pre-training, etc.

6.6 CONTINUOUS TIME ADAPTIVE CRITIC

Consider a non-linear control affine system given by

$$\dot{x} = f(x) + g(x)u \tag{6.59}$$

The task is to find a control input that minimizes the performance index given by

$$J(x(t_0), t_0) = \phi(x(T), T) + \int_{t_0}^{T} L(x(\tau), u(\tau))d\tau \tag{6.60}$$

along with the boundary conditions

$$x(t_0) = x_0 \text{ fixed and } x(t_f) \text{ free.} \tag{6.61}$$

The utility function L is given by

$$\psi(x, u) \triangleq \frac{1}{2}(x^T Q x + u^T R u) \tag{6.62}$$

Let us define a scalar function $J^*(x^*(t), t)$ as the *optimal* value of the performance index J for an initial state $x^*(t)$ at time t, i.e.,

$$J^*(x^*(t), t) = \phi(x(T), T) + \int_{t}^{T} L(x^*(\tau), u^*(\tau), \tau)d\tau \tag{6.63}$$

Consider a Hamiltonian given by

$$H(x, \lambda^*, u) = L(x, u) + \lambda^{*T}[f(x) + g(x)u] \tag{6.64}$$

where $\lambda^* = \frac{\partial J^*}{\partial x}$. The optimal control is obtained from the necessary condition given by

$$\frac{\partial H}{\partial u} = \frac{\partial L}{\partial u} + \lambda^{*T} \frac{\partial}{\partial u}[f(x) + g(x)u] = 0 \tag{6.65}$$

This gives the following optimal control equation for control affine system described in (6.59):

$$u = -R^{-1}g^T\lambda^* \tag{6.66}$$

Substituting the value of u into (6.64), we get

$$H(x^*, \lambda^*, u^*) = \frac{1}{2}x^{*T}Qx^* + \frac{1}{2}\lambda^{*T}gR^{-1}g^T\lambda + \lambda^{*T}[f - gR^{-1}g^T\lambda^*] \tag{6.67}$$

On simplification, we have the following optimal Hamiltonian:

$$H^* = \frac{1}{2}x^{*T}Qx^* - \frac{1}{2}\lambda^{*T}gR^{-1}g^T\lambda^* + \lambda^{*T}f$$

$$= \frac{1}{2}x^{*T}Qx^* - \frac{1}{2}\lambda^{*T}G\lambda + \lambda^{*T}f \tag{6.68}$$

where $G = gR^{-1}g^T$. We know that the optimal value function $J^*(x^*, t)$ must satisfy the HJB equation given by

$$\frac{\partial J^*}{\partial t} + \min_u H\left(x, \frac{\partial J^*}{\partial x}, u, t\right) = 0 \tag{6.69}$$

with the boundary condition given by

$$J^*(x^*(T), T) = \phi(x^*(T), T) \tag{6.70}$$

It provides the solution to the optimal control problem for general nonlinear dynamical systems. However, the analytical solution to the HJB equation is difficult to obtain in most cases. It is well known that the HJB equation is both necessary and sufficient conditions of optimality. Therefore, by combining (6.68) and (6.69), we can say that, in case of control affine systems (6.59), the optimal value function must satisfy the following nonlinear dynamic equation:

$$\frac{\partial J^*}{\partial t} + \frac{1}{2}x^{*T}Qx^* - \frac{1}{2}\left(\frac{\partial J^*}{\partial x}\right)^T G\frac{\partial J^*}{\partial x} + \left(\frac{\partial J^*}{\partial x}\right)^T f = 0 \tag{6.71}$$

Since, the analytical solution of the above equation is difficult, we take a different approach and approximate the optimal value function as follows:

$$V(x, t) = h(w, x) \tag{6.72}$$

where the approximating function $h(w, x)$ is selected so as to satisfy certain initial conditions stated in the next section. The parameter t has been put in $V(x, t)$ to show explicit dependence of the value function on time because of time varying parameters w in the approximating function $h(w, x)$.

For the value function given in (6.72) to be optimal, it must satisfy the HJB Eqn (6.71):

$$\frac{\partial V}{\partial t} + L(x, u) + \left(\frac{\partial V}{\partial x}\right)^T [f + gu] = 0 \tag{6.73}$$

$$\frac{\partial h}{\partial w}\dot{w} + \frac{1}{2}x^T Qx - \frac{1}{2}\left(\frac{\partial V}{\partial x}\right)^T G\frac{\partial V}{\partial x} + \left(\frac{\partial V}{\partial x}\right)^T f = 0 \tag{6.74}$$

This gives the following weight update law:

$$\frac{\partial h}{\partial w} \dot{w} = -\frac{1}{2}x^T Q x + \frac{1}{2}\frac{\partial h}{\partial x}^T G \frac{\partial h}{\partial x} - \left(\frac{\partial h}{\partial x}\right)^T f \qquad (6.75)$$

The task is to find $\dot{w}$ so that the above scalar equation is satisfied. This is an *under-determined* system of linear equations with the number of equations less than the number of variables to be estimated. Though, there are infinitely many solutions for which $\dot{w}$ would exactly satisfy the above equation, we seek the one which minimizes $\|\dot{w}\|_2$. The problem is referred to as finding a *minimum norm* solution to an *under-determined* system of linear equations. The pseudo-inverse method is used to solve this problem.

Eqn (6.75) may be written as

$$s\dot{w} = r \qquad (6.76)$$

where $s = \frac{\partial h}{\partial w}$ is a $1 \times N_w$ vector and $r = -\frac{1}{2}x^T Q x + \frac{1}{2}\frac{\partial h}{\partial x}^T G \frac{\partial h}{\partial x} - \left(\frac{\partial h}{\partial x}\right)^T f$ is a scalar quantity. The pseudo-inverse solution is given by

$$\dot{w} = s^T (ss^T)^{-1} r \qquad (6.77)$$

Note that the term ss^T is a scalar quantity and its inverse is easily computable. The control scheme is shown in Figure 6.4. The blocks are self-explanatory.

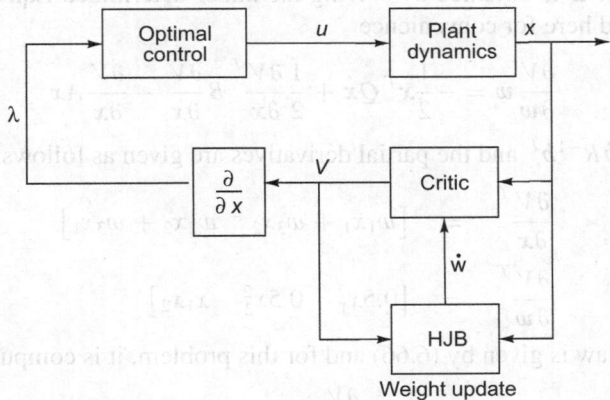

Figure 6.4 Continuous time single network adaptive critic scheme

Example 6.3 (Linear Systems) Consider a single input linear system of the form $\dot{x} = Ax + bu$ given by

$$\begin{bmatrix} \dot{x}_1 \\ \dot{x}_2 \end{bmatrix} = \begin{bmatrix} 0 & 1 \\ 0.4 & 0.1 \end{bmatrix} \begin{bmatrix} x_1 \\ x_2 \end{bmatrix} + \begin{bmatrix} 0 \\ 1 \end{bmatrix} u \qquad (6.78)$$

The task is to find a control law $u = c(x)$ that minimizes the cost function

$$J = \frac{1}{2} \int_0^\infty [x^T Q x + u^T R u]\, dt \qquad (6.79)$$

where

$$Q = \begin{bmatrix} 1 & 0 \\ 0 & 1 \end{bmatrix} \quad \text{and} \quad R = 1$$

We know that the optimal value function for a linear system is given by

$$V = \frac{1}{2} x^T P x \qquad (6.80)$$

where P is a symmetric positive definite matrix. It is trivial to show that the HJB Eqn (6.69) for this value function gives rise to the differential Riccati equation (DRE), given by

$$\dot{P} = -(PA + A^T P) - Q + P^T B R^{-1} B^T P \qquad (6.81)$$

and for an infinite time, $\dot{P} = 0$ and the above equation gives rise to the algebraic Riccati equation (ARE). In order to solve this problem using the above-described approach, we rewrite the optimal value function as

$$V = \frac{1}{2}(w_1 x_1^2 + w_2 x_2^2 + 2w_3 x_1 x_2) \qquad (6.82)$$

where the initial value of the weight vector $w = [w_1 \ w_2 \ w_3]^T$ is chosen so that V is at least positive semi-definite in the beginning. The derivative of the weight vector $\dot{w}$ is obtained by solving the under-determined Eqn (6.75) which is reproduced here for convenience

$$\frac{\partial V}{\partial w} \dot{w} = -\frac{1}{2} x^T Q x + \frac{1}{2} \frac{\partial V}{\partial x}^T \bar{B} \frac{\partial V}{\partial x} - \frac{\partial V}{\partial x} A x \qquad (6.83)$$

where $\bar{B} = b R^{-1} b^T$ and the partial derivatives are given as follows:

$$\frac{\partial V}{\partial x}^T = \begin{bmatrix} w_1 x_1 + w_3 x_2 & w_2 x_2 + w_3 x_1 \end{bmatrix} \qquad (6.84)$$

$$\frac{\partial V}{\partial w}^T = \begin{bmatrix} 0.5 x_1^2 & 0.5 x_2^2 & x_1 x_2 \end{bmatrix} \qquad (6.85)$$

The control law is given by (6.66) and for this problem, it is computed to be

$$u = -R^{-1} b^T \frac{\partial V}{\partial x} = -(w_2 x_2 + w_3 x_1) \qquad (6.86)$$

The weight vector w is updated as follows:

$$w(t + 1) = w(t) - \dot{w} dt \qquad (6.87)$$

The final values of weights after training are given below:

$$w = [2.10456 \ 2.09112 \ 1.4722]^T$$

The Eqn (6.82) may be written as

$$V = \frac{1}{2} x^T W x = \frac{1}{2} x^T \begin{bmatrix} w_1 & w_3 \\ w_3 & w_2 \end{bmatrix} x \qquad (6.88)$$

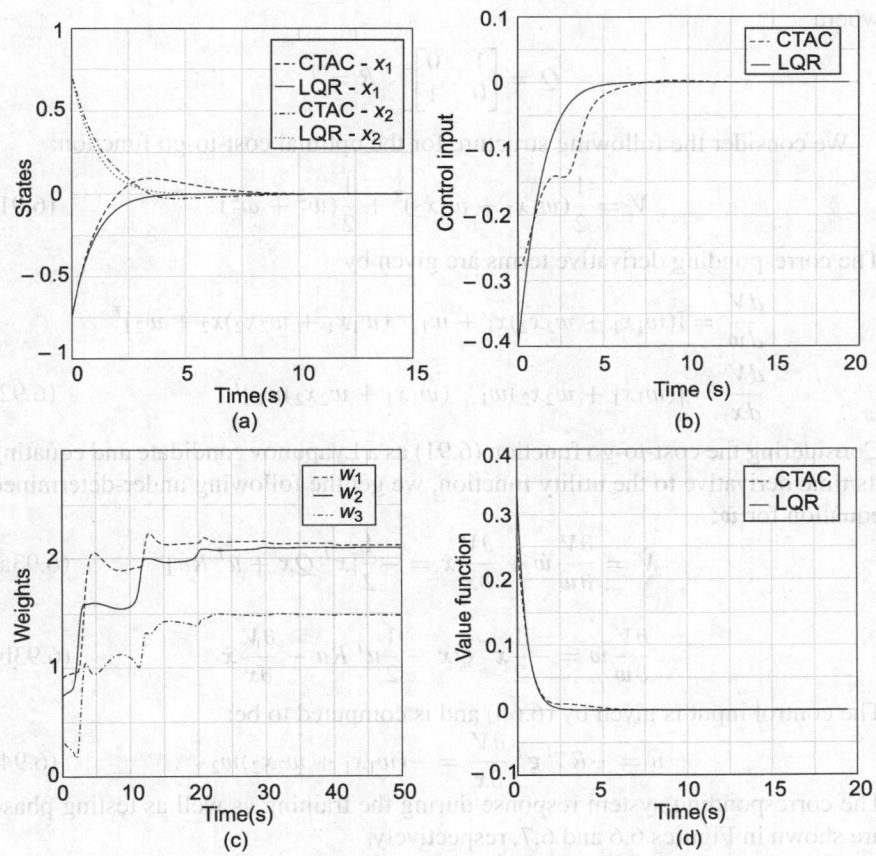

Figure 6.5 Linear system: comparison with LQR performance during training

It can be verified that the matrix W is same as the Riccati matrix P obtained by solving the ARE as shown below.

$$P = \begin{bmatrix} 2.10456 & 1.4722 \\ 1.4722 & 2.09112 \end{bmatrix}$$

Example 6.4 (Non-linear System) Consider the following single link manipulator system given by

$$\dot{x}_1 = x_2$$
$$\dot{x}_2 = -10 \sin x_1 + u \qquad (6.89)$$

We seek to find a controller that minimizes the following cost function:

$$J = \frac{1}{2} \int_0^\infty [x^T Q x + u^T R u] dt \qquad (6.90)$$

where

$$Q = \begin{bmatrix} 1 & 0 \\ 0 & 1 \end{bmatrix} \quad R = 1$$

We consider the following structure for the optimal cost-to-go function:

$$V = \frac{1}{2}(w_1 x_1 + w_2 x_2)^2 + \frac{1}{2}(w_1^2 + w_2^2) \qquad (6.91)$$

The corresponding derivative terms are given by

$$\frac{dV}{dw} = [(w_1 x_1 + w_2 x_2)x_1 + w_1 \quad (w_1 x_1 + w_2 x_2)x_2 + w_2]^T$$

$$\frac{dV}{dx} = [(w_1 x_1 + w_2 x_2)w_1 \quad (w_1 x_1 + w_2 x_2)w_2]^T \qquad (6.92)$$

Considering the cost-to-go function (6.91) as a Lyapunov candidate and equating its time derivative to the utility function, we get the following under-determined equation for $\dot{w}$:

$$\dot{V} = \frac{\partial V}{\partial w}\dot{w} + \frac{\partial V}{\partial x}\dot{x} = -\frac{1}{2}[x^T Q x + u^T R u] \qquad (6.93a)$$

$$\frac{\partial V}{\partial w}\dot{w} = -\frac{1}{2}x^T Q x - \frac{1}{2}u^T R u - \frac{\partial V}{\partial x}\dot{x} \qquad (6.93b)$$

The control input is given by (6.66) and is computed to be:

$$u = -R^{-1}g^T \frac{\partial V}{\partial x} = -(w_1 x_1 + w_2 x_2)w_2 \qquad (6.94)$$

The corresponding system response during the training as well as testing phase are shown in Figures 6.6 and 6.7, respectively.

6.7 ADAPTIVE CRITIC USING THE T–S FUZZY MODEL

Given a linear system, the optimal cost function is obtained as $x^T P x$, where P is the solution of the ARE. When a non-linear system is regulated in an operating point, one can expect that the optimal cost function may be close to the LQR solution of the corresponding linearized system.

This idea leads to the formation of a critic network that predicts the optimal cost function to be T–S fuzzy combination of the local cost functions as

$$\text{If } x_1 \text{ is } F_1^i \text{ and } \cdots \text{ and } x_n(k) \text{ is } F_n^i \text{ then}$$
$$V(x) = x^T W_i x \qquad (6.95)$$

where $F_j^i, j = 1, 2, \ldots, n$ is the jth fuzzy set of the ith rule. Let

$$\mu_i = \prod_{j=1}^{n} \mu_i^j(x_j) \qquad (6.96)$$

where $\mu_i^j(x_j)$ is the membership function of the fuzzy set $F_j^i, i = 1, 2, \ldots, N$.

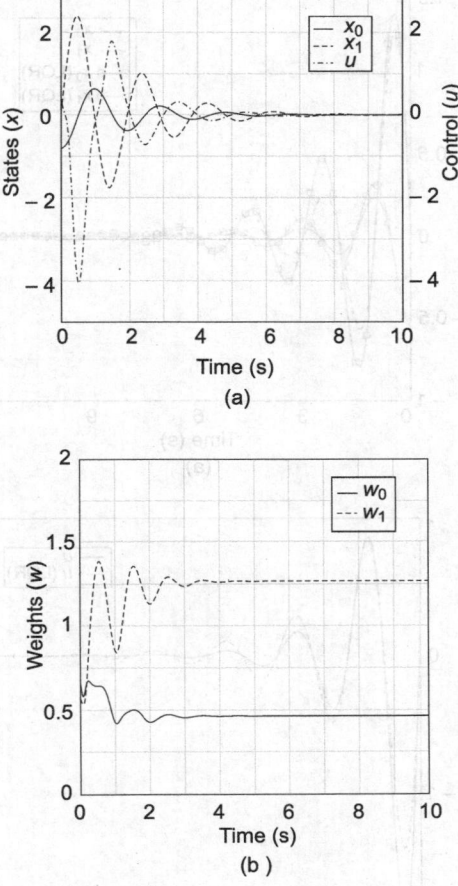

Figure 6.6 Single link manipulator: training phase

Then the critic output, the cost function predicted by the critic network is given as

$$V(x) = \sum_{i=0}^{N} \sigma_i x^T W_i x \qquad (6.97)$$

where $\sigma_i = \frac{\mu_i}{\sum_{i=0}^{N} \mu_i}$ is the normalized membership associated with i^{th} rule.

6.7.1 Continuous Time Adaptive Critic

Consider the input affine system as given in (6.59). The HJB formulation for this equation is given as

$$\frac{\partial V(x, t)}{\partial t} + \psi(x, u) + \lambda^{*T}[f(x) + g(x)u] = 0 \qquad (6.98a)$$

$$\frac{\partial V(x, t)}{\partial t} + \psi(x, u) + \frac{\partial V^T}{\partial x}[f(x) + g(x)u] = 0 \qquad (6.98b)$$

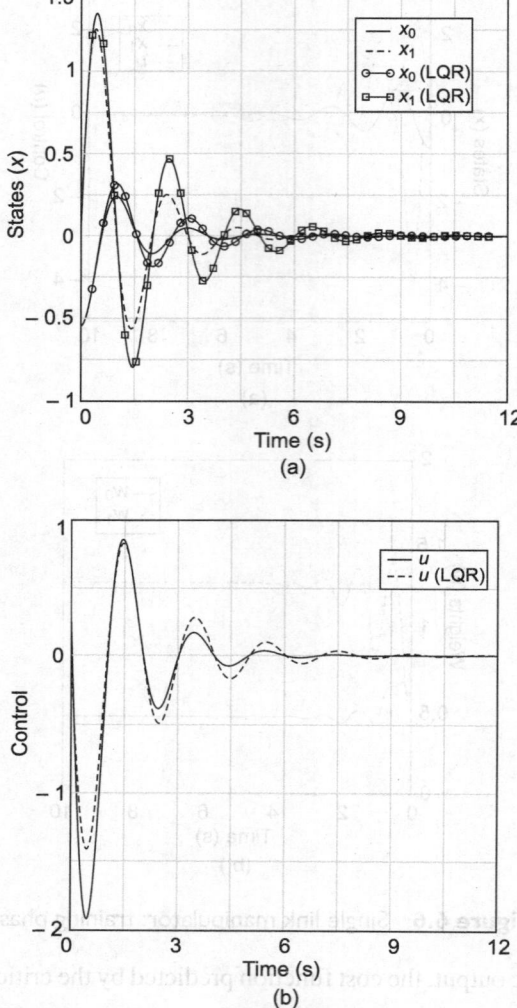

Figure 6.7 Single link manipulator: testing phase

where $\psi(x, u)$ is same as the instantaneous cost function L. This equation can be written as

$$\dot{V}(x, t) = -\psi(x, u) \tag{6.99}$$

where

$$\dot{V}(x, t) = \frac{\partial V(x, t)}{\partial t} + \frac{\partial V^T}{\partial x}[f(x) + g(x)u] \tag{6.100}$$

Eqn (6.100) gives the expression for change in the optimal cost function as the system moves along the optimal path. Since the critic should follow the HJB dynamics (6.100), the weight update law for the T–S fuzzy model based critic (6.97) should satisfy Eqn (6.100).

Though the system is continuous, the weight update is performed at each instant. The discrete time form of the above equation can be written as

$$\Delta V(x(k)) = \psi(x(k), u(k))\Delta T \qquad (6.101)$$

where ΔT is the sampling time.

The above equation can be rewritten as

$$V(x(k)) = V(x(k+1)) + \psi(x(k), u(k)\Delta T \qquad (6.102)$$

where $\Delta V(x(k)) = V(x(k)) - V(x(k+1))$. The forward differentiation is considered since we know from the dynamic programming that the optimal cost at the kth instant is given by

$$J^*(x(k), u^*(k)) = \psi(x(k), u^*(k)) + J^*(x(k+1), u^*(k+1)) \qquad (6.103)$$

At any instant, the T–S fuzzy model weights would satisfy (6.102). The weights are updated such that the above equation is satisfied. At any instant k, we have weight vector $w(k)$. Then, along the optimal path, we can compute

$$\hat{V}(w(k), x(k)) = V(w(k), x(k+1)) + \psi(x(k), u(k)) \qquad (6.104)$$

Then, weights are updated to minimize $\| \hat{V}(w(k), x(k)) - V(w(k), x(k)) \|$.

After learning the optimal cost, the costate vector $\frac{\partial J^*}{\partial x}$ can be computed from the network. The control scheme is shown in Figure 6.8.

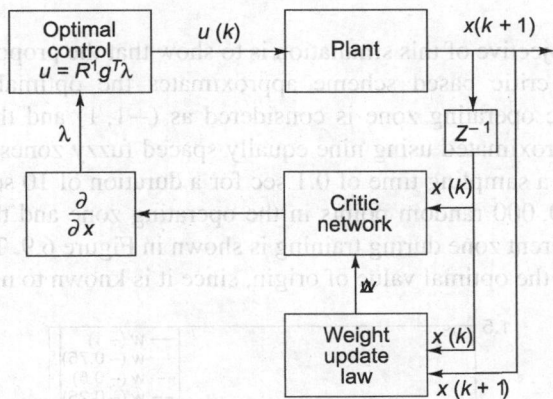

Figure 6.8 T–S fuzzy based critic

The network is trained as follows:

1. Initialize the states to a random point in the operating zone.
2. Compute the control input $u(k)$ from the T–S Fuzzy model, using Eqn (6.24).
3. Give the input $u(k)$ and obtain the next state $x(k+1)$.
4. Compute $\psi(x(k), u(k))$, $V(x(k+1))$, and $V(x(k))$, with current instant weight vector $w(k)$.
5. Compute $\hat{V}(w(k), x(k))$ from $V(w(k), x(k+1))$ and $\psi(x(k), u(k))$.
6. Update the weights to minimize, $\| \hat{V}(w(k), x(k)) - V(w(k), x(k)) \|$.
7. Repeat steps 2–6 for a fixed number of steps, say N.

8. Check for weight convergence. If it is not converged, repeat steps 1–7.

Since the T–S fuzzy model represents the optimal cost, it would be a positive definite function. Hence, the weight has to be properly initialized such that the the T–S fuzzy model is positive definite from the beginning of training. The simple way to initialize the weights in each zone is to assign the local weight matrix W_i associated with the ith zone with the solution obtained from the ARE of the linearized model in that zone.

Example 6.5 (First-Order Non-linear System) Consider the first-order non-linear system dynamics given by

$$\dot{x} = -x^3 + u \tag{6.105}$$

where x is the system state and u is the input to the system. The objective is to stabilize the system such that the input would minimize the cost function

$$J = \frac{1}{2} \int_0^\infty (x^2 + u^2)dt \tag{6.106}$$

The analytic expression of the optimal controller is known for this system and is given by

$$u^* = x^3 - \sqrt{x^6 + x^2} \tag{6.107}$$

The main objective of this simulation is to show that the proposed continuous time adaptive critic based scheme approximates the optimal cost function effectively. The operating zone is considered as $(-1, 1)$ and the optimal cost function is approximated using nine equally spaced fuzzy zones. The system is simulated with a sampling time of 0.1 sec for a duration of 10 sec. The critic is trained with 20, 000 random points in the operating zone and the evolution of weights at different zone during training is shown in Figure 6.9. The weights are initialized near the optimal value of origin, since it is known to us.

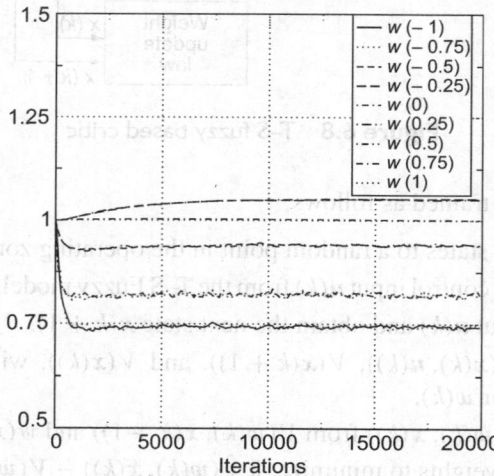

Figure 6.9 Fuzzy weights evolution during training phase

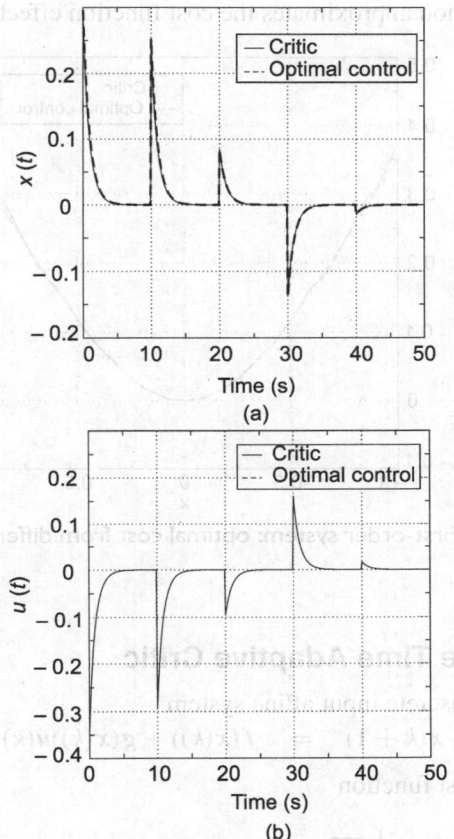

Figure 6.10 First-order system: controller performance. (a) State and (b) Input

After initial training, the system is controlled from different initial conditions and the control result is compared with the optimal control law. The performance comparison of both the proposed controller and optimal controller is shown in Figure 6.10. The corresponding optimal cost is given in Table 6.1. It is evident that, the proposed scheme approximates the optimal cost effectively and performs closer to the optimal controller.

Table 6.1 Computed cost at simulated points

x_0	Critic cost	Optimal cost
0.2915	0.0563	0.0540
0.2221	0.0415	0.0400
0.0762	0.0044	0.0046
−0.1237	0.01248	0.01244
−0.0149	0.00013	0.00014

To evaluate further, the optimal cost from various operating points is shown in Figure 6.11. The figure clearly shows that the proposed T–S fuzzy model based adaptive critic method approximates the cost function effectively.

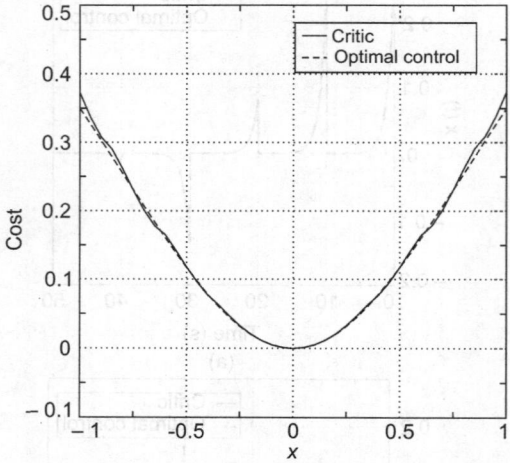

Figure 6.11 First-order system: optimal cost from different initial states

6.7.2 Discrete Time Adaptive Critic

Let us consider a discrete input affine system

$$x(k+1) \quad = \quad f(x(k)) + g(x(k))u(k) \tag{6.108}$$

with a quadratic cost function

$$J = \frac{1}{2} \sum_{k=0}^{\infty} (x^T(k)Qx(k) + u^T(k)Ru(k)) \tag{6.109}$$

where $x(k) \in R^n$ and $u(k) \in R^m$. In this case, the expression for control input is given by

$$u(k) = -R^{-1}g^T(x(k))\lambda(k+1) \tag{6.110}$$

where, $\lambda(k+1) = \frac{\partial J(k+1)}{\partial x(k+1)}$.

In [104], a critic network is proposed to learn $\lambda(k+1)$. In the case of linear systems, a critic network of the following architecture is considered:

$$\hat{\lambda}(k+1) = Wx(k) \tag{6.111}$$

It is shown that the above critic network approaches the optimum value of linear system which corresponds to

$$W = (I + PBR^{-1}B^T)^{-1}PA \tag{6.112}$$

where P is the solution of the ARE.

Instead of a neural network based critic, the T–S fuzzy based critic for such an input affine non-linear system can be used. It is well known that in a narrow operating region, the non-linear system behaves like a linear system. Hence, in a narrow operating range for a given quadratic cost function, the optimal cost can be approximated with a cost function $J^* = \frac{1}{2}x^T P x$. The critic network is given by

$$\text{IF } x_1(k) \text{ is } F_1^i \text{ AND } \cdots \text{ AND } x_n(k) \text{ is } F_n^i \text{ THEN}$$

$$\hat{\lambda}(k+1) = Wx(k)$$

where F_j^i, $j = 1, 2, \cdots, n$ is the jth fuzzy set of the ith rule. Let

$$\mu_i = \prod_{j=1}^{n} \mu_i^j(x_j) \tag{6.113}$$

where $\mu_i^j(x_j)$ is the membership function of the fuzzy set F_j^i, $i = 1, 2, \ldots, N$.

Given the current state vector $x(k)$, the fuzzy model around this operating point is constructed as the weighted average of the local models and has the form

$$\hat{\lambda}(k+1) = \frac{\sum_{i=0}^{N} \mu_i W_i x}{\sum_{i=0}^{N} \mu_i} \tag{6.114}$$

Since the weights of the critic network would vary smoothly from the optimum network at the origin, the weights of the critic network are always initialized with the optimal value at the origin which is known to us through the ARE.

Example 6.6 (First-order Non-linear System)   The discrete time representation of the first-order non-linear system discussed in 6.5 is given by

$$x(k+1) = x(k) + \Delta T(-x^3(k) + u(k)) \tag{6.115}$$

where $x(k)$ is the system state, $u(k)$ is the input to the system at the instant k, and ΔT is the sampling time. The objective is to stabilize the system such that the input would minimize the cost function:

$$J = \frac{1}{2} \sum_{0}^{\infty} (x^2 + u^2) \Delta T \tag{6.116}$$

As discussed earler, the analytic expression of the optimal controller is known for this system. The main objective of this example is to show that the TS Fuzzy model based discrete time single network adaptive critic approximates the costate vector effectively and results in optimal solution. The operating zone is considered as $(-1, 1)$ similar to the one given in 6.5 and the costate vector is approximated using nine equally spaced fuzzy zones. The system is simulated with a sampling time of 0.1 s for a duration of 10 s. The critic is trained with 5000 random points in the operating zone and the evolution of weights at different zones during training is shown in Figure 6.12. The weights are initialized near the costate vector at the origin, since it is known to us.

The controller performance is tested from different initial conditions and compared with the optimal control law. The performance comparison of the discrete time single network adaptive controller scheme and the optimal controller is shown in Figure 6.13. The corresponding optimal cost is given in Table 6.2. It is evident that the T–S fuzzy model approximates the costate vector better due to its universal approximation capability and performs closer to the optimal controller.

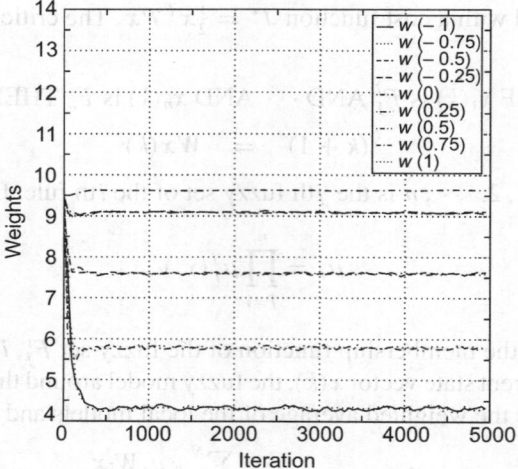

Figure 6.12 Fuzzy weights evolution during training phase

Table 6.2 Computed cost at simulated points

x_0	Critic cost	Optimal cost
−0.4248	0.1048	0.1046
0.464	0.1222	0.1222
−0.8348	0.3147	0.3147
0.5690	0.1730	0.1731
−0.5696	0.1733	0.1734

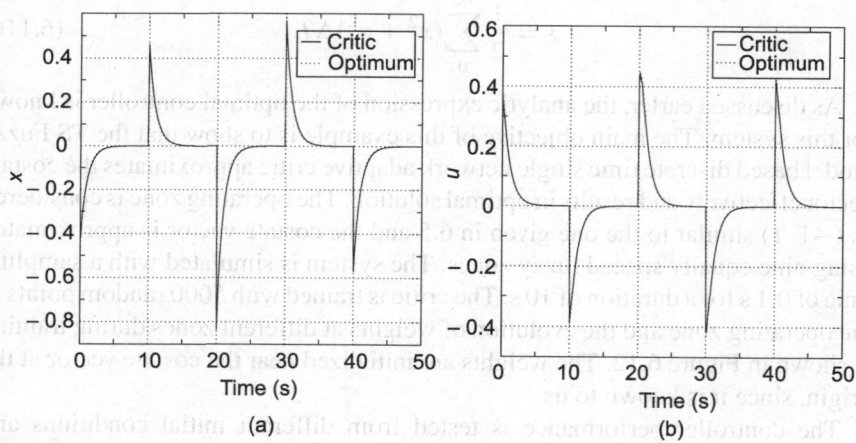

Figure 6.13 First order system: controller performance (a) state and (b) input

To evaluate further, the optimal cost from various operating points is shown in Figure 6.14. The figure clearly shows that the proposed T–S fuzzy model based adaptive critic method approximates the cost function effectively.

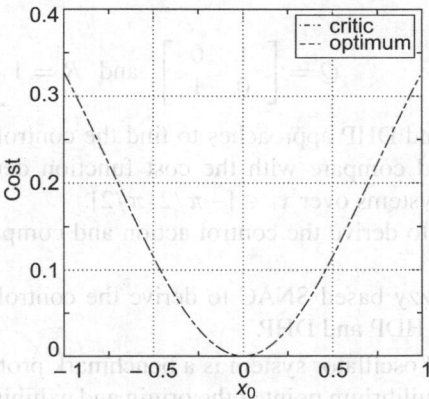

Figure 6.14 First order system: optimal cost from different initial states

SUMMARY

This chapter has presented an adaptive critic based optimal control design for a non-linear system. The HJB formulations for both continuous time and discrete time systems have been presented. A closed form solution for a scalar non-linear differential equation is presented as an example to show that the closed form solution for a generic non-linear system is very difficult to achieve. Thus, the motivation for the adaptive critic based controller design has been made clear. The chapter then discusses the design principles of popular critics such as HDP and DHP. The rest of the chapter deals with the critic design of an input affine non-linear system. The concept of SNAC has been presented. The weight update law for the continuous version of this SNAC has been derived. Two examples are presented to illustrate this concept. The chapter ends with the discussion regarding the T–S fuzzy based critic for SNAC. Readers should simulate this critic network for complex systems to appreciate the benefit of such a scheme. In recent years, researchers have taken great interests in adaptive critic design. There are many interesting works [109, 110, 111, 112, 113, 114, 115] in the literature which readers should refer to be able to appreciate the critic based controllers at a greater depth.

EXERCISES

1. The single link manipulator is expressed as

$$\dot{x}_1 = x_2$$

$$\text{venaya02} \dot{x}_2 = -10\sin(x_1) + u$$

It is desired to find a control law u that will minimize the cost function

$$J = \frac{1}{2}\int_0^\infty [x^T Q x + u^T R u]dt$$

where

$$Q = \begin{bmatrix} 1 & 0 \\ 0 & 1 \end{bmatrix} \quad \text{and} \quad R = 1$$

(a) Use HDP and DHP approaches to find the control action. Plot the cost function and compare with the cost function derived using LQR for linearized systems over $x_1 \in [-\pi/2, \pi/2]$.

(b) Use SNAC to derive the control action and compare results with HDP and DHP.

(c) Use T–S fuzzy based SNAC to derive the control action and compare results with HDP and DHP.

2. The Vanderpol oscillator system is a benchmark problem. The system has an unstable equilibrium point at the origin and exhibits the limit cycle. The dynamics is given by

$$\dot{x}_1 = x_2$$

$$\dot{x}_2 = \alpha(1 - x_1^2)x_2 - x_1 + (1 + x_1^2 + x_2^2)u$$

The control task is to compute the input u such that it will drive the states to $x = [0, 0]$ from any initial condition while minimizing the quadratic cost function

$$J = \frac{1}{2} \int_0^\infty [x^T Q x + u^T R u] dt$$

where

$$Q = \begin{bmatrix} 1 & 0 \\ 0 & 1 \end{bmatrix} \quad \text{and} \quad R = 1$$

Design a DHP based adaptive critic and compare with the T–S fuzzy based adaptive critic for the same problem. Compare the associated cost function.

3. The forward kinematics of CIRAS robot with rigid wrist [66] is given by

$$x = R\cos(\theta_1)$$

$$y = R\sin(\theta_1)$$

$$z = l_1 + l_2\sin(\theta_2) + l_3\sin(\theta_3)$$

where $R = l_2\cos(\theta_2) + l_3\cos(\theta_3) + t$ and l_1, l_2, l_3 are respective link lengths and t is the length of the wrist. The manipulator dimensions are given by $l_1 = l_2 = l_3 = 0.254\,\text{m}$ and $t = 0.05\,\text{m}$. Assuming $X(k + 1), X(k)$ are the end-effector position at the $(k + 1)$th and kth instants, respectively, and $\Delta\theta(k)$ is the change in the joint angle at the kth instant, the closed loop error dynamics which moves the end-effector from the current position X to the desired position X_d can be derived as

$$e(k + 1) = e(k) - J\Delta\theta(k)$$

$$= Ae(k) + Bu(k)$$

where, $e(k) = X_d(k) - X(k)$, $X_d(k+1) = X_d(k)$, $A = I$, $B = -J$, and $u(k) = \Delta\theta(k)$.

$$J = \begin{bmatrix} -R\sin(\theta_1) & -l_2\cos(\theta_1)\sin(\theta_2) & -l_3\cos(\theta_1)\sin(\theta_3) \\ R\cos(\theta_1) & -l_2\sin(\theta_1)\sin(\theta_2) & -l_3\sin(\theta_1)\sin(\theta_3) \\ 0 & l_2\cos(\theta_2) & l_3\cos(\theta_3) \end{bmatrix}$$

Design a T–S Fuzzy based adaptive critic, where the control action and joint angle input $\Delta\theta$ are given by

$$\Delta\theta(k) = R^{-1}J^T\lambda(k+1)$$

where $\lambda(k+1)$ is computed from the critic network. The control task is to compute the input $u(k) = \Delta\theta(k)$ which minimizes the cost function

$$J_c = \frac{1}{2}\sum_{k=0}^{\infty}(X^T(k)QX(k) + u^T(k)Ru(k))$$

where

$$Q = \begin{bmatrix} 1 & 0 & 0 \\ 0 & 1 & 0 \\ 0 & 0 & 1 \end{bmatrix} \quad \text{and} \quad R = \begin{bmatrix} 1 & 0 & 0 \\ 0 & 1 & 0 \\ 0 & 0 & 1 \end{bmatrix}$$

CHAPTER

7

Fuzzy Logic Control

Fuzzy logic controllers (FLCs)[116, 117] based on Zadeh's proposition [118] are being used successfully in an increasing number of application areas ranging from consumer electronics to the automobile industry. The usual decision making process to control a dynamical system in a fuzzy logic control (FLC) framework is structured as a set of $if < situation > then < action >$ rules where both *situation* and *action* have suitable fuzzy representations. In general, FLCs are mostly recommended for systems which are complex and mathematically ill-defined. Thus the performance of an FLC depends on human expertise about the system and the knowledge acquisition techniques to convert human expertise to appropriate fuzzy *if–then* rules as well as a proper fuzzy membership function for each fuzzy variable.

7.1   CONSTRUCTION OF AN FLC

As discussed in Chapter 3, one of the most popular fuzzy control systems is the Mamdani type fuzzy logic control where the controller is designed directly based on the consequences on error variables. Given any arbitrary plant, the control problem in the FLC paradigm is broken down into a set of IF–THEN rules that define the desired controller output response for a given system input conditions. The rules are either obtained from an expert or generated from various observations. The rules are local in nature, i.e., each rule tells us how we should control the system in a small region of the input space. Usually error and change in an error are input variables, and control action is the output variable for an FLC. A typical rule base in the Mamdani type FLC mimicking a PI type controller is of the following form:

Rule R_j: If error e is A_j^1 and change in error, $\dot{e}$, is A_j^2, then the change in control action Δu is B_j for $j = 1, 2, \cdots, r$, where A_j^1, A_j^2, and B_j are fuzzy linguistic variables (e.g. large, small).

Figure 7.1 shows the complete architecture of a Mamdani type FLC. The key issues in the design of an FLC are as follows.

(i) The rule base is usually obtained looking at the typical input–output response of a classical PID controller for the given plant.

(ii) The fuzzification of input variables error e and change in error $\dot{e}$ is done by critically ascertaining the values where the system is most sensitive. The range of these variables and sensitive zones is decided based on numerous PID responses obtained through real-time experiments or simulations.

(iii) The fuzzification of control action is usually most challenging in the design of an FLC. It is usually done through either heuristic tuning of a membership function over a repeated trial or a genetic algorithm (GA) based optimization technique.

The crisp value of the control input can be obtained from the fuzzy rule base and membership grade using the centre of gravity method which is explained in Chapter 3. Once the controller is designed, the system is tested through evaluation of results, tuning the rules and membership functions until satisfactory results are obtained.

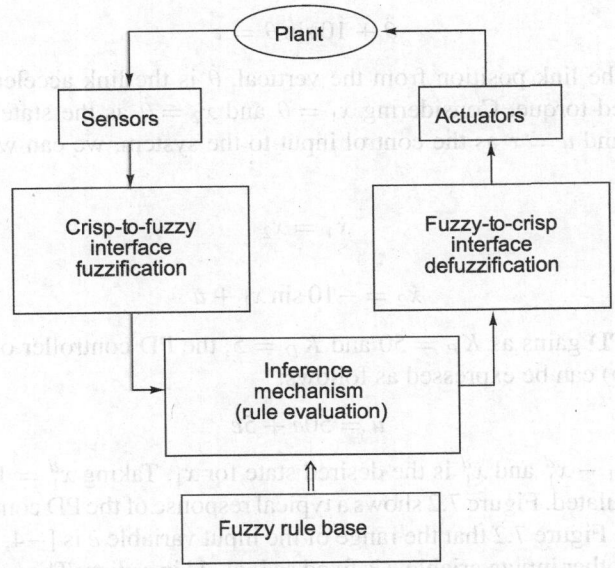

Figure 7.1 Complete architecture of a Mamdani type FLC

7.2 FUZZY PD CONTROLLER

A PD action is instantaneously dependent only on the input variables e and $\dot{e}$. Thus an FLC representing a fuzzy PD controller directly computes the control action unlike the change in the control action as in the case of a fuzzy PI controller.

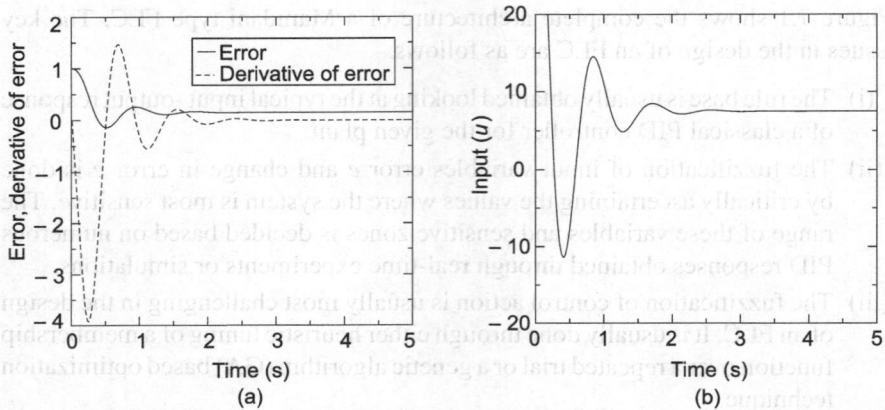

Figure 7.2   Single link manipulator: (a) error and derivative of error to track a set
point of 1 radian and (b) the output of a PD controller

PD controllers are suitable for multi-link robot manipulators. In this section,
the process of a fuzzy PD design will be explained using a single link manipulator.
The dynamics is expressed as

$$\ddot{\theta} + 10\sin\theta = \tau \tag{7.1}$$

where θ is the link position from the vertical, $\dot{\theta}$ is the link acceleration, and τ
is the applied torque. Considering $x_1 = \theta$ and $x_2 = \dot{\theta}$ as the state variables of
the system and $u = \tau$ as the control input to the system, we can write the state
equations as

$$\dot{x}_1 = x_2 \tag{7.2a}$$

$$\dot{x}_2 = -10\sin x_1 + u \tag{7.2b}$$

Setting the PD gains as $K_P = 50$ and $K_D = 5$, the PD controller output for the
system (7.2b) can be expressed as follows:

$$u = 50e + 5\dot{e} \tag{7.3}$$

where $e = x_1 - x_1^d$ and x_1^d is the desired state for x_1. Taking $x_1^d = 1$, the system
(7.2b) is simulated. Figure 7.2 shows a typical response of the PD controller. It can
be seen from Figure 7.2 that the range of the input variable $\dot{e}$ is $[-4, 4]$ in rad/s if
the range of other input variable e is fixed as $[-1, 1]$ in radian. The corresponding
range for the control action (torque) is found to be $[-20, 20]$ N m. If the absolute
value of the maximum torque is limited to 20 N m, then the above controller will
always lead to steady state error. These observations will be used to fuzzify input
variables and control action variables in the first phase of the fuzzy PD design.

Readers can generate these figures using the following MATLAB code:

MATLAB code
| ```
clear all;
x1(1,1)=0;
x2(1,1)=0;
x1d=1;
x2d=0;
dt=0.01;
for i=1:500
e(i,1)=(x1d-x1(i,1));
edot(i,1)=(x2d-x2(i,1));
u(i,1)=50*e(i,1)+5*edot(i,1);
if(u(i,1)>20)
u(i,1)=20;
elseif(u(i,1)<-20)
u(i,1)=-20;
end
x1(i+1,1)=x1(i,1)+dt*(x2(i,1));
x2(i+1,1)=x2(i,1)+dt*(-10*sin(x1(i,1))+u(i,1));
end
plot(e)
figure
plot(edot)
figure
plot(u)
``` |

The PD controller response u as in (7.3) is linear in terms of input variables e, and $\dot{e}$ while a fuzzy PD controller response is non-linear in terms of these input variables. Thus the fuzzy controller can indeed provide a better performance while controlling a non-linear plant.

7.2.1   The Rule Base

The construction of a fuzzy rule base is very simple. For the single link manipulator, you may like to fuzzify each of the input and output variables of the controller e, $\dot{e}$, and u into three fuzzy zones: Positive, zero, and negative. Since e is represented by three fuzzy regions and $\dot{e}$ is represented by three fuzzy regions, the rule base will consist of nine rules as shown in Table 7.1. Given that we have

Table 7.1   Fuzzy rule base for a single link manipulator

Rule	e	$\dot{e}$	u
1	Negative	Positive	Zero
2	Negative	Zero	Positive
3	Negative	Negative	Positive
4	Zero	Zero	Zero
5	Zero	Negative	Positive
6	Zero	Positive	Negative
7	Positive	Zero	Negative
8	Positive	Negative	Zero
9	Positive	Positive	Negative
```

constrained the action space u to only three fuzzy regions, we just have to fill in the fourth column of Table 7.1 with one of these fuzzy terms: Negative, zero, or positive. Of course you cannot do it blindly. You need to understand the system dynamics. Figure 7.2 will give you some idea regarding the relation between input and output variables. Applying a little bit of common sense, one can say that if the error e is negative, i.e., the manipulator angle is at the left of the desired angle and $\dot{e}$ is positive, i.e., the change in error is in the positive direction (in other words the manipulator is moving towards the right), there is no need for a control action, i.e., u is zero. You can use similar logic to formulate the complete rule base. The rule base shown in Table 7.1 is formed using this type of common sense.

7.2.2 Membership Function

Triangular membership functions as shown in Figure 7.3, are chosen for e, $\dot{e}$, and u. The operating regions are subdivided into three fuzzy zones each corresponding to a particular membership function. Looking at the response of a classical PD controller (refer to Figure 7.2), the three fuzzy centres for e are chosen as -1, 0, and 1, whereas for $\dot{e}$ and u they are -2, 0, 2, and -5, 0, 5, respectively.

After constructing the fuzzy rule base and membership grades for all the variables, the output of the fuzzy controller is obtained using the centre of gravity defuzzification method. The above-described fuzzy logic controller is used for set point and trajectory tracking of the single link manipulator. The results of set point tracking for three different set points are shown in Figure 7.4. We should note from Figure 7.4 that although the tracking is accurate and smooth for a desired angle of 0, there are large offsets for other values of x^d.

7.2.3 Fuzzy Parameter Optimization

In the previous section, it is shown how to construct a fuzzy logic controller for a single link manipulator. The rules are generated using a heuristic approach while fixing the fuzzy parameters *a priori*. In this section, we will optimize the fuzzy parameters using an optimization technique.

As shown in Figure 7.5, the operating zone of the variable e is expressed in terms of the following points: N_{min}, N_{max}, Z_{min}, Z_{cen}, Z_{max}, P_{min}, P_{max}, where

$$N_{min} < Z_{min} < N_{max} < Z_c < P_{min} < Z_{max} < P_{max} \qquad (7.4)$$

The centre of zero region, i.e., Z_c can safely be considered as 0. Thus the parameters which are to be optimized for e are N_{min}, N_{max}, Z_{min}, Z_{max}, P_{min}, , P_{max}. Similarly for $\dot{e}$ there are six parameters which are to be optimized. The operating zone of u is expressed in terms of the following points: NU_{min}, NU_{cen}, NU_{max}, ZU_{min}, ZU_{cen}, ZU_{max}, PU_{min}, PU_{cen}, PU_{max}, where

$$NU_{min} < NU_{cen} < NU_{max}$$
$$ZU_{min} < ZU_{cen} < ZU_{max}$$
$$PU_{min} < PU_{cen} < PU_{max} \qquad (7.5)$$

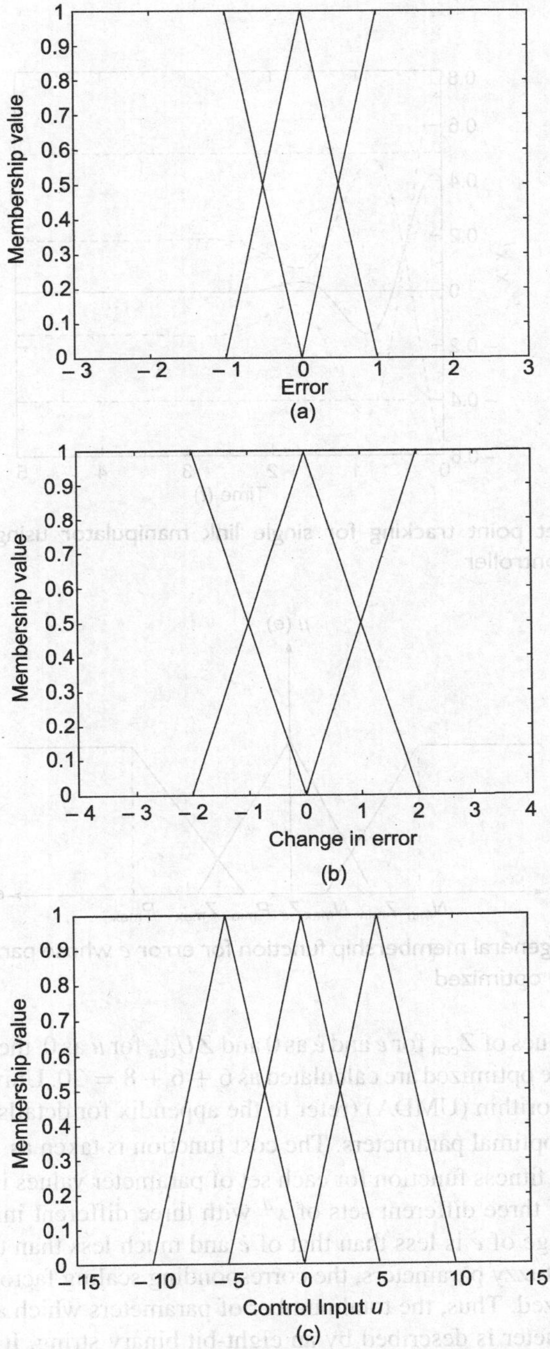

Figure 7.3 Membership functions for e, $\dot{e}$, and u for single link manipulator

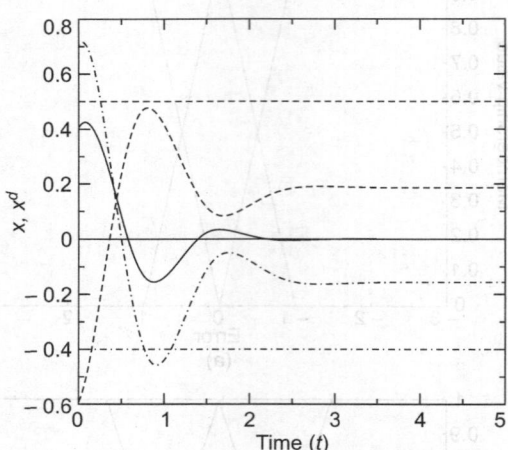

Figure 7.4 Set point tracking for single link manipulator using a fuzzy logic controller

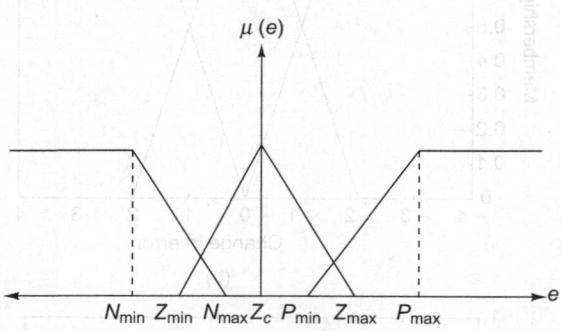

Figure 7.5 A general membership function for error e whose parameters are to be optimized

Keeping the values of Z_{cen} for e and $\dot{e}$ as 0 and ZU_{cen} for u as 0, the total number of parameters to be optimized are calculated as $6 + 6 + 8 = 20$. Univariate marginal distribution algorithm (UMDA) (refer to the appendix for details) has been used to find out the optimal parameters. The cost function is taken as $J = \frac{1}{T} \int_0^T (e^2 + 0.01u^2)dt$. The fitness function for each set of parameter values is assigned to an average cost of three different sets of x^d with three different initial conditions. Usually the range of e is less than that of $\dot{e}$ and much less than that of u. While optimizing the fuzzy parameters, the corresponding scaling factors of e, $\dot{e}$, and u are also optimized. Thus, the total number of parameters which are optimized is 23. Each parameter is described by an eight-bit binary string. It has been made sure during calculation of the parameters that the relationships, as described by expressions (7.4) and (7.5), are satisfied. Figure 7.6 shows a flowchart of the optimization algorithm.

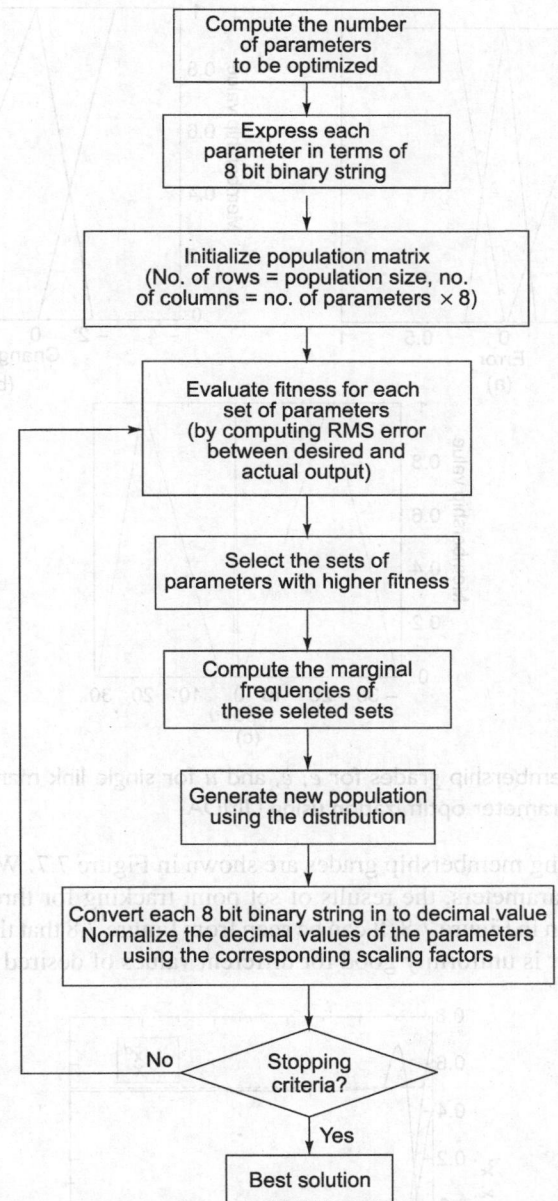

Figure 7.6 A flowchart of the optimization technique

The optimal parameters are found to be

$Z_{\min} = -0.19$, $Z_{\max} = 0.246$, $P_{\max} = 0.31$, $P_{\min} = 0.004$,

$N_{\max} = -0.012$, $N_{\min} = -0.623$ for e;

$Z_{\min} = -1.977$, $Z_{\max} = 3.834$, $P_{\max} = 5.729$, $P_{\min} = 1.917$,

$N_{\max} = -0.06$, $N_{\min} = -2.752$ for $\dot{e}$;

$ZU_{\min} = -0.462$, $ZU_{\max} = 0.615$, $PU_{\max} = 29.231$, $PU_{\min} = 3.307$,

$PU_{\text{cen}} = 19.61$, $NU_{\max} = -1.23$, $NU_{\min} = -34.23$, $NU_{\text{cen}} = -19.61$

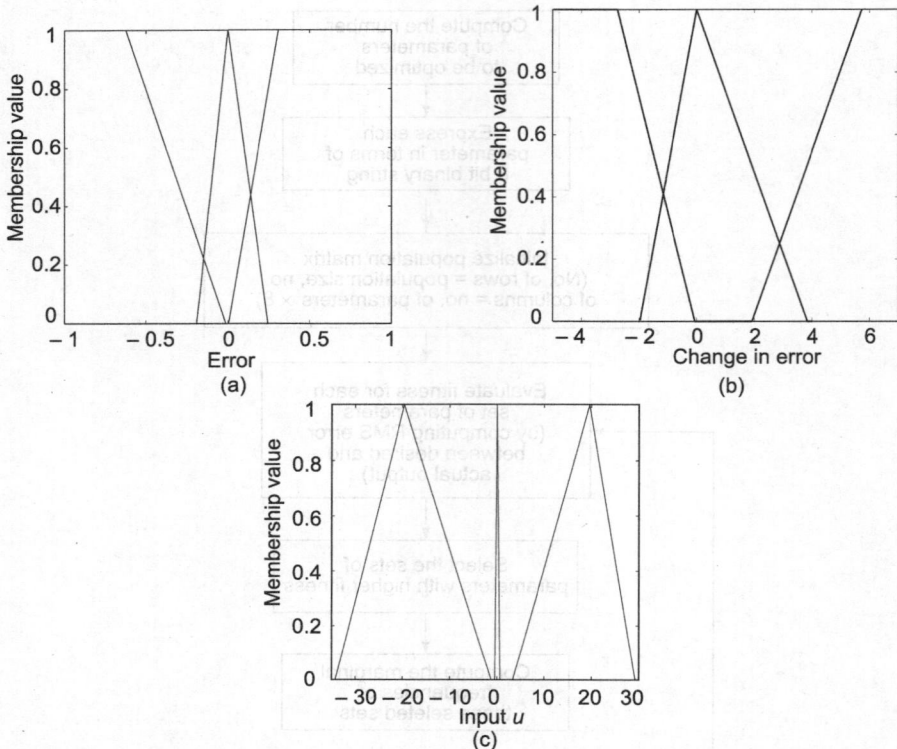

Figure 7.7 Membership grades for e, $\dot{e}$, and u for single link manipulator after parameter optimization using UMDA

Corresponding membership grades are shown in Figure 7.7. With the optimal values of the parameters, the results of set point tracking for three different set points are shown in Figure 7.8. It can be seen from Figure 7.8 that the performance of the controller is uniformly good for different values of desired angles.

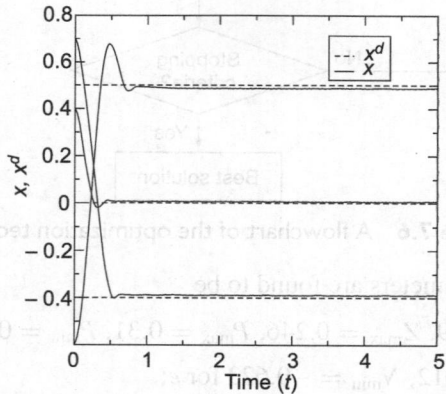

Figure 7.8 Set point tracking for single link manipulator after parameter optimization using UMDA

7.2.4 Rule Generation Using Optimization Technique

In this section, we will describe how to generate the rule base using the optimization technique. The rule base has a form as shown in Table 7.2 where the Rules from 1 to 9 are to be generated using the UMDA algorithm. Since there are only three values of u, i.e., positive, zero, and negative, each rule is represented with two binary bits. The addition of the bits will give us three distinct numerical values, i.e., 0, 1, and 2. We have assigned negative$_U$ = 0, positive$_U$ = 1, and Zero$_U$ = 2. Thus when the addition of the two bits assigned for Rule 1 equals 2, the corresponding u should be zero. The final rule after optimization is shown in Table 7.3. Using the optimized rule base and optimized fuzzy parameters, the fuzzy PD controller is applied to the single link manipulator for tracking a desired trajectory $x^d = \sin(3t)$.

Table 7.2 Rule base formation for single link manipulator

e	$\dot{e}$	u
Negative	Positive	Rule 1
Negative	Zero	Rule 2
Negative	Negative	Rule 3
Zero	Zero	Rule 4
Zero	Negative	Rule 5
Zero	Positive	Rule 6
Positive	Zero	Rule 7
Positive	Negative	Rule 8
Positive	Positive	Rule 9

Table 7.3 Optimized rule base for single link manipulator

e	$\dot{e}$	u
Negative	Positive	Positive
Negative	Zero	Zero
Negative	Negative	Positive
Zero	Zero	Zero
Zero	Negative	Positive
Zero	Positive	Negative
Positive	Zero	Zero
Positive	Negative	Zero
Positive	Positive	Positive

The tracking results are shown in Figure 7.9 with an RMS error of 0.004. It can be seen from Figure 7.9 that the tracking is highly satisfactory and the RMS error is much less than that of a general PD controller.

7.3 FUZZY PI CONTROLLER

In contrast to a fuzzy PD controller, instead of computing the control input u directly, fuzzy PI controller computes the incremental control input Δu as a function of the output error and derivative of the error which form the rule base

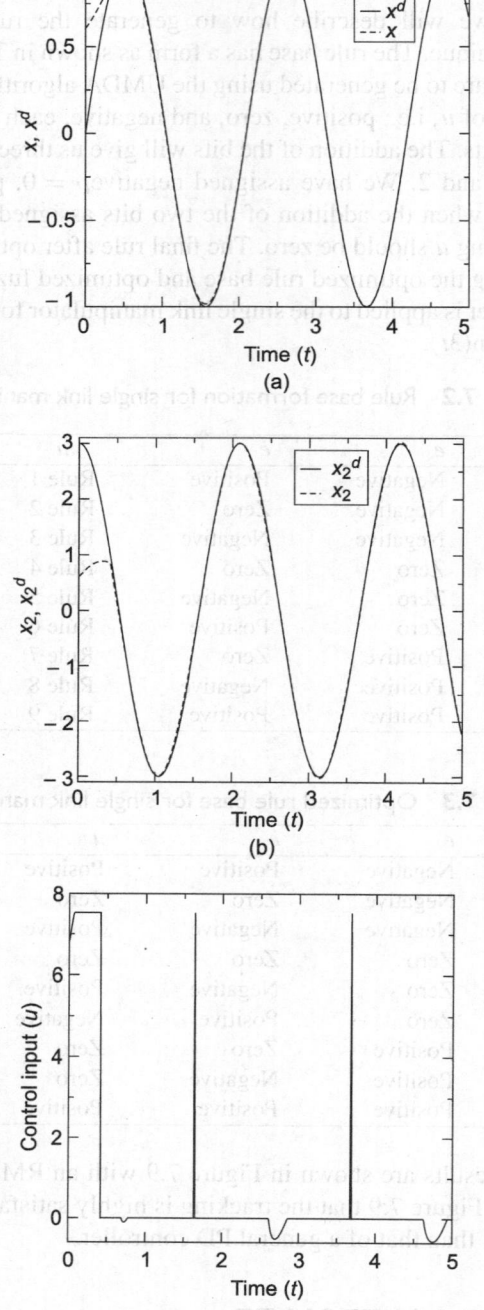

Figure 7.9 (a) The tracking of x_1, (b) tracking of x_2, and (c) the corresponding control input

of the fuzzy control system. To illustrate how to construct a fuzzy PI controller let us take the following linear system:

$$\ddot{y} + 12\dot{y} + 20y = 2u \tag{7.6}$$

where u is the input to the system and y is the output of the system. The above system can be controlled using a PI controller of the form

$$u(t) = 5\,e(t) + 25 \int_0^T e(t)\,dt$$

The MATLAB code to simulate the above system is as follows:

MATLAB code

```
s=tf('s');
g=2/(s*s+12*s+20);
c=5+25/s;
gc=feedback(g*c,1);
step(gc,5)
```

The output response of the system (7.6) with the PI controller is shown in Figure 7.10.

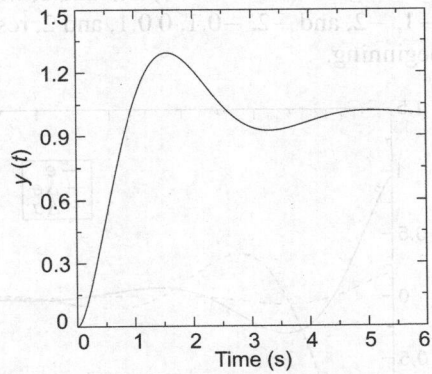

Figure 7.10 Response of system (7.6) using a classical PI controller

7.3.1 The Rule Base for the Fuzzy PI Controller

To construct the fuzzy rule base, the operating regions of e, Δe, and Δu are divided into five subparts, PL (positive large), PS (positive small), ZE (zero), NS negative small and NL (negative large). Thus the FLC will have 25 rules and each rule will produce an incremental control Δu from the set [PL, PS, ZE, NS, NL]. It is seen from Figure 7.10 that when e is PL and Δe is PL (i.e., change in y is NL), the error is likely to increase further in the next instant, thus one should give a large u to reduce the error, hence Δu should be PL in this case. Similarly when e is PL and Δe is PS, Δu is PS. Following a similar logic, the complete rule base is generated as shown in Figure 7.11.

e \ Δe	NL	NS	ZE	PS	PL
NL	NL	NL	NS	NS	ZE
NS	NL	NS	NS	ZE	PS
ZE	NS	NS	ZE	PS	PS
PS	NS	ZE	PS	PS	PL
PL	ZE	PS	PS	PL	PL

Figure 7.11   Rule base for a fuzzy PI controller

7.3.2   Membership Function

Triangular membership functions, as in the case of a fuzzy PD controller in Section 7.2, are chosen for e, Δe, and Δu. The operating regions are subdivided into five fuzzy zones each corresponding to a particular membership function. Looking at how e, Δe, and Δu change for a classical PI controller (Figure 7.12), the five fuzzy centres for e are chosen as -1, -0.5, 0, 0.5, and 1, whereas for Δe and Δu they are -2, -1, 0, -1, -2, and -2, -0.1, 0 0.1, and 2, respectively. Δu is very small except at the beginning.

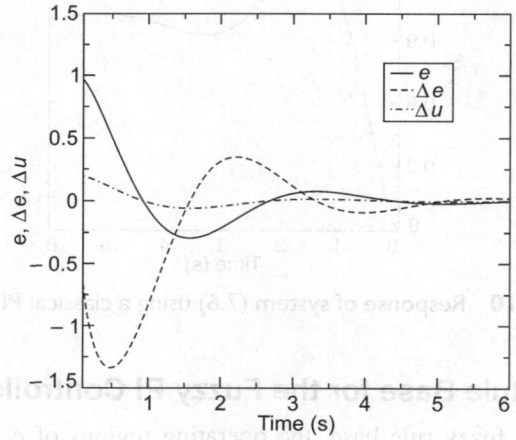

Figure 7.12   Response of system (7.6): e, Δe, and Δu are plotted against time

After constructing the fuzzy rule base and membership grades, the output of fuzzy controller is obtained using the centre of gravity defuzzification method. The output response for a set point of 1 with the above-described fuzzy PI controller is shown in Figure 7.13. The steady state error is found to be 0.00002. However, it is noticed that the control input u fluctuates between its maximum and minimum values.

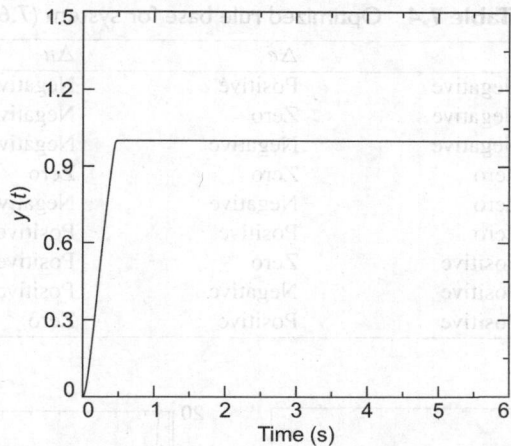

Figure 7.13 Response of system (7.6) using a fuzzy PI controller

7.3.3 Parameter Optimization and Rule Generation Using UMDA

In the previous section, it is shown how to construct a fuzzy PI controller for a linear dynamical system. The rules are generated using a heuristic approach while fixing the fuzzy parameters *a priori*. In this section, we will optimize the fuzzy parameters as well as the rules using the procedure as described in Section 7.2. The operating regions of e and Δe are divided into three subparts, positive, zero and negative, i.e., the number of fuzzy rules is reduced to 9. As in Section 7.2, including the scaling factors, the total number of parameters to be optimized is found to be 23 and two binary bits are assigned to find the rule. The cost function is given as

$$J = \frac{1}{T} \int_0^T (e^2 + 0.005 \Delta u^2) dt$$

The optimized parameters are found to be

$Z_{min} = -0.97, Z_{max} = 0.985, P_{max} = 2.22, P_{min} = 0.046, N_{max} = -0.11;$

$N_{min} = -2.2, VZ_{min} = -1.85, VZ_{max} = 0.8, VP_{max} = 1.35;$

$VP_{min} = 0.15, VN_{max} = -0.93, VN_{min} = -2.38, ZU_{min} = -0.61;$

$ZU_{max} = 1.52, PU_{max} = 2.22, PU_{min} = 0;$

$PU_{cen} = 1.53, NU_{max} = 0, NU_{min} = -2.355, NU_{cen} = -0.87$

The optimized rule base is shown in Table 7.4.

The output response of the system to a reference of 1 is shown in Figure 7.14(a), whereas Figure 7.14(b) shows the corresponding control input, which is smooth.

Table 7.4 Optimized rule base for system (7.6)

e	Δe	Δu
Negative	Positive	Negative
Negative	Zero	Negative
Negative	Negative	Negative
Zero	Zero	Zero
Zero	Negative	Negative
Zero	Positive	Positive
Positive	Zero	Positive
Positive	Negative	Positive
Positive	Positive	Zero

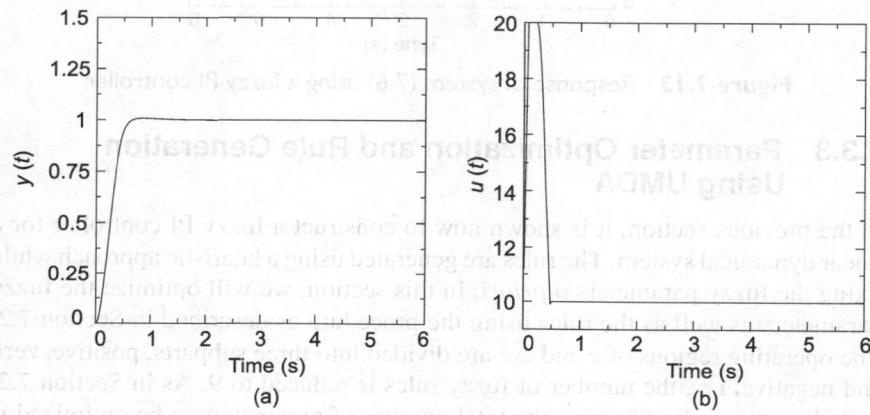

Figure 7.14 Response of system (7.6) using a fuzzy PI controller with optimized rule base and parameters: (a) the system output and (b) the control input

7.4 FUZZY PI CONTROLLER FOR A SERIES DC MOTOR

The field circuit of a series DC motor [69] is connected in series with its armature circuit. The torque produced by this motor is proportional to the square of the current. Thus these kinds of motors are used in applications where a high torque is required at a low speed. The equations of a series DC motor with $i_a - i_f = i$, where i_a is armature current and i_f is field current, are described by

$$J\frac{d\omega}{dt} = -D\omega + K_m L_f i^2 - \tau_L \tag{7.7a}$$

$$L\frac{di}{dt} = -K_m L_f \omega i - Ri + V \tag{7.7b}$$

where ω is the speed to be controlled, V is the input voltage, τ_L is the load torque, J is the moment of inertia of the motor, D is the viscous friction coefficient, K_m is the motor torque constant, L_f is the field inductance, and $L = L_a + L_f$,

$R = R_a + R_f$, where L_a and R_a are the armature inductance and resistance, respectively. The parameters used for the simulation are given in Table 7.5.

Table 7.5 Series DC motor: list of parameters

L	0.0917 H
R	7.2 Ω
D	0.0004 Nm/rad/s
$K_m L_f$	0.1236 Nm/Wb − A
J	0.000704 kgm^2

The objective is to design a fuzzy PI controller so that the motor achieves a desired value of 100 rad/s. Considering the state variables as $x_1 = \omega$ and $x_2 = i$ and input variable as $u = V$, the state space model of the system can be written as

$$\dot{x}_1 = -\frac{D}{J}x_1 + \frac{K_m L_f}{J}x_2^2 - \tau_L \tag{7.8a}$$

$$\dot{x}_2 = -\frac{K_m L_f}{L}x_1 x_2 - \frac{R}{L}x_2 + \frac{1}{L}u \tag{7.8b}$$

7.4.1 Parameter Optimization and Rule Generation

Let us denote the error between the actual and desired speed by e and change in error by Δe. The operating regions of e, Δe, and Δu are divided into three subparts, positive, zero and negative. Thus the FLC will have nine rules and each rule will produce an incremental control Δu from the set (positive, zero, negative). Triangular membership functions are chosen for e, Δe, and Δu. The fuzzy centres as well as the rules will be optimized using the procedure as described in Section 7.2. As in Section 7.2, including the scaling factors, total number of parameters to be optimized is found to be 23 and two binary bits are assigned to find the rule. The cost function is given as

$$J = \frac{1}{T} \int_0^T (e^2 + \Delta u^2)dt$$

The optimized parameters are found to be

$Z_{min} = -1.17$, $Z_{max} = 1.77$, $P_{max} = 2.645$, $P_{min} = 0.59$, $N_{max} = -0.009$;

$N_{min} = -1.207$, $VZ_{min} = -27.889$, $VZ_{max} = 55$, $VP_{max} = 75.444$;

$VP_{min} = 26.667$, $VN_{max} = -0.444$, $VN_{min} = -43.78$, $ZU_{min} = -11$;

$ZU_{max} = 12.048$, $PU_{max} = 12.381$, $PU_{min} = 0$;

$PU_{cen} = 12.143$, $NU_{max} = 0$, $NU_{min} = -10.857$, $NU_{cen} = -9.81$

The optimized rule base is shown in Table 7.6.

Figure 7.15 shows the simulation results where (a) shows how the actual speed tracks the desired speed and (b) shows the corresponding control input. It can be noticed from the figure that the settling time is only about 0.4 s. A voltage limiter has been added in the simulation program to make sure that the input voltage remains within the maximum allowable range, i.e., ± 40 V.

Table 7.6 Optimized rule base for series DC motor

e	Δe	Δu
Negative	Positive	Zero
Negative	Zero	Zero
Negative	Negative	Negative
Zero	Zero	Zero
Zero	Negative	Zero
Zero	Positive	Positive
Positive	Zero	Positive
Positive	Negative	Positive
Positive	Positive	Positive

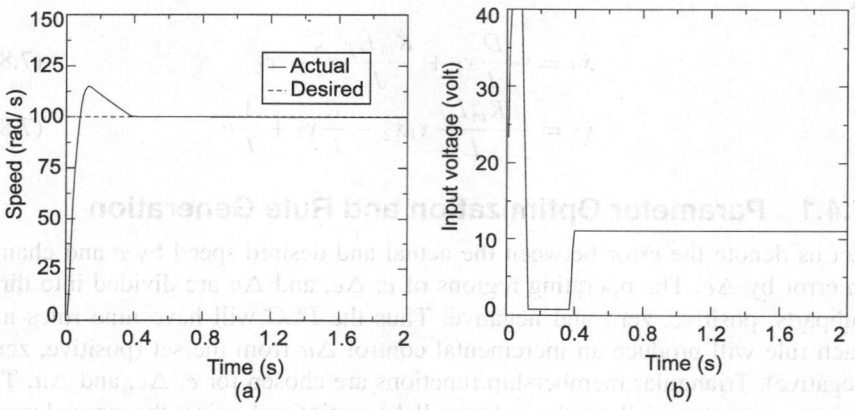

Figure 7.15 Speed control of a series DC motor using a fuzzy PI controller: (a) the speed of the motor and (b) the corresponding input voltage

7.5 FLC USING LYAPUNOV SYNTHESIS

The rule base derived in Mamdani type FLC is based on heuristics. Although rules and parameters are optimized to provide a stable operation around an approximate model of the plant, the stability is not intrinsically embedded into the rule base. The rule base for the controller is obtained using the response behaviour of a classical PD or a PID controller. Margaliot and Langholz [119] introduced the idea of rule generation using the Lyapunov synthesis approach. This work demonstrated that the classical Lyapunov stability analysis can be extended to the domain of computing with words. In this approach, a Lyapunov function candidate V is selected for a given system. The rate derivative of the Lyapunov function $\dot{V}$ is expressed in terms of states and control input. The fuzzy IF–THEN rule is proposed, so that $\dot{V}$ is negative definite.

Consider a scalar non-linear system

$$\dot{x} = -x^3 + u \qquad (7.9)$$

We are required to design an FLC, so that the plant is stabilized around the origin, $x = 0$. Let the Lyapunov function for this plant be

$$V = \frac{1}{2}x^2 \qquad (7.10)$$

The rate derivative of the Lyapunov function is given as

$$\dot{V} = x\dot{x} \qquad (7.11)$$

$$= x(-x^3 + u) \quad [\text{using (7.9)}] \qquad (7.12)$$

$$= -x^4 + xu \qquad (7.13)$$

Here are two rules that will make the term $\dot{V}$ negative definite:

(i) IF x is negative, THEN u is positive

(ii) IF x is positive, THEN u is negative

We can notice that each rule results in a negative quantity for the second term of $\dot{V}$ which yields $\dot{V}$ as negative definite as the other term $-x^4$ is always negative. This argument is valid as long as the fuzzy term negative (positive) is defined over negative (positive) values of both x and u. This simple example should make readers clear about the derivation of rules that can guarantee the stability of the system. However, it is still required to optimize other parameters of FLC to optimize the performance.

FLC design using the Lyapunov synthesis approach will be described in detail for two different dynamic systems: rotational–translational proof mass actuator (RTAC) and two-link robot manipulator.

7.5.1 Rotational–Translational Proof Mass Actuator

Consider the example of a rotational–translational proof mass actuator RTAC [119] as shown in Figure 7.16. The RTAC system combines a translational oscillator with a rotational proof mass actuator. The oscillator consists of a cart connected to a fixed wall by a linear spring. The cart is constrained to move in the x-direction only. The proof mass actuator is attached to the cart and is controlled by an applied torque N. F is a disturbance force that perturbs the cart.

Table 7.7 defines different parameter values of the system. The cart position and the actuator angle are x_c and θ, respectively. Define the normalized variable as

$$\xi = x_c\sqrt{\frac{M+m}{I+mL^2}} \qquad \tau = t\sqrt{\frac{k}{M+m}}$$

$$u = N\frac{M+m}{k(I+mL^2)} \qquad w = \frac{1}{k}\sqrt{\frac{M+m}{I+mL^2}}F$$

Table 7.7 RTAC parameters

Mass of the cart	M=1.36 Kg
Proof mass	m=0.096 Kg
Distance	L=0.059 m
Moment of inertia(proof mass)	I=0.00022 Kg m^2
Spring constant	k=186.3 N/m

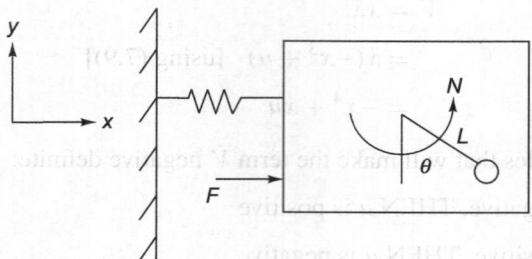

Figure 7.16 RTAC system

RTAC Dynamics

Considering the state variable as $x = [x_1\ x_2\ x_3\ x_4]^T = [\xi\ \dot{\xi}\ \theta\ \dot{\theta}]^T$, control variable as u, and disturbance as w, the dynamics can be written as

$$\dot{x} = \begin{bmatrix} x_2 \\ \frac{-x_1+\epsilon x_4^2 \sin x_3}{1-\epsilon^2 \cos^2 x_3} \\ x_4 \\ \frac{\epsilon(x_1-\epsilon x_4^2 \sin x_3)\cos x_3}{1-\epsilon^2 \cos^2 x_3} \end{bmatrix} + \begin{bmatrix} 0 \\ \frac{-\epsilon \cos x_3}{1-\epsilon^2 \cos^2 x_3} \\ 0 \\ \frac{1}{1-\epsilon^2 \cos^2 x_3} \end{bmatrix} u + \begin{bmatrix} 0 \\ \frac{1}{1-\epsilon^2 \cos^2 x_3} \\ 0 \\ \frac{-\epsilon \cos x_3}{1-\epsilon^2 \cos^2 x_3} \end{bmatrix} w \quad (7.14)$$

where $\epsilon = \dfrac{mL}{\sqrt{(I+mL^2)(M+m)}}$ is a very small number.

The system dynamics can be written as

$$\dot{x} = f(x) + g(x)u + d(x)w \quad (7.15)$$

Without knowing $f(x)$, $g(x)$, and $d(x)$, the following linguistic statements can be formulated:

- S1: The state of the system is described by x_1 (cart's position), $x_2 = \dot{x}_1$, x_3 (actuator's angle), and $x_4 = \dot{x}_3$
- S2: $\dot{x}_4$ is proportional to the control input u
- S3: $\dot{x}_2$ is proportional to $-x_1$

Fuzzy Controller: Regulation Problem

The control objective is to find a u such that the system is stable around $x = [x_1\ x_2\ x_3\ x_4]^T = 0$.

Let $V = \frac{1}{2}(x_1^2 + x_2^2 + x_3^2 + x_4^2)$ be the Lyapunov function candidate. Then, using S1,

$$\dot{V} = x_1 x_2 + x_2 \dot{x}_2 + x_3 x_4 + x_4 \dot{x}_4 \quad (7.16)$$

using S3 and S2,

$$\dot{V} \approx x_3 x_4 + x_4 u \tag{7.17}$$

The classical Lyapunov synthesis suggests the design of u that will guarantee $\dot{V} < 0$. Same idea is applicable here, but in the domain of computing with words.

Rule Base: Fuzzy PD Controller

$\dot{V} \approx x_3 x_4 + x_4 u$ can be made negative definite if the rule base is formed as follows:

- IF x_3 is negative AND x_4 is negative, THEN u is positive
- IF x_3 is positive AND x_4 is positive, THEN u is negative
- IF x_3 is negative AND x_4 is positive, THEN u is zero
- IF x_3 is positive AND x_4 is negative, THEN u is zero

Inference Mechanism and Defuzzification

Gaussian membership functions are used to define the linguistic terms in the rule base as

$$\mu_p(x) = e^{-(x-a_x)^2}$$

$$\mu_n(x) = e^{-(x+a_x)^2}$$

$$\mu_z(x) = e^{-x^2}$$

$a_x > 0$, x stands for the state variable, and p, n, and z denote positive, negative, and zero, respectively. Using 'product for AND' inference and 'centre of gravity' method for defuzzification, the fuzzy controller is given as

$$u = \frac{\mu_n(x_3)\mu_n(x_4)(a_u) + \mu_p(x_3)\mu_p(x_4)(-a_u) + \mu_n(x_3)\mu_p(x_4)(0) + \mu_p(x_3)\mu_n(x_4)(0)}{\mu_n(x_3)\mu_n(x_4) + \mu_p(x_3)\mu_p(x_4) + \mu_n(x_3)\mu_p(x_4) + \mu_p(x_3)\mu_n(x_4)}$$

Fuzzy Control Law

Since $\mu_p(x) = e^{-(x-a_x)^2}$ and $\mu_n(x) = e^{-(x+a_x)^2}$, we can write

$$
\begin{aligned}
u &= \frac{e^{-(x_3+a_{x_3})^2} e^{-(x_4+a_{x_4})^2}(a_u) - e^{-(x_3-a_{x_3})^2} e^{-(x_4-a_{x_4})^2}(a_u)}{e^{-(x_3+a_{x_3})^2}(e^{-(x_4+a_{x_4})^2} + e^{-(x_4-a_{x_4})^2}) + e^{-(x_3-a_{x_3})^2}(e^{-(x_4+a_{x_4})^2} + e^{-(x_4-a_{x_4})^2})} \\[2mm]
&= \frac{(a_u)[e^{-(x_3^2+a_{x_3}^2+x_4^2+a_{x_4}^2)}][e^{-2a_{x_3}x_3} e^{-2a_{x_4}x_4} - e^{2a_{x_3}x_3} e^{2a_{x_4}x_4}]}{e^{-(x_3^2+a_{x_3}^2+x_4^2+a_{x_4}^2)}[e^{-2a_{x_3}x_3}(e^{-2a_{x_4}x_4} + e^{2a_{x_4}x_4}) + e^{2a_{x_3}x_3}(e^{-2a_{x_4}x_4} + e^{2a_{x_4}x_4})]} \\[2mm]
&= \frac{(a_u)\left[e^{-(2a_{x_3}x_3+2a_{x_4}x_4)} - e^{(2a_{x_3}x_3+2a_{x_4}x_4)}\right]}{(e^{-2a_{x_3}x_3} + e^{2a_{x_3}x_3})(e^{-2a_{x_4}x_4} + e^{2a_{x_4}x_4})} \\[2mm]
&= -\frac{a_u}{2}\left[\frac{e^{-(x_3-a_{x_3})^2} - e^{-(x_3+a_{x_3})^2}}{e^{-(x_3-a_{x_3})^2} + e^{-(x_3+a_{x_3})^2}} + \frac{e^{-(x_4-a_{x_4})^2} - e^{-(x_4+a_{x_4})^2}}{e^{-(x_4-a_{x_4})^2} + e^{-(x_4+a_{x_4})^2}}\right]
\end{aligned}
$$

$$= -\frac{a_u}{2}\left[\frac{e^{2a_{x_3}x_3} - e^{-2a_{x_3}x_3}}{e^{2a_{x_3}x_3} + e^{-2a_{x_3}x_3}} + \frac{e^{2a_{x_4}x_4} - e^{-2a_{x_4}x_4}}{e^{2a_{x_4}x_4} + e^{-2a_{x_4}x_4}}\right]$$

$$= -\frac{a_u}{2}\left[\tanh(2a_{x_3}x_3) + \tanh(2a_{x_4}x_4)\right] \qquad (7.18)$$

The parameters a_{x_3}, a_{x_4}, and a_u can be found out using any optimization technique.

Optimal Parameters

To find out the optimal controller parameters, the following cost function is used:

$$J = \int_0^{t_f} (\mathbf{x}^T(\tau)\mathbf{x}(\tau) + u^2(\tau))d\tau \qquad (7.19)$$

The univariate marginal distribution algorithm is used to find the optimal parameters. The optimal parameters are found off-line as $a_{x_3} = 0.73$, $a_{x_4} = 0.25$, $a_u = 1.4$, yielding the cost function $J = 41.139$.

Rule Base: Fuzzy State Feedback Controller

Assume that u has a form $u = -k_1 x_3 - k_2 x_4$. Putting the expression of u in $\dot{V} \approx x_3 x_4 + x_4 u$,

$$\dot{V} \approx x_3 x_4 + x_4(-k_1 x_3 - k_2 x_4) = (1 - k_1)x_3 x_4 - k_2 x_4^2 \qquad (7.20)$$

To make $\dot{V}$ negative, the following rule base is formed

- IF x_3 is negative AND x_4 is negative, THEN u is $u = -k_{1_{nn}}x_3 - k_{2_{nn}}x_4$ where $k_{1_{nn}} \geq 1, k_{2_{nn}} > 0$
- IF x_3 is positive AND x_4 is positive, THEN u is $u = -k_{1_{pp}}x_3 - k_{2_{pp}}x_4$ where $k_{1_{pp}} \geq 1, k_{2_{pp}} > 0$
- IF x_3 is negative AND x_4 is positive, THEN u is $u = -k_{1_{np}}x_3 - k_{2_{np}}x_4$ where $k_{1_{np}} \leq 1, k_{2_{np}} > 0$
- IF x_3 is positive AND x_4 is negative, THEN u is $u = -k_{1_{pn}}x_3 - k_{2_{pn}}x_4$ where $k_{1_{pn}} \leq 1, k_{2_{pn}} > 0$

Inference Mechanism and Defuzzification

Considering the same membership functions and using 'product for AND' inference and 'centre of gravity' method for defuzzification the fuzzy controller is given as

$$u = \frac{\mu_n(x_3)\mu_n(x_4)(-k_{1_{nn}}x_3 - k_{2_{nn}}x_4) + \mu_p(x_3)\mu_p(x_4)(-k_{1_{pp}}x_3 - k_{2_{pp}}x_4)}{\mu_n(x_3)\mu_n(x_4) + \mu_p(x_3)\mu_p(x_4) + \mu_n(x_3)\mu_p(x_4) + \mu_p(x_3)\mu_n(x_4)}$$

$$+ \frac{\mu_n(x_3)\mu_p(x_4)(-k_{1_{np}}x_3 - k_{2_{np}}x_4) + \mu_p(x_3)\mu_n(x_4)(-k_{1_{pn}}x_3 - k_{2_{pn}}x_4)}{\mu_n(x_3)\mu_n(x_4) + \mu_p(x_3)\mu_p(x_4) + \mu_n(x_3)\mu_p(x_4) + \mu_p(x_3)\mu_n(x_4)}$$

Since $\mu_p(x) = e^{-(x-a_x)^2}$ and $\mu_n(x) = e^{-(x+a_x)^2}$, we can write

$$u = \frac{e^{-(x_3+a_{x_3})^2}e^{-(x_4+a_{x_4})^2}(-k_{1_{nn}}x_3 - k_{2_{nn}}x_4) + e^{-(x_3-a_{x_3})^2}e^{-(x_4-a_{x_4})^2}(-k_{1_{pp}}x_3 - k_{2_{pp}}x_4)}{e^{-(x_3+a_{x_3})^2}(e^{-(x_4+a_{x_4})^2} + e^{-(x_4-a_{x_4})^2}) + e^{-(x_3-a_{x_3})^2}(e^{-(x_4+a_{x_4})^2} + e^{-(x_4-a_{x_4})^2})}$$

$$+ \frac{e^{-(x_3+a_{x_3})^2}e^{-(x_4-a_{x_4})^2}(-k_{1_{np}}x_3 - k_{2_{np}}x_4) + e^{-(x_3-a_{x_3})^2}e^{-(x_4+a_{x_4})^2}(-k_{1_{pn}}x_3 - k_{2_{pn}}x_4)}{e^{-(x_3+a_{x_3})^2}(e^{-(x_4+a_{x_4})^2} + e^{-(x_4-a_{x_4})^2}) + e^{-(x_3-a_{x_3})^2}(e^{-(x_4+a_{x_4})^2} + e^{-(x_4-a_{x_4})^2})}$$

Fuzzy Control Law

After simplification, the fuzzy control law becomes

$$u = \frac{(-k_{1_{nn}}x_3 - k_{2_{nn}}x_4)[e^{-(2a_{x_3}x_3 + 2a_{x_4}x_4)}] + (-k_{1_{pp}}x_3 - k_{2_{pp}}x_4)[e^{(2a_{x_3}x_3 + 2a_{x_4}x_4)}]}{(e^{-2a_{x_3}x_3} + e^{2a_{x_3}x_3})(e^{-2a_{x_4}x_4} + e^{2a_{x_4}x_4})}$$
$$+ \frac{(-k_{1_{np}}x_3 - k_{2_{np}}x_4)[e^{(-2a_{x_3}x_3 + 2a_{x_4}x_4)}] + (-k_{1_{pn}}x_3 - k_{2_{pn}}x_4)[e^{(2a_{x_3}x_3 - 2a_{x_4}x_4)}]}{(e^{-2a_{x_3}x_3} + e^{2a_{x_3}x_3})(e^{-2a_{x_4}x_4} + e^{2a_{x_4}x_4})}$$

(7.21)

The optimal parameters are found as

$$a_{x_3} = 0.25 \quad a_{x_4} = 0.12$$
$$k_{1_{nn}} = 1.2 \quad k_{2_{nn}} = 0.35$$
$$k_{1_{pp}} = 1.4 \quad k_{2_{pp}} = 0.2$$
$$k_{1_{np}} = 0.6 \quad k_{2_{np}} = 0.4$$
$$k_{1_{pn}} = 0.8 \quad k_{2_{pn}} = 0.25$$

The system (7.14) is simulated for the fuzzy PD and state feedback controllers (7.18) and (7.21), and the results are shown in Figure 8.7. It is clear from this figure that $u \approx x_4 = \dot{x}_3$, which was our assumption at the beginning.

Tracking Controller

The control objective is to find a u such that the actuator angle will follow any given desired trajectory.

Let $V = \frac{1}{2}(e_1^2 + e_2^2 + e_3^2 + e_4^2)$ be the Lyapunov function candidate, where $e_i = x_i^d - x_i$, x_i is ith state and x_i^d is the corresponding desired response. Since $\dot{e}_i = \dot{x}_i^d - \dot{x}_i = x_{i+1}^d - x_{i+1} = e_{i+1}$ for $i = 1, 3$, we can write

$$\dot{V} = e_1 e_2 + e_2 \dot{e}_2 + e_3 e_4 + e_4 \dot{e}_4$$

(7.22)

Again $\dot{x}_2 \approx -x_1$ and $\dot{x}_4 \approx u$. Thus one can simplify the above expression to

$$\dot{V} \approx e_3 e_4 + e_4 v$$

(7.23)

where $v = \dot{x}_4^d - u$.

- Regulation: $\dot{V} \approx x_3 x_4 + x_4 u$
- Tracking: $\dot{V} \approx e_3 e_4 + e_4 v$

Thus v can be designed in the same way as u in the regulation problem.

- Tracking controller–PD: $u = \dot{x}_4^d - v$ where
$$v = \frac{(a_v)\left[e^{-(2a_{e_3}e_3 + 2a_{e_4}e_4)} - e^{(2a_{e_3}e_3 + 2a_{e_4}e_4)}\right]}{(e^{-2a_{e_3}e_3} + e^{2a_{e_3}e_3})(e^{-2a_{e_4}e_4} + e^{2a_{e_4}e_4})}$$
- Tracking controller–state feedback: $u = \dot{x}_4^d - v$ where
$$v = \frac{(-k_{1_{nn}}e_3 - k_{2_{nn}}e_4)\left[e^{-(2a_{e_3}e_3 + 2a_{e_4}e_4)}\right] + (-k_{1_{pp}}e_3 - k_{2_{pp}}e_4)\left[e^{(2a_{e_3}e_3 + 2a_{e_4}e_4)}\right]}{(e^{-2a_{e_3}e_3} + e^{2a_{e_3}e_3})(e^{-2a_{e_4}e_4} + e^{2a_{e_4}e_4})}$$
$$+ \frac{(-k_{1_{np}}e_3 - k_{2_{np}}e_4)\left[e^{(-2a_{e_3}e_3 + 2a_{e_4}e_4)}\right] + (-k_{1_{pn}}e_3 - k_{2_{pn}}e_4)\left[e^{(2a_{e_3}e_3 - 2a_{e_4}e_4)}\right]}{(e^{-2a_{e_3}e_3} + e^{2a_{e_3}e_3})(e^{-2a_{e_4}e_4} + e^{2a_{e_4}e_4})}$$

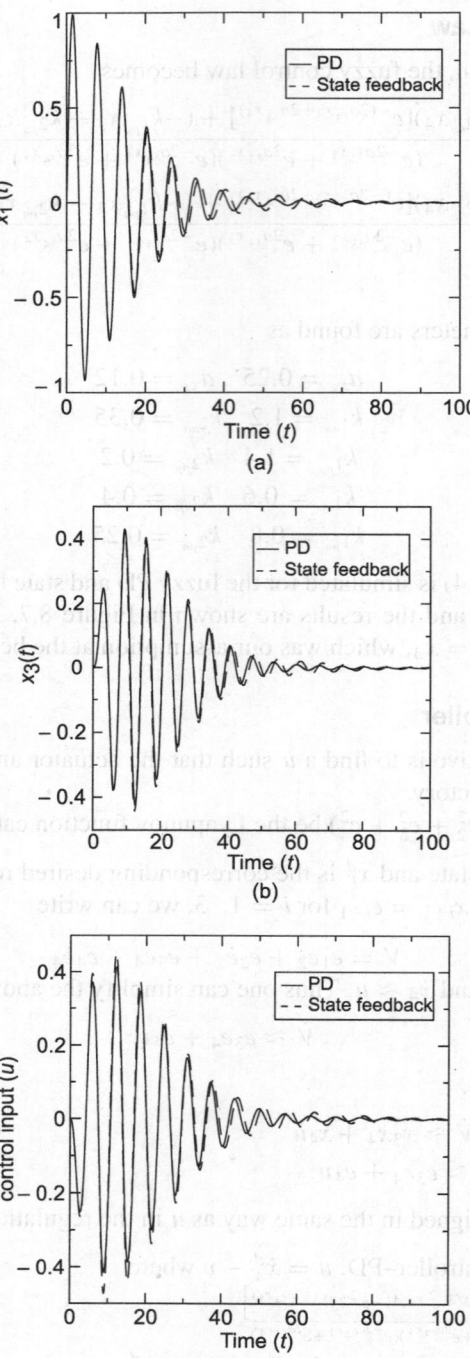

Figure 7.17 RTAC–fuzzy controllers: (a) x_1 converges to 0 with time, (b) the convergence of x_3, and (c) the corresponding control input u

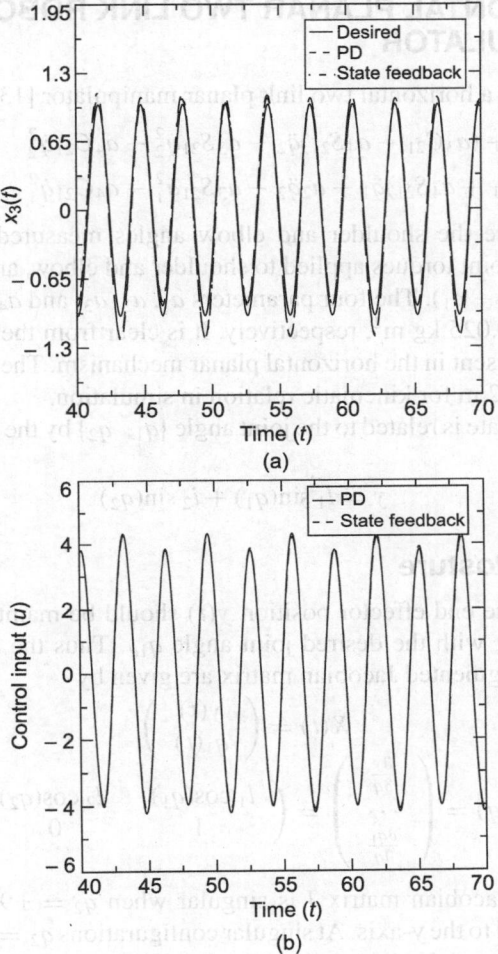

Figure 7.18 RTAC tracking results: (a) x_3 plotted against time and (b) the corresponding control input

The optimal parameters for both PD and state feedback type tracking controllers are found as:

- PD controller:
  $a_v = 1.5 \quad a_{e_3} = 1.23 \quad a_{e_4} = 0.55$
- State feedback controller: $a_{e_3} = 0.37 \quad a_{e_4} = 0.26$

  $k_{1_{nn}} = 1.3 \quad k_{2_{nn}} = 0.6 \quad k_{1_{pp}} = 1.1 \quad k_{2_{pp}} = 0.5$

  $k_{1_{np}} = 0.9 \quad k_{2_{np}} = 0.45 \quad k_{1_{pn}} = 0.7 \quad k_{2_{pn}} = 0.55$

The tracking results are shown in Figure 7.18. It is clear from this figure that $u \approx x_4 = \dot{x}_3$. The RMS for the fuzzy PD Controller is found to be $= 0.06$, whereas for fuzzy state feedback controller it is $= 0.08$.

7.6   HORIZONTAL PLANAR TWO LINK ROBOT MANIPULATOR

The dynamics of a horizontal two link planar manipulator [13] can be written as

$$a_1\ddot{q}_1 + (a_3C_{21} + a_4S_{21})\ddot{q}_2 - a_3S_{21}\dot{q}_2^2 + a_4C_{21}\dot{q}_2^2 = \tau_1$$
$$(a_3C_{21} + a_4S_{21})\ddot{q}_1 + a_2\ddot{q}_2 + a_3S_{21}\dot{q}_1^2 - a_4C_{21}\dot{q}_1^2 = \tau_2 \qquad (7.24)$$

where $\{q_1\ q_2\}$ are the shoulder and elbow angles measured from the x-axis, $\{\tau_1,\ \tau_2\}$ are the joint torques applied to shoulder and elbow, and $C_{21} = \cos(q_2 - q_1)$, $S_{21} = \sin(q_2 - q_1)$. The four parameters a_1, a_2, a_3, and a_4 are taken as 0.15, 0.04, 0.03, and 0.025 kg m^2, respectively. It is clear from the dynamics that the gravity term is absent in the horizontal planar mechanism. The length of the links are taken as 0.432 m for kinematic relation in simulation.

The y-coordinate is related to the joint angle $\{q_1,\ q_2\}$ by the forward kinematic equation,

$$y = l_1 \sin(q_1) + l_2 \sin(q_2) \qquad (7.25)$$

7.6.1   Arm Posture

In arm posture the end effector position $y(t)$ should be maintained at a desired position y_d along with the desired joint angle q_{1d}. Thus the augmented output vector and the augmented Jacobian matrix are given by

$$X(t) = \begin{pmatrix} y(t) \\ q_1(t) \end{pmatrix} \qquad (7.26)$$

$$J(q) = \begin{pmatrix} \frac{\partial y}{\partial q} \\ \cdots \\ \frac{\partial q_1}{\partial q} \end{pmatrix} = \begin{pmatrix} l_1 \cos(q_1) & l_2 \cos(q_2) \\ 1 & 0 \end{pmatrix} \qquad (7.27)$$

The augmented Jacobian matrix J is singular when $q_2 = \pm 90°$, i.e., when the forearm is parallel to the y-axis. At singular configurations $q_2 = \pm 90^o$, the change in q_1 and y must satisfy the relationship $(l_1 \cos(q_1))\Delta q_1 - \Delta y = 0$ and hence q_1 and y cannot change arbitrarily. In our simulation we have taken the initial condition as $q_1 = 45^o$ and $q_2 = 0°$, yielding $y(0) = 0.3054$ m.

7.6.2   Elbow Control

In elbow control, we wish to move the elbow to a final horizontal position (x_e) along a desired trajectory $x_{de}(t)$ where $x_e = l_1 \cos(q_1)$. The augmented output vector and augmented Jacobian matrix in this case are given as

$$X(t) = \begin{pmatrix} y(t) \\ x_e(t) \end{pmatrix} \qquad (7.28)$$

$$J(q) = \begin{pmatrix} \frac{\partial y}{\partial q} \\ \cdots \\ \frac{\partial x_e}{\partial q} \end{pmatrix} = \begin{pmatrix} l_1 \cos(q_1) & l_2 \cos(q_2) \\ -l_1 \sin(q_1) & 0 \end{pmatrix} \qquad (7.29)$$

The augmented Jacobian matrix J is singular, when $q_1 = 0°$ or $180°$, and $q_2 = \pm 90°$, i.e., when the shoulder is along the x-axis and the elbow is parallel to the y-axis. At singular configurations $q_2 = \pm 90°$, the change in x_e and y must satisfy the relationship $\Delta x_e + \tan(q_1)\Delta y = 0$ and at $q_1 = 0°$ or $180°$, we have $\Delta x_e = 0$. Hence x_e and y cannot be changed arbitrarily at singular configuration.

7.6.3 Controller Design

The dynamics of a horizontal plane two link manipulator can be expressed as

$$\begin{bmatrix} \ddot{q}_1 \\ \ddot{q}_2 \end{bmatrix} = \frac{1}{D} \begin{bmatrix} a_2 & -a_3 C_{21} - a_4 S_{21} \\ -a_3 C_{21} - a_4 S_{21} & a_1 \end{bmatrix} \begin{bmatrix} \tau_1 - (a_4 C_{21} - a_3 S_{21})\dot{q}_2^2 \\ \tau_2 + (a_4 C_{21} - a_3 S_{21})\dot{q}_1^2 \end{bmatrix}$$

$$(7.30)$$

where $D = a_1 a_2 - (a_3 C_{21} + a_4 S_{21})^2$. The following partial information about plant dynamics will be used for deriving the fuzzy rule base.

1. $\ddot{q}_1$ is proportional to both shoulder and elbow torques τ_1, τ_2. The shoulder torque accelerates the q_1, while the elbow torque retards q_1.

2. $\ddot{q}_2$ is proportional to both shoulder and elbow torques τ_1, τ_2. The elbow torque accelerates the q_2, while the shoulder torque retards q_2.

3. The joint angles are more influenced by their own actuators, i.e., τ_1 influences the dynamics of q_1 and τ_2 influences q_2.

The first two assumptions are made by neglecting the Coriolis term in dynamics and considering the sign of the torque coefficients. The third assumption comes from the values of the manipulator parameters as $a_1 > a_2 > a_3 > a_4$.

Let us consider the states of the system as $x(t) = [x_1 \ x_2 \ x_3 \ x_4]^T = [q_1 \ \dot{q}_1 \ q_2 \ \dot{q}_2]^T$. Consider the quadratic Lyapunov candidate V as

$$V = \frac{1}{2} x^T(t) x(t) \tag{7.31}$$

The time derivative of the Lyapunov candidate is

$$\dot{V} = x^T(t)\dot{x}(t) = x_1\dot{x}_1 + \dot{x}_1\ddot{x}_1 + x_3\dot{x}_3 + \dot{x}_3\ddot{x}_3$$

$$= \dot{x}_1\{x_1 + \ddot{x}_1\} + \dot{x}_3\{x_3 + \ddot{x}_3\} \tag{7.32}$$

Now, V will be a valid Lyapunov function if $\dot{V} < 0$. This can by ensured by giving the input torques such that $\dot{x}_1$ and $\{x_1 + \ddot{x}_1\}$ are of opposite signs and also $\dot{x}_3$ and $\{x_3 + \ddot{x}_3\}$ are be of opposite signs. The rule base can be easily derived with the approximate derivative of the Lyapunov candidate:

$$\dot{V} \approx \dot{x}_1\{x_1 + a_2\tau_1 - (a_3 C_{21} + a_4 S_{21})\tau_2\} + \dot{x}_3\{x_3 - (a_3 C_{21} + a_4 S_{21})\tau_1 + a_1\tau_2\}$$

$$(7.33)$$

The entire state space is fuzzified into the fuzzy subspaces positive (P), negative (N), and zero (Z). The output space is fuzzified into seven linguistic variables. The linguistic variables and their corresponding crisp values of input are given in Table 7.8. The ith rule base of the fuzzy controller is of the form

IF x_1 is F_{1_i} and $\dot{x}_1$ is F_{2_i} and x_2 is F_{3_i} and $\dot{x}_2$ is F_{4_i} THEN

τ_1 is M_{1_i} and τ_2 is M_{2_i}

Table 7.8 Crisp values of input fuzzy space

Fuzzy Subspace		Crisp Value
Large Positive	LP	400
Medium Positive	MP	300
Small Positive	SP	200
Zero	Z	0
Small Negative	SN	−200
Medium Negative	MN	−300
Large Negative	LN	−400

where F_{j_i} is the ith fuzzy set of jth state and M_{j_i} is the ith fuzzy set of jth input. As it is not feasible to show the entire rule base construction, two cases are presented here for better understanding.

Consider the case that $\dot{x}_1$ is positive, and $\dot{x}_3$ is positive. Four generic cases will occur based on the sign of x_1 and x_3. Consider that x_1 and x_3 are positive. With zero torques, $\dot{V}$ in (7.33) will be positive definite rendering the system unstable. Torques $\{\tau_1 \ \tau_2\}$ should be selected in such a way that their actuation will be more negative than the products of q_1 and $\dot{q}_2$ and also q_2 and $\dot{q}_2$. Hence the joint inputs are selected as $\tau_1 = \text{LN}$ and $\tau_2 = \text{LN}$. If x_1 and x_3 are negative, then $\dot{V}$ is negative with zero input. To incorporate the unknown dynamics, the torques are selected as $\tau_1 = \text{SN}$ and $\tau_2 = \text{SN}$.

The constructed rule base is given in Table 7.9. In simulation, the gains k_1, k_2 are taken as 50. The rule base shown in Table 7.9 is constructed for stabilization. The same rule base is applicable for tracking also, where the Lyapunov function would be a quadratic function of error vector $e(t) = x_d(t) - x(t)$ and the input to fuzzification module would be the error vector $e(t)$.

Table 7.9 Horizontal two link manipulator: Rule base using Lyapunov approach

x_1	P	N	P	N	P	N	P	N	P	N	P	N	P
x_2	P	P	P	P	N	N	N	N	P	P	P	P	N
x_3	P	N	N	P	P	N	N	P	P	N	N	P	P
x_4	P	P	P	P	N	N	N	N	N	N	N	N	P
τ_1	LN	SN	MN	SP	SP	LP	SP	LP	LN	SN	LN	MN	MP
τ_2	LN	SN	SP	MN	SP	MP	MP	SN	SP	LP	LP	MP	LN
x_1	N	P	N	P	N	P	N	–	–	–	–	–	
x_2	N	N	N	N	N	N	N	Z	Z	Z	Z	N	
x_3	N	N	P	–	–	–	–	P	N	P	N		
x_4	P	P	P	Z	Z	Z	Z	P	P	N	N	Z	
τ_1	LP	MP	LP	MN	SN	SN	MP	SP	SP	SN	SN		$-k_1x_1$
τ_2	MN	MN	LN	SP	SP	SP	SN	MN	SN	MP	SP		$-k_2x_2$

The desired end effector position is given as

$$y_d(t) = 0.6108 + 0.9162\exp\left(\frac{-t}{0.3}\right) - 1.2216\exp\left(\frac{-t}{0.4}\right) \quad (7.34)$$

For arm posture, the angle q_1 is maintained at $45°$ and for elbow control the desired trajectory of horizontal position is taken as

$$x_{\text{de}}(t) = 0.2160 + 0.0894\exp\left(\frac{-t}{0.1}\right) - 0.1788\exp\left(\frac{-t}{0.2}\right) \quad (7.35)$$

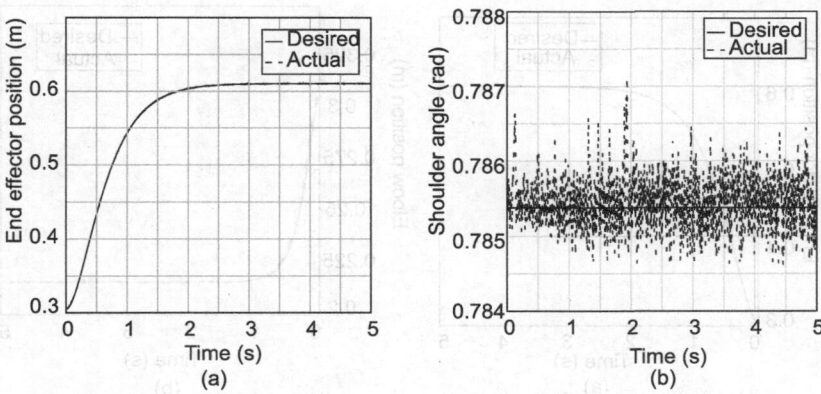

Figure 7.19 Horizontal manipulator response: arm posture–(a) end effector position $y(t)$ and (b) joint angle

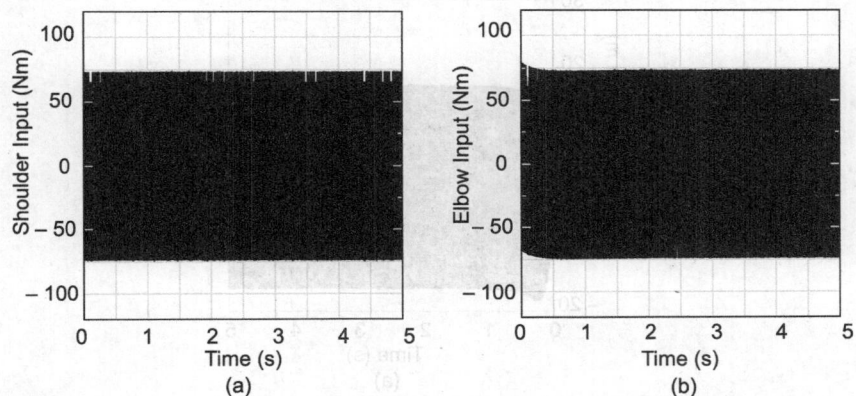

Figure 7.20 Horizontal manipulator response: arm posture–(a) shoulder input and (b) elbow input

Tracking results (trajectory tracking, joint angles, and control inputs) for arm posture control are shown in Figures 7.19 and 7.20 and for elbow control are shown in Figures 7.21 and 7.22. The tracking performance is good, but it is achieved with high control action. The control action is highly fluctuating similar to that of a sliding mode control and bang–bang control. The rule base of the controller is constructed with partial system dynamics, and the fuzzy subspaces are mutually exclusive. The system will be stable only if the control action can cancel the effect of the neglected dynamics. In this context, the Lyapunov fuzzy controller is similar to a sliding mode control law where the control action is discontinuous on the sliding surface due to model uncertainty and results in chattering.

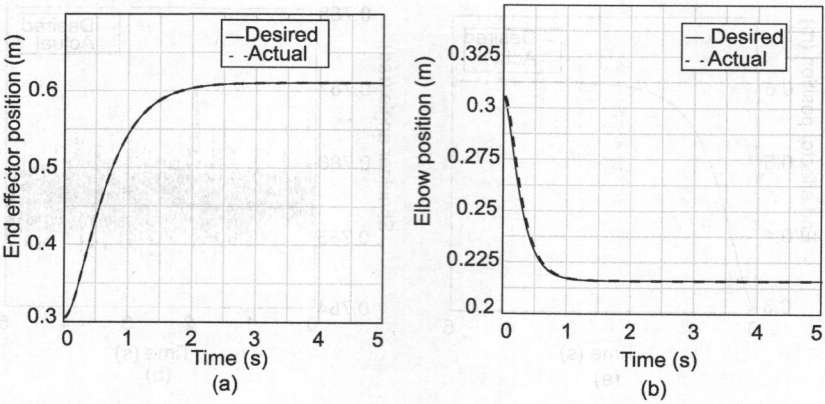

Figure 7.21   Horizontal manipulator response: elbow control–(a) end effector position $y(t)$ and (b) elbow position $x(t)$

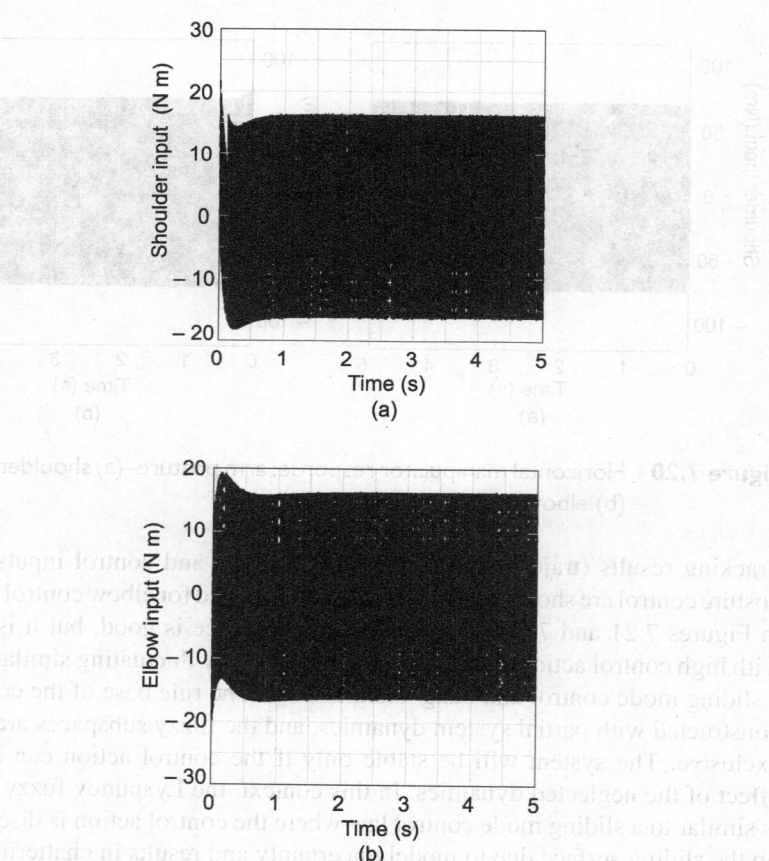

Figure 7.22   Horizontal manipulator response: elbow control–(a) shoulder input and (b) elbow input

SUMMARY

This chapter presents systematic design of fuzzy PD and fuzzy PI controllers. The FLC parameters have been optimized by a genetic algorithm. The design processes have been illustrated through examples such as single link manipulator and DC motor. It is discussed that traditional design of FLC consists of rule formation using heuristics that has little concern for stability. However, rule formation can be achieved while ensuring the stability using the Lyapunov synthesis method. The systematic construction of a fuzzy rule base using the Lyapunov synthesis method described in a tutorial fashion for two different systems: rotational translational proof-mass actuator and two-link manipulator. In a sense, this chapter provides a comprehensive approach to the design of a classical fuzzy logic controller. Those who are interested in recent applications of FLC parameter optimization using genetic algorithm (GA) can refer to works such as [120, 121]. Some interesting industrial applications of fuzzy logic control can also be found in [122, 123, 124].

APPENDIX

Genetic Algorithm

Genetic Algorithm (GA) is an adaptive heuristic search algorithm premised on the evolutionary ideas of natural selection and genetic variation. GA ideas are motivated by Charles Darwin's seminal thesis of 'survival of the fittest'. As such they represent an intelligent exploitation of a random search within a defined search space to solve a problem. GA algorithm consists of four principal concepts:

   (i) Representation
   (ii) Reproduction using Crossover and Mutation
   (iii) Selection
   (iv) Fitness Function

GA algorithm starts with an initial population. Each individual in a population represents a random solution for a problem. Through crossover and mutation, these solutions are given variation. Finally a selection process is adopted based on a fitness function, so that better performing individuals with diversity in solution space are retained for the next generation processing. The flow chart of a typical *genetic algorithm* is shown in Figure 7.23.

The Univariate Marginal Distribution Algorithm

The univariate marginal distribution algorithm (UMDA) estimates the distribution of gene frequencies using mean-field approximation. Each string in the population is represented by a binary vector x. The algorithm generates new points according

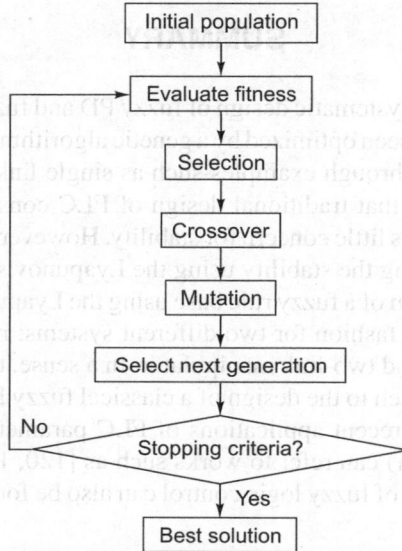

Figure 7.23 Basic flowchart of genetic algorithm

to the following distribution:

$$p(\boldsymbol{x}, t) = \prod_{i=1}^{n} p_i^s(x_i, t) \qquad (7.36)$$

The UMDA algorithm is given as follows:

- Step 1: Set $t = 1$, generate $N(>> 0)$ binary strings randomly.
- Step 2: Select $M < N$ strings according to a selection method.
- Step 3: Compute the marginal frequencies $p_i^s(x_i, t)$ from the selected strings.
- Step 4: Generate new N points according to the distribution $p(\boldsymbol{x}, t) = \prod_{i=1}^{n} p_i^s(x_i, t)$.
- Set $t = t + 1$. If the termination criteria are not met, go to Step 2.

For infinite populations and proportionate selection, it has been shown [125] that average fitness never decreases for maximization problem (increases for minimization problem).

EXERCISES

1. Consider the dynamics of a surge tank as represented by the following differential equation:

$$\frac{dh(t)}{dt} = \frac{-\sqrt{2gh(t)}}{A_r(h(t))} + \frac{1}{A_r(h(t))} u(t)$$

where $u(t)$ is the input flow, which can be positive or negative, $h(t)$ is the liquid level of the tank, and $A_r(h(t)) = \sqrt{ah^2(t) + b}$ is the cross-sectional area of the tank, where $a = 3$ and $b = 1$. Design a fuzzy PD controller for the above system to track a square wave whose magnitude flips between 1 and 4 and time period is 20 s. Optimize the rules and parameters of the controller using a optimization technique. Plot the desired and actual output trajectories of the system.

2. Assuming that the change in level is proportional to the input flow, generate the rule base of a fuzzy logic controller using the Lyapunov analysis to track the same desired trajectory as given in Exercise 1. Optimize the parameters involved in the controller for a cost function of $J = \int_0^{10} (e^2 + 0.01u^2)dt$.

3. Consider the following dynamics of a magnetically levitated system

$$\dot{x} = v$$

$$\dot{v} = g - \frac{k_m k_a^2 V^2}{mx^2}$$

where x is the position and v is the velocity of the system. $g = 9.81$ m/s^2 is the acceleration due to gravity, $k_m = 0.0008$ N $-$ (m/A)^2 is the magnetic constant, $k_a = 0.397$ f is the conductance $m = 0.02$ kg is the mass, and V is the input voltage. Assuming $u = V^2$, design a fuzzy PD controller u such that x converges to 0.

4. Consider the dynamics of an automobile as given by

$$\dot{v}(t) = \frac{1}{m}(-A_\rho v^2(t) - d + f(t))$$

$$\dot{f}(t) = \frac{1}{\tau}(-f(t) + u(t))$$

where u is the control input. $u > 0$ represents a throttle input and $u < 0$ represents a brake input. m is the mass of the vehicle, A_ρ is the aerodynamic drag, d is a constant frictional force, f is the driving or braking force, and τ is the engine/brake time constant. Assume that $u \in [-1000, 1000]$. Take the system parameters as $m = 1300$ kg, $A_\rho = 0.3$ N s^2/m^2, $d = 100$ N, and $\tau = 0.2$ s. Design a fuzzy PI controller for the system, so that the speed of the automobile achieves a desired speed of 60 kmph. Optimize the rules and parameters of the controller, simulate the system response.

5. Construct the rule base of a fuzzy PD controller for the single link manipulator as given in Eqn (7.2b) using the Lyapunov analysis. Optimize the parameters and simulate the system response.

8

Takagi–Sugeno Fuzzy Model Based Control

Takagi and Sugeno [63, 126] proposed a framework to deal with fuzzy systems with the traditional mathematical rigor. In this framework, a non-linear system is represented as the fuzzy average of local linear models which is popularly known as T–S fuzzy model. Given a non-linear system in terms of a T–S fuzzy model, various control schemes can be designed using traditional control techniques such as linear matrix inequality (LMI), robust control, or adaptive control technique. The subject matter discussed in this chapter has a very strong connection with existing works [68, 127, 128].

8.1   T–S FUZZY MODEL

In a T–S fuzzy model, a non-linear system is approximated by a fuzzy cluster of locally valid linear systems. Consider a general non-linear system of the form

$$\dot{x}(t) = f(x(t), u(t)) \tag{8.1}$$

where $x(t) \in R^n$, $u(t) \in R^m$. The T–S model of the above system is expressed in terms of r fuzzy rules where the jth fuzzy rule has the following form:

IF $x_1(t)$ is F_1^j and $\cdots$ and $x_n(t)$ is F_n^j THEN

$$\dot{x}(t) = A_j x(t) + B_j u(t)$$

where F_i^j, $i = 1, 2, \ldots, n$, is the ith fuzzy term of the jth rule corresponding to the state x_i. Let

$$\mu_j = \prod_{i=1}^{n} \mu_j^i(x_i) \tag{8.2}$$

where $\mu_j^i(x_i)$ is the membership function of the fuzzy term F_i^j, $j = 1, 2, \cdots, r$. Given an input–output pair $(x(t), u(t))$, the fuzzy model around this operating point is constructed as the weighted average of the local models and has the form

$$\dot{x}(t) = \frac{\sum_{j=1}^r \mu_j (A_j x(t) + B_j u(t))}{\sum_{j=1}^r \mu_j} \tag{8.3}$$

We should note that the state space of a system at a given operating point may not span the entire rule base consisting of r rules. In general, the number of rules, r_1, which are active at an operating point is less than r. Thus, the T–S model representation (10.25) of the non-linear system (8.1) is more informative in terms of local dynamics. The fuzzy system (10.25) can be rewritten as

$$\dot{x}(t) = \sum_{j=1}^r \sigma_j (A_j x(t) + B_j u(t)) \tag{8.4}$$

where $\sigma_j = \dfrac{\mu_j}{\sum_{j=1}^r \mu_j}$, $\sum_{j=1}^r \sigma_j = 1$. Once a non-linear system is identified in terms of a T–S fuzzy model, the following design techniques can be applied to compute a suitable controller for the system.

8.2 LINEAR MATRIX INEQUALITY TECHNIQUE

In this design technique, first a set of inequalities is formed based on the Lyapunov sufficient condition to ensure stability of the T–S fuzzy model using a parallel distributor fuzzy compensator [128], and the compensator gains are found by solving the set of inequalities.

8.2.1 Common Lyapunov Matrix Criterion for Stability of the T–S Model

The unforced T–S fuzzy system is described by the following equation:

$$\dot{x}(t) = \sum_{j=1}^r \sigma_j A_j \, x(t) \tag{8.5}$$

The equilibrium point of the T–S fuzzy system described by (8.5) is asymptotically stable if there exists a common positive definite matrix P such that

$$A_j^T P + P A_j < 0 \tag{8.6}$$

for $j = 1, 2, \ldots, r$. The above inequality gives a sufficient condition for ensuring stability of (8.5), but it is conservative in the sense that it searches for a common Lyapunov function for all subsystems, which may not even exist.

8.2.2 Parallel Distributed Fuzzy Compensator

In parallel distributed compensation (PDC) [128], the control rule in each fuzzy region is designed based on the corresponding linear subsystem of the T–S fuzzy model. The designed fuzzy controller uses the same fuzzy sets of the model. For the fuzzy model (8.4), a fuzzy regulator can be designed as follows:

Regulator rule j:

IF $x_1(t)$ is F_1^j and $x_2(t)$ is $F_2^j \cdots$ and $x_n(t)$ is F_n^j THEN

$$\boldsymbol{u}(t) = -K_j\boldsymbol{x}(t) \quad j = 1, 2, \ldots, r \tag{8.7}$$

The overall fuzzy regulator is represented by

$$\boldsymbol{u}(t) = -\sum_{j=1}^{r} \sigma_j K_j \boldsymbol{x}(t) \tag{8.8}$$

The design problem simplifies to designing the local feedback gains K_j while establishing the stability of the overall system using the control input (8.8). Substituting the control law (8.8) in Eqn (8.4), the closed loop system becomes

$$\dot{\boldsymbol{x}}(t) = \sum_{i=1}^{r} \sigma_i(A_i - B_i \sum_{j=1}^{r} \sigma_j K_j \boldsymbol{x}(t))$$

$$= \sum_{i=1}^{r}\sum_{j=1}^{r} \sigma_i\sigma_j(A_i - B_iK_j)\boldsymbol{x}(t) \tag{8.9}$$

The above system will be asymptotically stable if there exists a common P for all the subsystems such that

$$H_{ij}^T P + H_{ij} P < 0 \tag{8.10}$$

where $H_{ij} = A_i - B_iK_j$.

Proof: Let us take the following Lyapunov function candidate for the T–S fuzzy system (8.4).

$$V = \boldsymbol{x}^T(t)P\boldsymbol{x}(t)$$

where P is the common Lyapunov matrix for all linear systems. Taking the time derivative of the Lyapunov function candidate, we get

$$\dot{V} = \dot{\boldsymbol{x}}^T(t)P\boldsymbol{x}(t) + \boldsymbol{x}^T(t)P\dot{\boldsymbol{x}}(t)$$

$$= \sum_{i=1}^{r}\sum_{j=1}^{r} \sigma_i\sigma_j\boldsymbol{x}^T(t)H_{ij}^T P\boldsymbol{x}(t) + \boldsymbol{x}^T(t)P\sum_{i=1}^{r}\sum_{j=1}^{r} \sigma_i\sigma_j H_{ij}\boldsymbol{x}(t)$$

$$= \sum_{i=1}^{r}\sum_{j=1}^{r} \sigma_i\sigma_j\boldsymbol{x}^T(t)(H_{ij}^T P + P H_{ij})\boldsymbol{x}(t)$$

If $H_{ij}^T P + H_{ij} P < 0$ for all i, j, $\dot{V}$ will be negative definite, thus the system (8.4) will be asymptotically stable. Hence the proof. Although the controller design becomes simpler in terms of linear subsystems, inequality (8.10) gives a constraint that the linear controllers should stabilize not only the corresponding subsystems but also the other subsystems. Tanaka, *et. al.* [128] relaxed the above stability condition for the closed loop T–S fuzzy system and, based on the relaxed stability condition, proposed a controller design scheme where the controller gains K_js are computed using the LMI technique. Using the relaxed stability condition, the following problem can be formulated to determine the gains K_js.

Find $X > 0$, $Y \geq 0$ and M_j ($j = 1, \ldots, r$) satisfying

$$-XA_j^T - A_j X + M_j^T B_j^T + B_j M_j - (s - 1)Y > 0 \tag{8.11}$$

$$2Y - XA_j^T - A_j X - XA_i^T - A_i X + M_j^T B_j^T + B_j M_j +$$
$$M_i^T B_i^T + B_i M_i \geq 0, \quad \text{for } j < i \tag{8.12}$$

where $X = P^{-1}$, $M_j = K_j X$, and $Y = XQX$. The above conditions are LMIs with respect to the variable X. We can find a positive definite matrix X, a positive semidefinite matrix Y, and M_j satisfying the LMIs or determine that no such solutions exist. This is a convex feasibility problem for which one of the most popular techniques is interior-point methods [128].

Example 8.1 Consider the dynamics of a single-link manipulator as

$$\dot{x}_1 = x_2$$

$$\dot{x}_2 = -g \, \sin(x_1) + u \tag{8.13}$$

Find out the continuous time T–S fuzzy model of the above system. Design a parallel distributed fuzzy compensator, as described by Eqn (8.8) by solving Eqns (8.11) and (8.12). Simulate the system response.

Solution
To find out the T–S fuzzy model of the system, manipulator angle x_1 is fuzzified in the operating region $[-\pi/2, \pi/2]$, and x_2 is always considered as 0. The system is approximated with four fuzzy regions in positive half with centres 0, $\frac{\pi}{6}$, $\frac{\pi}{3}$, and $\frac{\pi}{2}$. The Gaussian function is chosen as membership function. Around the equilibrium point $(0, 0)$, the system is linearized using the standard Taylor series expansion. At other operating points, linear models are obtained by the technique described in Chapter 1. Similar approximation is done in the negative half of the operating region. Thus, the T–S fuzzy model of the system (8.64) is described by the following four rules:

Rule 1: If $x(t)$ is around $[0 \; 0]^T$

Then $\dot{x}(t) = \begin{bmatrix} 0 & 1 \\ -9.81 & 0 \end{bmatrix} x(t) + \begin{bmatrix} 0 \\ 1 \end{bmatrix} u(t)$

Rule 2: If $x(t)$ is around $[\pm\frac{\pi}{6} \; 0]^T$

Then $\dot{x}(t) = \begin{bmatrix} 0 & 1 \\ -9.37 & 0 \end{bmatrix} x(t) + \begin{bmatrix} 0 \\ 1 \end{bmatrix} u(t)$

Rule 3: If $x(t)$ is around $[\pm\frac{\pi}{3} \; 0]^T$

Then $\dot{x}(t) = \begin{bmatrix} 0 & 1 \\ -8.11 & 0 \end{bmatrix} x(t) + \begin{bmatrix} 0 \\ 1 \end{bmatrix} u(t)$

Rule 4: If $x(t)$ is around $[\pm\frac{\pi}{2} \; 0]^T$

Then $\dot{x}(t) = \begin{bmatrix} 0 & 1 \\ -6.24 & 0 \end{bmatrix} x(t) + \begin{bmatrix} 0 \\ 1 \end{bmatrix} u(t)$

where $x(t) = [x_1(t) \; x_2(t)]^T$. The MATLAB code to find the above rules is given as follows:

```
function out = LinSlm(x);
g=9.81;
f = [x(2);-g*sin(x(1))];
gradf = [0 -g*cos(x(1));1 0];
if(norm(x)==0)
A=[0 1;-g 0];
else
for i=1:2
A(i,:)=(gradf(:,i)+((f(i)-x'*gradf(:,i))/(norm(x)*
norm(x)))*x)';
end
end
A
B=[0;1]
```

Once the T–S fuzzy model of the system is found, a fuzzy regulator of the form (8.8) can be designed using the LMI technique. We have used the MATLAB toolbox to solve the feasibility problem. After solving the LMIs (8.11) and (8.12), the controller gains for the four fuzzy rules are found as

Rule 1: If $x(t)$ is around $[0\ 0]^T$ *Rule* 2: If $x(t)$ is around $[\pm\frac{\pi}{6}\ 0]^T$
Then $K_1 = [-8.6228\ 0.6422]$ Then $K_2 = [-8.1828\ 0.6422]$
Rule 3: If $x(t)$ is around $[\pm\frac{\pi}{3}\ 0]^T$ *Rule* 4: If $x(t)$ is around $[\pm\frac{\pi}{2}\ 0]^T$
Then $K_3 = [-6.9228\ 0.6422]$ Then $K_4 = [-5.0578\ 0.6422]$

with a common Lyapunov matrix $P = \begin{bmatrix} 1.6201 & 0.4722 \\ 0.4722 & 1.6201 \end{bmatrix}$. The corresponding MATLAB code is given as follows:

```
clear all
A(:,:,1)=[0 1;-9.81 0];   B(:,:,1)=[0;1];
A(:,:,2)=[0 1;-9.37 0];   B(:,:,2)=[0;1];
A(:,:,3)=[0 1;-8.11 0];   B(:,:,3)=[0;1];
A(:,:,4)=[0 1;-7.42 0];   B(:,:,4)=[0;1];
s=2; setlmis([]); X=lmivar(1,[2 1]); Y=lmivar(1,[2 1]);
for i=1:4  M(:,:,i)=lmivar(2,[1,2]);  end
for i=1:4
T1=newlmi;
lmiterm([T1 1 1 X],1,A(:,:,i)');  lmiterm([T1 1 1
Y],s-1,1);
lmiterm([T1 1 1 X],A(:,:,i),1);   lmiterm([T1 1 1
M(:,:,i)],B(:,:,i),-1);
lmiterm([T1 1 1 -M(:,:,i)],-1,B(:,:,i)');
if(i>1)  for j=1:i-1
T1=newlmi;
lmiterm([T1 1 1 X],1,A(:,:,i)');   lmiterm([T1 1 1
Y],-2,1);
lmiterm([T1 1 1 X],A(:,:,i),1);   lmiterm([T1 1 1
X],1,A(:,:,j)');
```

```
lmiterm([T1 1 1 X],A(:,:,j),1);    lmiterm([T1 1 1
-M(:,:,j)],-1,B(:,:,i)');
lmiterm([T1 1 1 -M(:,:,i)],-1,B(:,:,j)');    lmiterm([T1
1 1 M(:,:,i)],B(:,:,j),-1);
lmiterm([T1 1 1 M(:,:,j)],B(:,:,i),-1);
end
end
end
T1=newlmi;    lmiterm([T1 1 1 X],-1,1);
T1=newlmi;    lmiterm([T1 1 1 Y],-1,1);
lmisys=getlmis;    [tmin,xfeas]=feasp(lmisys);
Xf = dec2mat(lmisys,xfeas,X)
Yf = dec2mat(lmisys,xfeas,Y)
for i=1:4    Mf(:,:,i)=dec2mat(lmisys,xfeas,M(:,:,i))    end
```

Simulation results are shown in Figure 8.1 where (a) shows how the system states converge to the equilibrium (0, 0) and (b) shows the corresponding control input.

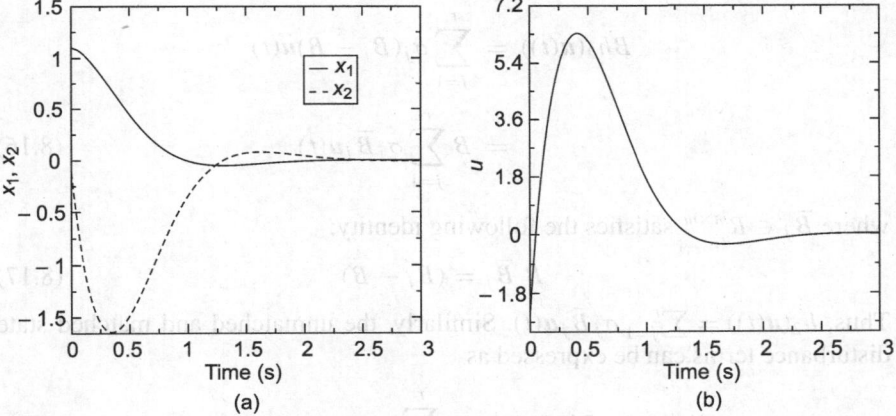

Figure 8.1 (a) System states and (b) control input

8.3 FIXED GAIN STATE FEEDBACK CONTROLLER DESIGN TECHNIQUE

In 1999, Stanislaw Zak [68] proposed a fixed gain state feedback controller for the global T–S fuzzy system using the concept of robust control technique. The T–S fuzzy model is rewritten in terms of a single nominal plant and rest of the plants are expressed as a disturbance to the nominal plant. The controller is designed such that the overall system becomes stable in the presence of the disturbance terms. However, the implementation of the controller is constrained by the norm bound of the disturbance term.

The T–S fuzzy model (8.4) can be rewritten as

$$\dot{x}(t) = Ax(t) + Bu(t) + \sum_{j=1}^{r} \sigma_j(A_j - A)x(t) + \sum_{j=1}^{r} \sigma_j(B_j - B)u(t)$$

$$= Ax(t) + Bu(t) + F(x(t), u(t)) \tag{8.14}$$

The above equation is a representation of the fuzzy dynamics (8.4) in terms of a linear nominal plant $\dot{x}(t) = Ax(t) + Bu(t)$ and a non-linear disturbance term $F(x(t), u(t))$. The matrices A and B can be selected from the set $\{A_j\}$ and $\{B_j\}$, respectively. Further the disturbance $F(x(t), u(t))$ can be expressed as

$$F(x(t), u(t)) = f(x(t)) + Bh_1(x(t)) + Bh_2(u(t)) \tag{8.15}$$

We should note that $h_1(x(t))$ and $h_2(u(t))$ in (8.15) affect the system dynamics via matrix B in the same way as the control input $u(t)$ does. In other words, these uncertainties *match the input*. The unmatched part of the disturbance is modelled as $f(x(t))$. The matched input disturbance term in (8.15) can be computed from Eqn (8.14) as

$$Bh_2(u(t)) = \sum_{j=1}^{r} \sigma_j(B_j - B)u(t)$$

$$= B \sum_{j=1}^{r} \sigma_j \overline{B}_j u(t) \tag{8.16}$$

where $\overline{B}_j \in R^{m \times m}$ satisfies the following identity:

$$B\,\overline{B}_j = (B_j - B) \tag{8.17}$$

Thus, $h_2(u(t)) = \sum_{j=1}^{r} \sigma_j \overline{B}_j u(t)$. Similarly, the unmatched and matched state disturbance terms can be expressed as

$$f(x(t)) + Bh_1(x(t)) = \sum_{j=1}^{r} \sigma_j(A_j - A)x(t)$$

$$= \sum_{j=1}^{r} \sigma_j(A_{1_j} + BA_{2_j})x(t) \tag{8.18}$$

where $(A_j - A) = (A_{1_j} + BA_{2_j})$. Here $A_{1_j} \in R^{n \times n}$ represents the unmatched term and $A_{2_j} \in R^{m \times n}$ represents the matched term. Thus, $f(x(t)) = \sum_{j=1}^{r} \sigma_j A_{1_j} x(t)$ and $h_1(x(t)) = \sum_{j=1}^{r} \sigma_j A_{2_j} x(t)$. In general, B matrix is of rank m which is less than n and the rank of the column space of B matrix is m. Hence, the n-dimensional columns of $(A_j - A)$ will not be spanned by the column space of B. The components of $(A_j - A)$ which are in the column space of B form matrix A_{2_j} and the remaining components are expressed as unmatched term A_{1_j}.

The control problem is to design a state feedback controller using the robust control theory, such that the nominal plant $\dot{x}(t) = Ax(t) + Bu(t)$ becomes stable in the presence of the disturbance term $F(x(t), u(t))$. Before proceeding we first need to compute the norm bounds of the disturbance terms.

The norm bound of the matched state disturbance is given by

$$\| h_1(\boldsymbol{x}(t)) \| = \| \sum_{j=1}^{r} \sigma_j A_{2_j} \boldsymbol{x}(t) \|$$

$$\leq \| \sum_{j=1}^{r} \sigma_j A_{2_j} \| \| \boldsymbol{x}(t) \|$$

$$\leq \sum_{j=1}^{r} \sigma_j \| A_{2_j} \| \| \boldsymbol{x}(t) \|$$

$$\leq \alpha_{hx} \| \boldsymbol{x}(t) \| \tag{8.19}$$

where $\alpha_{hx_j} = \max_j \| A_{2_j} \|$. Similarly, the norm bound of the matched input disturbance is

$$\| h_2(\boldsymbol{u}(t)) \| = \| \sum_{j=1}^{r} \sigma_j \bar{B}_j \boldsymbol{u}(t) \|$$

$$\leq \left(\sum_{j=1}^{r} \sigma_j \| \bar{B}_j \| \right) \| \boldsymbol{u}(t) \|$$

$$= \alpha_u \| \boldsymbol{u}(t) \| \tag{8.20}$$

where $\alpha_u = \max_j \| \bar{B}_j \|$. The norm bound of the unmatched state disturbance is

$$\| f(\boldsymbol{x}(t)) \| = \| \sum_{j=1}^{r} \sigma_j A_{1_j} \boldsymbol{x}(t) \|$$

$$\leq \left(\sum_{j=1}^{r} \sigma_j \| A_{1_j} \| \right) \| \boldsymbol{x}(t) \|$$

$$\leq \max_j \| A_{1_j} \| \| \boldsymbol{x}(t) \|$$

$$= \alpha_f \| \boldsymbol{x}(t) \| \tag{8.21}$$

where $\alpha_f = \max_j \| A_{1_j} \|$.

8.3.1 Fixed Gain State Feedback Controller

The nominal plant (A, B) is so selected that A is asymptotically stable. If A is not asymptotically stable, a preliminary state feedback can be used to stabilize it before designing the controller. In such a case, the disturbance term will include the preliminary feedback as well. The overall control law in such case is a summation of both preliminary feedback and fixed gain controller. In further discussion, we will use the term $\lambda_{\min}(.)$ and $\lambda_{\max}(.)$ to denote the minimum and maximum eigenvalues of a matrix.

Theorem 8.1 Suppose that A is asymptotically stable and that P is a positive definite matrix satisfying $A^T P + P A = -2Q$ for some symmetric positive definite Q. Suppose also that

$$\alpha_f < \frac{\lambda_{\min}(Q)}{\lambda_{\max}(P)} \tag{8.22a}$$

$$\alpha_u < 1 \tag{8.22b}$$

Then, the state feedback controller

$$u(t) = -\gamma B^T P x(t) \tag{8.23}$$

where

$$\gamma > \frac{\alpha_{hx}^2}{4(1 - \alpha_u)(\lambda_{\min}(Q) - \alpha_f \lambda_{\max}(P))} \tag{8.24}$$

asymptotically stabilizes the fuzzy model (8.14).

Proof: The time derivative of the Lyapunov candidate $V = x^T P x$ evaluated along a state trajectory is

$$\dot{V} = 2x^T P \dot{x}$$

$$= -2x^T Q x - 2(\sum_{k=1}^{r} \sigma_k \gamma_k) x^T P B B^T P x$$

$$+2x^T P f(x) + 2x^T P B h_1(x) + 2x^T P B h_2(u)$$

To proceed further, we will use the following properties of matrices:

$$\lambda_{\min}(Q) \|x\|^2 \leq x^T Q x \leq \lambda_{\max}(Q) \|x\|^2 \tag{8.25}$$

and therefore,

$$-x^T Q x \leq -\lambda_{\min}(Q) \|x\|^2 \tag{8.26}$$

For a symmetric positive definite matrix P, its induced 2-norm is

$$\|P\| = \lambda_{\max}(P) \tag{8.27}$$

Furthermore,

$$x^T P B B^T P x = x^T P B (x^T P B)^T = \|x^T P B\|^2 \tag{8.28}$$

Taking all the above relations into account, and using norm bounds on uncertain elements, we get

$$\dot{V} \leq -2\lambda_{\min}(Q) \|x\|^2 - 2\gamma \|x^T P B\|^2 + 2\alpha_f \lambda_{\max}(P) \|x\|^2$$

$$+2\alpha_{hx} \|x^T P B\| \|x\| + 2\alpha_u \gamma \|x^T P B\|^2$$

$$= -2\bar{x}^T \bar{Q} \bar{x}$$

where $\bar{x} = \begin{bmatrix} \|x\| & \|x^T P B\| \end{bmatrix}^T$ and $\bar{Q} = \begin{bmatrix} q_{11} & q_{12} \\ q_{21} & q_{22} \end{bmatrix}$

Here, $q_{11} = \lambda_{\min}(Q) - \alpha_f \lambda_{\max}(P)$

$$q_{12} = q_{21} = -\frac{1}{2}\alpha_{hx}$$

$$q_{22} = \gamma(1 - \alpha_u)$$

We can find the positive definiteness of $\overline{Q}$ using the Sylvester criterion i.e., all principal minors should be positive. Therefore, we have

$$\lambda_{\min}(Q) - \alpha_f \lambda_{\max}(P) > 0 \tag{8.29a}$$

$$| \overline{Q} | > 0 \tag{8.29b}$$

Condition (8.29a) gives constraint (8.22a).

$$| \overline{Q} | = -\frac{1}{4}\alpha_{hx}^2 + \gamma(1 - \alpha_u)(\lambda_{\min}(Q) - \alpha_f \lambda_{\max}(P)) \tag{8.30}$$

Equation (8.29b) is satisfied if

$$\gamma > \frac{\alpha_{hx}^2}{4(1 - \alpha_u)(\lambda_{\min}(Q) - \alpha_f \lambda_{\max}(P))} \tag{8.31}$$

Hence the proof.

8.4 VARIABLE GAIN CONTROLLER DESIGN USING SINGLE LINEAR NOMINAL PLANT

As described in the previous section, the T–S fuzzy model (8.4) can be rewritten as a single plant with a disturbance term as given in Eqn (8.14). The nominal is given as $\dot{x}(t) = Ax(t) + Bu(t)$, and the non-linear disturbance term $F(x(t), u(t))$ is given in Eqn (8.15).

8.4.1 The Control Problem

In the T–S fuzzy model (8.4), the non-linear system (8.1) is approximated as a fuzzy cluster of r linear subsystems. Since each subsystem is linear, classical linear control theory can be applied to design either PID or state feedback type of fixed gain controller for each subsystem. As the desired system output traverses a specific trajectory, system states traverse across different fuzzy zones. It is thus expected that a PID or a state feedback controller will be characterized by variable gains instead of fixed gains. Thus the control problem is defined as follows:

Given a T–S fuzzy model (8.4) represented as a linear nominal plant with non-linear disturbance term as given in (8.14), design a variable gain state feedback controller $u(t) = -Kx(t)$ such that the T–S fuzzy model is Lyapunov stable. Here the matrix K represents variable state feedback gain.

Since in this case, the disturbance term is considered to be variable we need to recalculate the norm bounds. The norm bound of matched state disturbance is given by

$$\| h_1(x(t)) \| = \| \sum_{j=1}^{r} \sigma_j A_{2j} x(t) \|$$

$$\leq \sum_{j=1}^{r} \sigma_j \| A_{2j} \| \| x(t) \|$$

$$= \sum_{j=1}^{r} \sigma_j \alpha_{hx_j} \|x(t)\| \tag{8.32}$$

where $\alpha_{hx_j} = \|A_{2_j}\|$. Similarly, the norm bound of the matched input disturbance is

$$\| h_2(\boldsymbol{u}(t)) \| = \left\| \sum_{j=1}^{r} \sigma_j \overline{B}_j \boldsymbol{u}(t) \right\|$$

$$\leq \left(\sum_{j=1}^{r} \sigma_j \|\overline{B}_j\| \right) \| \boldsymbol{u}(t) \|$$

$$= \sum_{j=1}^{r} \sigma_j \alpha_{u_j} \|\boldsymbol{u}(t)\| \qquad (8.33)$$

where $\alpha_{u_j} = \|\overline{B}_j\|$. The norm bound of the unmatched state disturbance remains the same as that of (8.21).

We can infer from (8.32), (8.33), and (8.21) that the disturbance measure at each operating point is governed only by the subsystems which have non-zero membership values, and this measure is continuously varying.

8.4.2 Variable Gain Controller I

The nominal plant (A, B) is so selected that A is asymptotically stable. If A is not asymptotically stable, a preliminary state feedback can be used to stabilize it before designing the controller. In such a case, the disturbance term will include the preliminary feedback as well. The overall control law in such case is a summation of both preliminary feedback and variable gain controller. In further discussion, we will use the term $\lambda_{\min}(.)$ and $\lambda_{\max}(.)$ to denote the minimum and maximum eigenvalues of a matrix.

Theorem 8.2   Suppose that A is asymptotically stable and that P is a positive definite matrix satisfying $A^T P + P A = -2Q$ for some symmetric positive definite Q. Suppose also that

$$\alpha_f < \frac{\lambda_{\min}(Q)}{\lambda_{\max}(P)} \qquad (8.34a)$$

$$\sum_{j=1}^{r} \sigma_j \alpha_{u_j} < 1 \qquad (8.34b)$$

Then, the state feedback controller

$$\boldsymbol{u}(t) = - \left(\sum_{k=1}^{r} \sigma_k \gamma_k \right) B^T P \boldsymbol{x}(t) \qquad (8.35)$$

where

$$\gamma_k > \frac{\alpha_{hx_k} \sum_{j=1}^{r} \sigma_j \alpha_{hx_j}}{4(1 - \sum_{j=1}^{r} \sigma_j \alpha_{u_j})(\lambda_{\min}(Q) - \alpha_f \lambda_{\max}(P))} \qquad (8.36)$$

asymptotically stabilizes the fuzzy model (8.14).

Proof: The time derivative of the Lyapunov candidate $V = x^T P x$ evaluated along a state trajectory is

$$\dot{V} = 2x^T P \dot{x}$$

$$= -2x^T Q x - 2(\sum_{k=1}^{r} \sigma_k \gamma_k) x^T P B B^T P x$$

$$+ 2x^T P f(x) + 2x^T P B h_1(x) + 2x^T P B h_2(u)$$

To proceed further, we will use the following properties of matrices:

$$\lambda_{\min}(Q) \|x\|^2 \le x^T Q x \le \lambda_{\max}(Q) \|x\|^2 \tag{8.37}$$

and therefore

$$-x^T Q x \le -\lambda_{\min}(Q) \|x\|^2 \tag{8.38}$$

For a symmetric positive definite matrix P, its induced 2-norm is

$$\|P\| = \lambda_{\max}(P) \tag{8.39}$$

Furthermore,

$$x^T P B B^T P x = x^T P B (x^T P B)^T = \|x^T P B\|^2 \tag{8.40}$$

Taking all the above relations into account, and using norm bounds on uncertain elements, we get

$$\dot{V} \le -2\lambda_{\min}(Q) \|x\|^2 - 2\sum_{k=1}^{r} \sigma_k \gamma_k \|x^T P B\|^2 + 2\alpha_f \lambda_{\max}(P) \|x\|^2$$

$$+ 2\|x^T P B\| \sum_{j=1}^{r} \sigma_j \alpha_{hx_j} \|x\| + 2\sum_{j=1}^{r} \sigma_j \alpha_{u_j} (\sum_{k=1}^{r} \sigma_k \gamma_k) \|x^T P B\|^2$$

$$= -2\bar{x}^T \bar{Q} \bar{x}$$

where $\bar{x} = \begin{bmatrix} \|x\| & \|x^T P B\| \end{bmatrix}^T$ and $\bar{Q} = \begin{bmatrix} q_{11} & q_{12} \\ q_{21} & q_{22} \end{bmatrix}$

Here, $q_{11} = \lambda_{\min}(Q) - \alpha_f \lambda_{\max}(P)$

$$q_{12} = q_{21} = -\frac{1}{2} \sum_{j=1}^{r} \sigma_j \alpha_{hx_j}$$

$$q_{22} = (1 - \sum_{j=1}^{r} \sigma_j \alpha_{u_j})(\sum_{k=1}^{r} \sigma_k \gamma_k)$$

We can find the positive definiteness of $\bar{Q}$ using the Sylvester criterion, i.e., all principal minors should be positive. Therefore, we have

$$\lambda_{\min}(Q) - \alpha_f \lambda_{\max}(P) > 0 \tag{8.41a}$$

$$|\bar{Q}| > 0 \tag{8.41b}$$

Condition (8.41a) gives constraint (8.34a).

$$| \bar{Q} | = -(\frac{1}{2} \sum_{k=1}^{r} \sigma_k \alpha_{hx_k})(\frac{1}{2} \sum_{j=1}^{r} \sigma_j \alpha_{hx_j}) +$$

$$\left[\sum_{k=1}^{r} \sigma_k \gamma_k (1 - \sum_{j=1}^{r} \sigma_j \alpha_{u_j}) \right] (\lambda_{\min}(Q) - \alpha_f \lambda_{\max}(P)) \quad (8.42)$$

Equation (8.41b) is satisfied if

$$\sum_{k=1}^{r} \sigma_k \gamma_k > \frac{\sum_{k=1}^{r} \sigma_k \alpha_{hx_k} \sum_{j=1}^{r} \sigma_j \alpha_{hx_j}}{4(1 - \sum_{j=1}^{r} \sigma_j \alpha_{u_j})(\lambda_{\min}(Q) - \alpha_f \lambda_{\max}(P))} \quad (8.43)$$

comparing the coefficients on both sides, we can rewrite the constraint as

$$\gamma_k > \frac{\alpha_{hx_k} \sum_{j=1}^{r} \sigma_j \alpha_{hx_j}}{4(1 - \sum_{j=1}^{r} \sigma_j \alpha_{u_j})(\lambda_{\min}(Q) - \alpha_f \lambda_{\max}(P))} \quad \forall k \quad (8.44)$$

The above equation gives the constraint on the controller parameter for the kth subsystem. The terms in the denominator give constraint (8.34). As the controller parameter is a positive one, it results in the constraint (8.34b) from (8.41a). Hence the proof.

Since the derivative of the Lyapunov function V consists of a term $+2x^T P f(x)$, we need to find out the worst case scenario to check the negative definiteness of $\dot{V}$. Thus, the upper bound of $f(x)$ is calculated in terms of $\|x\|$ as given in Eqn (8.21).

The controller parameter γ_k is being computed considering the kth subsystem as disturbance. As discussed earlier, each fuzzy subspace will not span the entire state space. The value of γ_k will depend on the local disturbances. As γ_k is varying while the system is traversing from one operating zone to another operating zone, the controller gain is also varying accordingly. Thus, the variable gain controller is expected to achieve a better tracking accuracy compared to the fixed gain controller because the variation in system parameters along the trajectory is taken into account by a variable state feedback gain.

8.5 VARIABLE GAIN CONTROLLER DESIGN USING EACH LINEAR SUBSYSTEM AS NOMINAL PLANT

The T–S fuzzy model (8.4) is a cluster of r linear subsystems. Thus, model (8.4) may be represented as (8.14) in r different forms by considering each subsystem as the nominal plant. Each representation can be used to compute a control action. The final control action can be computed using the fuzzy blending concept which will be explained later in this section. The advantage of such an approach will become clear subsequently.

Considering the kth subsystem to be the nominal plant, the T–S fuzzy system (8.4) can be written as

$$\dot{x}(t) = \sum_{j=1}^{r} \sigma_j (A_j x(t) + B_j u(t))$$

$$= \sum_{j=1}^{r} \sigma_j A_j x(t) + Bu(t) + \sum_{j=1}^{r} \sigma_j (B_j - B) u(t)$$

$$= A_k x(t) + \sum_{j=1, j \neq k}^{r} \sigma_j (A_j - A_k) x(t) + Bu(t) + \sum_{j=1}^{r} \sigma_j (B_j - B) u(t)$$

$$= A_k x(t) + Bu(t) + F_k(x(t), u(t)) \tag{8.45}$$

where B is the common input matrix for all such representations so that the pair (A_k, B) is controllable $\forall k$. The motivation for introducing a common B matrix for all the subsystems will be explained by the following theorem.

Theorem 8.3 For a class of the T–S fuzzy system with common input matrix in all fuzzy zones, the global control action $u(t)$ given by

$$u(t) = \sum_{j=1}^{r} \sigma_j u_j(t) \tag{8.46}$$

will imply that the jth subsystem is excited by the control action $u_j(t)$ for all j.

Proof: The fuzzy dynamics with a common input matrix can be expressed as

$$\dot{x}(t) = \sum_{j=1}^{r} \sigma_i A_j x(t) + \sum_{j=1}^{r} \sigma_j Bu(t)$$

$$= \sum_{j=1}^{r} \sigma_j A_j x(t) + Bu(t) \tag{8.47}$$

Substituting the control input (8.46) in (8.47), we get

$$\dot{x}(t) = \left(\sum_{j=1}^{r} \sigma_j A_j \right) x(t) + B \sum_{j=1}^{r} \sigma_j u_j(t)$$

$$= \left(\sum_{j=1}^{r} \sigma_j A_j \right) x(t) + \left(\sum_{j=1}^{r} \sigma_j Bu_j(t) \right)$$

$$= \sum_{j=1}^{r} \sigma_j (A_j x(t) + Bu_j(t)) \tag{8.48}$$

It is clear from (8.48) that the jth subsystem is excited by $u_j(t)$ with the control action $u(t)$ defined in (8.46). Hence the proof.

This theorem allows us to compute the overall control action from individual control actions. The procedure to compute the overall control action (8.46) is termed as the *fuzzy blending* of individual control actions. Since individual subsystems are linear, one has freedom to design linear controllers for each fuzzy zone. However, in practice, input matrix B_j is not the same for each fuzzy zone. To accommodate this generic case, a common B matrix is introduced in (8.45), and the deviation in input matrices is expressed as disturbance.

$F_k(x(t), u(t))$ incorporates the fuzzy dynamics as disturbance from the kth fuzzy region

$$F_k(x(t), u(t)) = \sum_{j=1, j \neq k}^{r} \sigma_j(A_j - A_k)x(t) + \sum_{j=1}^{r} \sigma_j(B_j - B)u(t) \qquad (8.49)$$

Similar to Section 8.4, the disturbance term $F_k(x(t), u(t))$ is expressed as

$$F_k(x(t), u(t) = f_k(x(t)) + Bh_{1_k}(x(t)) + Bh_2(u(t)) \qquad (8.50)$$

The input disturbance is expressed as

$$Bh_2(u(t)) = \sum_{j=1}^{r} \sigma_j(B_j - B)u(t)$$

$$= B \sum_{j=1}^{r} \sigma_j \overline{B}_j u(t) \qquad (8.51)$$

where $\overline{B}_j \in R^{m \times m}$ satisfies the identity (8.17). Thus, $h_2(u(t)) = \sum_{j=1}^{r} \sigma_j \overline{B}_j u(t)$. In a similar manner, the unmatched and matched terms in state disturbance are expressed as

$$f_k(x(t)) + Bh_{1_k}(x(t)) = \sum_{j=1}^{r} \sigma_j(A_j - A_k)x(t)$$

$$= \sum_{j=1}^{r} \sigma_j(A_{1k_j} + BA_{2k_j})x(t) \qquad (8.52)$$

where $A_j - A_k = A_{1k_j} + BA_{2k_j}$, $A_{1k_j} \in R^{n \times n}$, and $A_{2k_j} \in R^{m \times n}$. Thus, $f_k(x(t)) = \sum_{j=1}^{r} \sigma_j A_{1k_j} x(t)$ and $h_{1_k}(x(t)) = \sum_{j=1}^{r} \sigma_j A_{2k_j} x(t)$. It is clear from (8.50) and (8.51) that the input disturbance is common for all systems fired at each operating point.

8.5.1 The Control Problem

Given a set of r representative dynamics (8.45), compute u_k such that each representative dynamics is locally stable. Show that fuzzy-blending of these individual control actions defined as $u = \sum_{k=1}^{r} \sigma_k u_k$ makes the T–S fuzzy model (8.4) Lyapunov stable.

The norm bound of the disturbances can be computed as

$$\| h_{1_k}(x(t)) \| = \| \sum_{j=1}^{r} \sigma_j A_{2k_j} x(t) \|$$

$$\leq \| \sum_{j=1}^{r} \sigma_j A_{2k_j} \| \| x(t) \|$$

$$\leq \sum_{j=1}^{r} \sigma_j \| A_{2k_j} \| \| x(t) \|$$

$$\leq \max_j \| A_{2k_j} \| \| x(t) \|$$

$$= \alpha_{hxk} \| x(t) \| \tag{8.53}$$

where $\alpha_{hxk} = \max_j \| A_{2k_j} \|$ gives the maximum matched state disturbance acting on the kth subsystem from vicinity. The overall maximum of local state disturbance is

$$\alpha_{hx} = \max_k \alpha_{hxk} \tag{8.54}$$

Similarly, the norm bound of the input disturbance is

$$\| h_2(u(t)) \| = \| \sum_{j=1}^{r} \sigma_j \overline{B}_j u(t) \|$$

$$\leq \left(\sum_{j=1}^{r} \sigma_j \| \overline{B}_j \| \right) \| u(t) \|$$

$$\leq \max_j \alpha_{u_j} \| u(t) \| \tag{8.55}$$

where $\alpha_{u_j} = \| \overline{B}_j \|$. The disturbance due to the jth fuzzy zone input matrix is α_{hu_j}, and the overall maximum of input disturbance is computed as

$$\alpha_u = \max_j \alpha_{u_j} \tag{8.56}$$

The norm bound of the unmatched state disturbance on the kth subsystem is

$$\| f_k(x(t)) \| = \| \sum_{j=1}^{r} \sigma_j A_{1k_j} x(t) \|$$

$$\leq \left(\sum_{j=1}^{r} \sigma_j \| A_{1k_j} \| \right) \| x(t) \|$$

$$\leq \max_j \| A_{1k_j} \| \| x(t) \|$$

$$= \alpha_{f_k} \| x(t) \| \tag{8.57}$$

where $\alpha_{f_k} = \max_j \| A_{1k_j} \|$.

8.5.2 Variable Gain Controller II

In this approach, each of the r subsystems is considered as a nominal plant. Hence, it is assumed that A_k is asymptotically stable $\forall\, k$. Even if they are not stable, we can give a preliminary feedback

$$u_p(t) = -\sum_{k=1}^{r} \sigma_k K_k x(t) \tag{8.58}$$

in such a way that the individual controller, $K_k x(t)$, stabilizes the kth subsystem. Further explanation is given in the appendix.

It should be noted that no assumption has been made on the stable pole locations. So there is flexibility in selecting the pole locations for each subsystem. By choosing the pole locations properly, the bounds on a non-linear disturbance may be reduced.

Theorem 8.4 Suppose that A_k is asymptotically stable and that P_k is a positive definite matrix satisfying $A_k^T P_k + P_k A_k = -2Q_k$ for some symmetric positive definite Q_k. Suppose also that

$$\alpha_{f_k} < \frac{\lambda_{\min}(Q_k)}{\lambda_{\max}(P_k)} \tag{8.59a}$$

$$\alpha_u < 1 \tag{8.59b}$$

Then, the state feedback controller

$$u(t) = -\gamma \sum_{k=1}^{r} \sigma_k B^T P_k x(t) \tag{8.60}$$

where

$$\gamma > \frac{\alpha_{hx}^2}{4(1 - \alpha_u)[\sum_{k=1}^{r} \sigma_k(\lambda_{\min}(Q_k) - \alpha_{f_k}\lambda_{\max}(P_k))]} \tag{8.61}$$

asymptotically stabilizes the fuzzy model (8.45).

Proof: Let us consider the Lyapunov function candidate as

$$V = x^T \sum_{k=1}^{r} \sigma_k P_k \, x \tag{8.62}$$

The time derivative of the Lyapunov candidate V evaluated along a state trajectory is

$$\dot{V} = 2x^T \sum_{k=1}^{r} \sigma_k P_k \, \dot{x}$$

$$= 2x^T \sum_{k=1}^{r} \sigma_k P_k (A_k x + Bu + f_k(x) + Bh_{1_k}(x) + Bh_2(u))$$

The derivative can be further expanded as

$$\dot{V} = 2\left(-x^T \sum_{k=1}^{r} \sigma_k Q_k x + x^T \sum_{k=1}^{r} \sigma_k P_k Bu + x^T \sum_{k=1}^{r} \sigma_k P_k f_k(x)\right.$$

$$\left. + x^T \sum_{k=1}^{r} \sigma_k P_k Bh_{1_k}(x) + x^T \sum_{k=1}^{r} \sigma_k P_k Bh_2(u)\right)$$

Using the properties of matrices (8.37)–(8.40) and using the norm bounds on the uncertain elements, we get

$$\dot{V} \leq -2 \sum_{k=1}^{r} \sigma_k \lambda_{\min}(Q_k) \|x\|^2 + 2 \sum_{k=1}^{r} \sigma_k \alpha_{f_k} \lambda_{\max}(P_k) \|x\|^2$$

$$-2\gamma \sum_{k=1}^{r} \sigma_k \|x^T P_k B\|^2 + 2\alpha_{hx} \sum_{k=1}^{r} \sigma_k \|x^T P_k B\| \|x\| +$$

$$2\gamma\alpha_u (\sum_{k=1}^{r} \sigma_k \|x^T P_k B\|)^2 = -2\overline{x}^T \overline{Q}\overline{x}$$

where $\overline{x} = \begin{bmatrix} \|x\| & \sum_{k=1}^{r} \sigma_k \|x^T P_k B\| \end{bmatrix}^T$ and

$$\overline{Q} = \begin{bmatrix} \sum_{k=1}^{r}(\lambda_{\min}(Q_k) - \alpha_{f_k}\lambda_{\max}(P_k)) & -\frac{\alpha_{hx}}{2} \\ -\frac{\alpha_{hx}}{2} & \gamma(1-\alpha_u) \end{bmatrix} \tag{8.63}$$

$\overline{Q}$ will be positive definite if the constraints (8.59) and (8.61) are satisfied. Hence the proof.

The control law (8.60) can be written as

$$u(t) = -\sum_{k=1}^{r} \sigma_k(\gamma B^T P_k x(t)) = \sum_{k=1}^{r} \sigma_k u_k(t)$$

where $u_k(t) = -\gamma B^T P_k x(t)$ The individual control law $u_k(t)$ stabilizes the kth fuzzy zone by considering the disturbance acting on it.

Discussion:
In all methods, the constraint given in Table 8.1 should be satisfied before

Table 8.1 Controller parameters—comparison

	Control action	Constraint
Fixed gain controller [68]	$-\gamma B^T P x$	$\alpha_f < \frac{\lambda_{\min}(Q)}{\lambda_{\max}(P)}$
Variable gain controller I [127]	$-\gamma B^T P x,$ $\gamma = \sum_{k=1}^{r} \sigma_k \gamma_k$	$\alpha_f < \frac{\lambda_{\min}(Q)}{\lambda_{\max}(P)}$
Variable gain controller II [127]	$-\gamma B^T \sum_{k=1}^{r} \sigma_k P_k x$	$\alpha_{f_k} < \frac{\lambda_{\min}(Q_k)}{\lambda_{\max}(P_k)} \ \forall k$

designing the controller. In the fixed gain controller proposed by Zak [68] and in the variable gain controller I [127], one of the subsystems is selected as the nominal plant while in variable gain controller II each subsystem is considered as nominal plant.

In variable gain controller I, the disturbance measure changes with the operating point, resulting in a variable controller gain. The upper bound of the disturbance and controller parameter γ are the same as that of the fixed gain controller for a given nominal plant. Hence, the value of constraint α_f is the same in both cases for a particular application. It may happen that for some applications, the condition on α_f is not satisfied in both cases. In such applications, both controllers will fail.

In variable gain controller II, each subsystem is considered as nominal plant. At each operating point, the disturbance is calculated by considering only active subsystems (subsystems that are fired). This relaxes the constraint on α_{f_k} for each nominal plant. Thus, variable gain controller II may be useful even in those cases where the variable controller I and the fixed gain controller fail, violating the constraint on the unmatched state disturbance α_f.

Example 8.2 (Cart–pole system) Consider the dynamics of pendulum subsystem of a cart–pole system, which is written as

$$\dot{x}_1 = x_2$$
$$\dot{x}_2 = \frac{g \, \sin(x_1) - amlx_2^2 \, \sin(2x_1)/2 - a \, \cos(x_1)u}{4l/3 - aml \, \cos^2(x_1)} \qquad (8.64)$$

where x_1 is the angle of the pendulum from vertical, x_2 is the angular velocity of the pendulum, u is the control input applied to the cart, m is mass of the pendulum, M is mass of the cart, $2l$ is length of the pendulum, g is the acceleration due to gravity and $a = 1/(m + M)$. Take the system parameters as $m = 2.0$ kg, $M = 8.0$ kg, $2l = 1.0$ m, and $g = 9.8$ m/s^2 for simulation. Express the system using a T–S fuzzy model. Design regulators based on the T–S fuzzy model of the system using the design schemes as described in Sections 8.3, 8.4, and 8.5. Compare the performances of the three controllers.

Solution
Pendulum angle x_1 is fuzzified in the operating region $[-\pi/2, \pi/2]$, and x_2 is always considered as 0. The system is approximated with four fuzzy regions in the positive half with centres 0, $\frac{\pi}{6}$, $\frac{\pi}{3}$, and $\frac{88\pi}{180}$. The gaussian function is chosen as the fuzzy membership function and the spread of the Gaussian function is chosen in such a way that at each operating point only two fuzzy subsystems will be fired. Around the equilibrium point (0,0), we have linearized the system using the standard Taylor series expansion. At other operating points, linear models are obtained by the technique as described in Chapter 1. Similar approximation is done in the negative half of the operating region. Controllers are designed assuming that the approximation is fair enough. Thus, the T–S fuzzy model of the system (8.64) is described by the following four rules.

Rule 1: If $x(t)$ is around $[0 \ 0]^T$

Then $\dot{x}(t) = \begin{bmatrix} 0 & 1 \\ 17.2941 & 0 \end{bmatrix} x(t) + \begin{bmatrix} 0 \\ -0.1765 \end{bmatrix} u(t)$

Rule 2: If $x(t)$ is around $[\pm\frac{\pi}{6} \ 0]^T$

Then $\dot{x}(t) = \begin{bmatrix} 0 & 1 \\ 15.8169 & 0 \end{bmatrix} x(t) + \begin{bmatrix} 0 \\ -0.1464 \end{bmatrix} u(t)$

Rule 3: If $x(t)$ is around $[\pm\frac{\pi}{3}\ 0]^T$

Then $\dot{x}(t) = \begin{bmatrix} 0 & 1 \\ 12.6304 & 0 \end{bmatrix} x(t) + \begin{bmatrix} 0 \\ -0.0779 \end{bmatrix} u(t)$

Rule 4: If $x(t)$ is around $[\pm\frac{88\pi}{180}\ 0]^T$

Then $\dot{x}(t) = \begin{bmatrix} 0 & 1 \\ 9.6193 & 0 \end{bmatrix} x(t) + \begin{bmatrix} 0 \\ -0.0065 \end{bmatrix} u(t)$

where $x(t) = [x_1(t)\ x_2(t)]^T$.

For the fixed gain controller and variable gain controller I, the linearized system at the origin (*Rule* 1) is taken as the nominal system. As the nominal system is unstable, a preliminary feedback is applied to achieve the subsystem asymptotic stability. The closed loop pole locations of preliminary feedback are chosen as $[-2, -2]$. The preliminary feedback gain is found as $K = [-120.6667 - 22.6667]$ with $P = \begin{bmatrix} 2.2500 & 0.2500 \\ 0.2500 & 0.3125 \end{bmatrix}$. The disturbance is modelled as (8.15). There is no unmatched disturbance as $A_j - A$ matrix in (8.18) is in the column space of B (*Rule* 1), resulting the value of α_f as zero. Hence, the condition on α_f is satisfied here. The upper bound of the norm of matched state disturbance α_{hx} is found to be 76.5277. The upper bound of the norm of input disturbance α_u is found to be 0.9703. It is seen that the conditions on the norm bounds of the input and state disturbances are satisfied for all the three controllers, and these controllers can be applied for this particular example. For variable gain controller I, γ at different operating regions are found according to (8.36) and are given in Table 8.2.

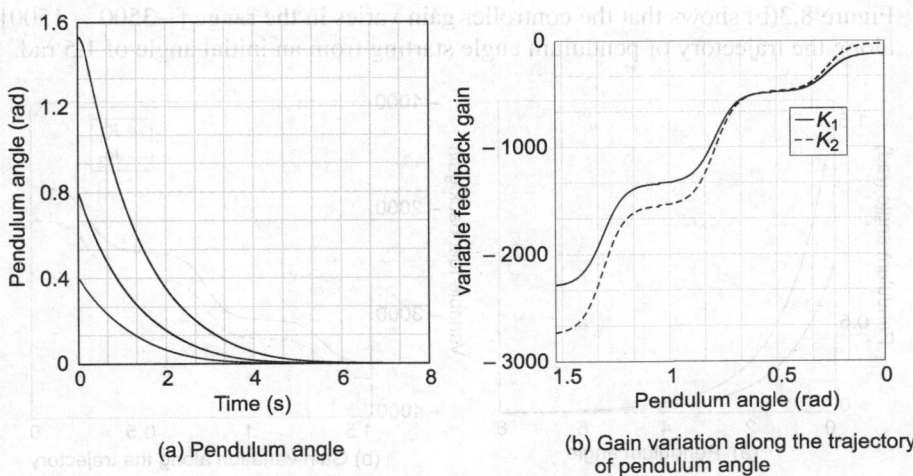

(a) Pendulum angle

(b) Gain variation along the trajectory of pendulum angle

Figure 8.2 Inverted pendulum mounted on cart: controller I

The controller gain at the ith region is thus $-(K + \gamma_i B^T P)$. Simulation results for different initial pendulum angles are shown in Figure 8.2(a). This figure shows that starting from different initial angles, the pendulum finally goes to the vertical upward position characterized by angle 0. It is also noticed that when the initial

Table 8.2 Variation of γ in different operating regions

Region	Centre	γ
1	0	10
2	$\pi/6$	8300
3	$\pi/3$	27700
4	$(88\pi)/180$	49400

condition is close to the nominal plant (*Rule* 1), the time required to settle down the pendulum at vertical upward position is less. Figure 8.2 shows the variation in the controller gain along the trajectory of pendulum angle for the initial condition 1.5 rad. It is seen from this figure that the gain varies in the range $[-3000,\ 0]$ where the maximum absolute gain occurs at the starting point 1.5 rad. This maximum value is the same as the fixed gain controller described in [68].

For variable gain controller II, at a specific trajectory point all the rules fired are considered as nominal plants. In this example, the maximum number of rules fired at any instant will be 2. The input matrix at the origin (*Rule* 1) is taken as the common actuation term B (refer to Section 8.5). As the nominal subsystems are unstable, preliminary feedback is necessary to make them stable. The four subsystems of the T–S fuzzy model are stabilized at $[-2\ -2]$, $[-1.7\ -1.7]$, $[-1.5\ -1.5]$ and $[-1.4\ -1.4]$, respectively using preliminary feedback. There is no unmatched disturbance and the maximum bound of the norm of matched state disturbance α_{hx} is found to be 67.0925. The controller parameter γ is computed from (8.61) as 40000. Simulation results are shown in Figure 8.3, where (a) shows pendulum angle and (b) shows the gain variation. It is seen from Figure 8.3(a) that the pendulum angle goes to 0 starting from different initial values. Figure 8.3(b) shows that the controller gain varies in the range $[-3500\ -1500]$ along the trajectory of pendulum angle starting from an initial angle of 1.5 rad.

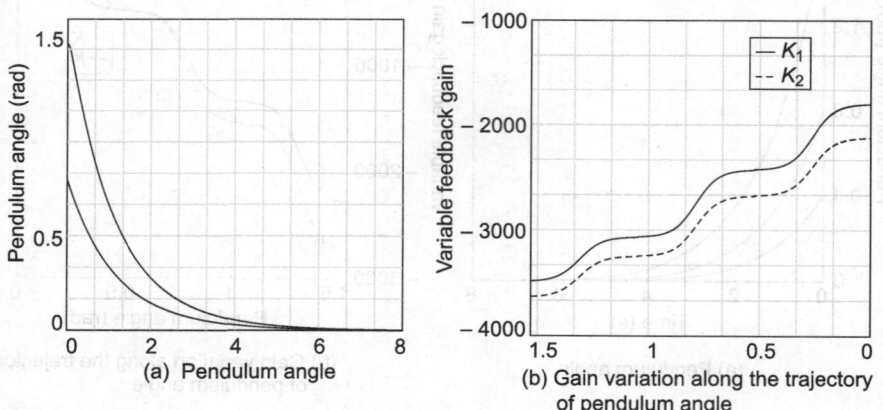

(a) Pendulum angle

(b) Gain variation along the trajectory of pendulum angle

Figure 8.3 Inverted pendulum mounted on cart: controller II

A comparative performance analysis of the fixed gain controller and the variable gain controllers I and II is given in Table 8.3 in terms of settling time. The values of K, P, and upper bounds on α_{hx} and α_u are the same for both controller I and Zak [68]. The fixed gain controller parameter is given as $K + \gamma B^T P = [2297.9 \ 2744.2]$, which is the upper bound of the variable gain controller parameter given in Figure 8.3(a). Lower settling time for controller I compared to the fixed gain controller is due to the effect of the variable nature of the gain of controller I. The settling time also varies considerably with the initial conditions for controller I. The dynamics become faster in variable controller I as we start nearer to the nominal plant. This is due to the fact that the state feedback gain is varying and it considers only the systems being fired instead of the entire fuzzy system as in the case of a fixed gain controller.

Table 8.3 Pendulum—Performance

Initial condition (rad)	Settling time (variable gain controller 1) (s)	Settling time (variable gain controller 2) (s)	Settling time (fixed gain controller) (s)
0.4 r	3.9 s	6.6 s	6.7 s
0.8 r	4.7 s	6.5 s	6.8 s
1.53 r	5.6 s	6.4 s	6.8 s

Example 8.3 The dynamics of an automobile are given by

$$\dot{v}(t) = \frac{1}{m}(-A_\rho v^2(t) - d + f(t)) \tag{8.65}$$

$$\dot{f}(t) = \frac{1}{\tau}(-f(t) + u(t)) \tag{8.66}$$

where $v(t)$ is the automobile speed and u is the control input. $u > 0$ represents a throttle input and $u < 0$ represents a brake input. m is the mass of the vehicle, A_ρ is the aerodynamic drag, d is a constant frictional force, f is the driving or braking force, and τ is the engine/brake time constant. Assume that $u \in [-1000, \ 1000]$. Take the system parameters as $m = 1300$ kg, $A_\rho = 0.3$ N s^2/m^2, $d = 100$ N, and $\tau = 0.2$ s. Derive the T–S fuzzy model from the state–space representation of the above system. Design a controller based on the T–S fuzzy model of the system using the design schemes as described in Sections 8.4 and 8.5 such that the automobile achieves a speed of $v_d = 54$ kmph.

Solution
Considering $x_1 = v(t)$ and $x_2 = f(t)$ as the system states, the state–space model of the system can be written as

$$\dot{x}_1 = \frac{1}{m}(-A_\rho x_1^2 - d + x_2)$$

$$\dot{x}_2 = \frac{1}{\tau}(-x_2 + u)$$

The desired speed is $v_d = 54$ kmph = 15 m/s. The system is linearized at nine equally spaced fuzzy zones in the operating range of $x_1 = (0, 30)$, with centre at $x = [x_1, \ 0]$. The membership function is chosen as Gaussian. The linearized system dynamics around $x_2 = 0$ has the form:

IF x_1 is around X_{1i} THEN

$$\dot{x} = \begin{bmatrix} a_{1i} & 0.0008 \\ 0 & -5 \end{bmatrix} x + \begin{bmatrix} 0 \\ 5 \end{bmatrix} u \tag{8.67}$$

The values of a_{1i} in matrix A for all fuzzy zones are given in Table 8.4. The input matrix, $B = [0\ 5]^T$ is same for all linear subsystems. Hence, the input disturbance $B_j - B$ will vanish while computing disturbance norms in (8.16) and (8.51).

Table 8.4 Automobile system: identified subsystem parameter

Fuzzy zone	x_1	a_{1i}
1	3	−0.0263
2	6	−0.0142
3	9	−0.0106
4	12	−0.0092
5	15	−0.0086
6	18	−0.0084
7	21	−0.0085
8	24	−0.0087
9	27	−0.0091

Table 8.5 Automobile system: disturbance tolerance limit—controller II

Fuzzy zone	α_{f_i}	$\dfrac{\lambda_{\min}(Q_i)}{\lambda_{\max}(P_i)}$
1	0.012128	0.026333
2	0.012128	0.014205
3	0.003581	0.010624
4	0.001444	0.009179
5	0.000590	0.008590
6	0.000162	0.008427
7	0.000234	0.008509
8	0.000336	0.008744
9	0.000336	0.009080

The condition (8.34a) for implementing controller I is satisfied for the linear systems in regions 1 and 2. Thus, the linear system in region 2 is chosen as the nominal plant for which α_f and $\frac{\lambda_{\min}(Q)}{\lambda_{\max}(P)}$ come out to be 0.0121 and 0.0142, respectively. Similarly, condition (8.59a) must satisfy for implementing Controller II. α_{f_i} and the corresponding $\frac{\lambda_{\min}(Q_i)}{\lambda_{\max}(P_i)}$ in Eqn (8.59a) for all the subsystems are given in Table 8.5. It is seen from Table 8.5 that each subsystem satisfies condition (8.59a). However, the matched state disturbance α_{hx} is found as 0 for both the controllers. Since the variable nature of controller I depends on the matched state disturbance, controller I becomes a fixed gain controller in this case. However, the variable nature of controller II is preserved due to the varying Lyapunov matrix P_i. The value of γ is chosen as $\gamma = 9.5$ and the feedback gains as well as feed-forward gains are found for all the regions. The simulation result for controller II is shown in Figure 8.4 which shows the desired and actual automobile speed. The variable gains for controller II are given in Table 8.6.

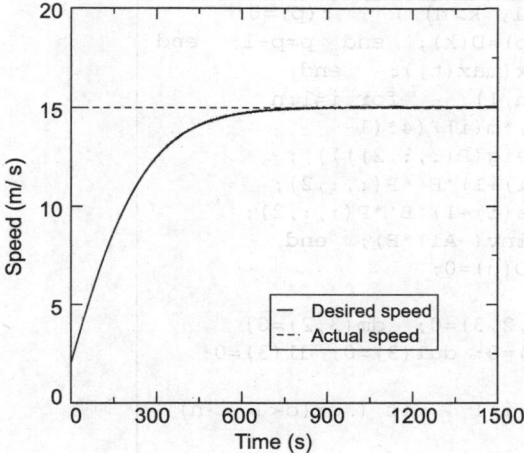

Figure 8.4 Automobile system: controller II—speed tracking

The MATLAB codes to find the linear systems, disturbance measures, and feedback gains are as follows.

```
clear all; n=9;
x1(n)=0; x2(n)=0;
for j=1:n   x1(j) = 3*(j);
end
B=[0;5];  Bi=pinv(B);  c=[1
0];
a(2,2,n)=0; P(2,2,n)=0;
la(n)=0;
for i=1:n   x=[x1(i);0];
a(:,:,i)=LinAutomobile(x);
P(:,:,i) =
lyap(a(:,:,i)',2*eye(2));
la(i) =
1/max(eig(P(:,:,i)));   end
d(2,2,n)=0;
dm(n,2)=0; dum(2,2,n)=0;
Du(n)=0; D(n)=0;
A=a(:,:,2);
for i=1:n   d(:,:,i) =
A-a(:,:,i);
dm(i,:) = Bi*d(:,:,i);
dum(:,:,i) = d(:,:,i) -
B*dm(i,:);

Du(i)=norm(dum(:,:,i));
D(i)=norm(dm(i,:));   end
m=zeros(n,1);
t=zeros(3,1);
for i=1:n   p=1;   for
k=i-1:i+1
```

```
LinAutomobile.m

function
a=LinAutomobile(x);
Ap=0.3; m=1300; d=100;
tau=0.2;
    f = [-Ap/m*x(1)*x(1)-
d/m+x(2)/m;
-x(2)/tau];
gradf = [-(Ap/m)*2*x(1) 0;
1/m -1/tau];
if(norm(x)==0)

a=gradf';

else

for i=1:2

a(i,:)=(gradf(:,i) +
((f(i)-
x'*gradf(:,i))/(norm(x)^2))
*x)';
end
```

```
if( or(k<1, k>n) )      t(p)=0;
else   t(p)=D(k);    end  p=p+1;  end
m(i) = max(max(t));     end
ga=zeros(n,1);     for i=1:n
ga(i)=D(i)*m(i)/(4*(1-
0.01*max(eig(P(:,:,2)))))));
K1 = (ga(i)+1)*B'*P(:,:,2);
A1=A-B*(ga(i)+1)*B'*P(:,:,2);
F= inv(c*inv(-A1)*B);   end
Du(n)=0;  D(n)=0;
for i=1:n
l=1;  d(2,2,3)=0;   dm(3,2)=0;
dum(2,2,3)=0;  du1(3)=0;  d1(3)=0;
e=1;
for o=i-1:i+1     if ( or(o<1,o>n)
)
A=a(:,:,i);      else
A=a(:,:,o);    end
d(:,:,e)=(A-a(:,:,i)); dm(e,:) =
Bi*(d(:,:,e));
dum(:,:,e) = d(:,:,e) - B*dm(e,:);
du1(e) = norm(dum(:,:,e));
d1(e) = norm(dm(e,:));
e=e+1;    end
Du(i)=max(du1); D(i)=max(d1); end
c1(n)=0;    for i=1:n
c1(i) = (1/la(i) ) *Du(i);    end
n1=max(D); cm=1-max(c1);
ga1=n1*n1/(4*cm);     for i=1:n
K1 = (ga1+9.5)*B'*P(:,:,i);
A1=A-B*(ga1+9.5)*B'*P(:,:,i);
F= inv(c*inv(-A1)*B);   end
```

Table 8.6 Automobile system: variable gains—controller II

Fuzzy zone	K_{1i}	K_{2i}	Feed-forward gain F_i
1	0.276053	9.500042	124.215444
2	0.512983	9.500079	124.452803
3	0.686394	9.500106	124.626529
4	0.794631	9.500122	124.734962
5	0.849287	9.500131	124.789718
6	0.865681	9.500133	124.806142
7	0.857344	9.500132	124.797790
8	0.834318	9.500128	124.774722
9	0.803373	9.500124	124.743721

Example 8.4 Consider the same example of a single link manipulator as given in Example 8.1. Design tracking controllers based on the T–S fuzzy model of the system using the design schemes as described in Sections 8.3, 8.4, and 8.5 such that the manipulator angle tracks the following trajectory.

$$\theta_d = 0.2\sin(t)$$

Simulate the system response.

Solution

The T–S fuzzy model of the single link manipulator system along with the corresponding MATLAB code is given in Example 8.1 where the system (8.64) is described by four rules in the operating region $[0, \pm\frac{\pi}{6}, \pm\frac{\pi}{3}, \pm\frac{\pi}{2}]$. The MATLAB code to find out the gains is the same as that of Example 8.3. The number of regions n is 4 in this case. Since the subsystems are unstable, a preliminary feedback of gain $[-3.81\ 5]$ is given to stabilize the subsystems dynamics. The unmatched disturbance for all the control schemes is found to be 0; thus, all the control schemes are applicable for this system. Using the maximum norm bound α_{hx}, the feedback gain of the fixed gain controller is computed as $K = [0.8845\ 1.2383]$. The maximum bound of the norm of the matched state disturbance α_{hx} is found to be 0.87 for controller II. The parameter γ for controller II is thus chosen as 3 (> 0.87). The variable gains for controllers I and II are given in Table 8.7.

Table 8.7 Single Link manipulator: variable gains—controllers I and II

Type of controller	x_1	K_i
Variable gain controller I	0	[0.3550 0.4970]
	$\pm\frac{\pi}{6}$	[0.3863 0.5408]
	$\pm\frac{\pi}{3}$	[0.6071 0.8499]
	$\pm\frac{\pi}{2}$	[0.8845 1.2383]
Variable gain controller II	0	[0.5003 0.7005]
	$\pm\frac{\pi}{6}$	[0.5401 0.7084]
	$\pm\frac{\pi}{3}$	[0.6976 0.7399]
	$\pm\frac{\pi}{2}$	[1.2327 0.8469]

Since the objective is to track a desired trajectory, we have to design the controller for tracking. The overall control input is given as

$$u = -Kx + v \tag{8.68}$$

where $-Kx$ is the stabilizing control input, computed using (8.60), and v is the tracking controller, yet to be designed. The subsystem dynamics after stabilization is of the form

$$\dot{x} = \begin{bmatrix} 0 & 1 \\ a_{1i} & a_{2i} \end{bmatrix} x + \begin{bmatrix} 0 \\ 1 \end{bmatrix} v$$

$$y = \begin{bmatrix} 1 & 0 \end{bmatrix} x \tag{8.69}$$

where y is the output of the system. The input–output dynamics of the subsystem is obtained by finding the derivatives of output equation as

$$\ddot{y} = v + \sum_i \sigma_i(a_{1i}x_1 + a_{2i}x_2)$$

Let us define the output tracking error as $e = y - y_d$. If we choose $v = -\sum_i \sigma_i(a_{1i}x_1 + a_{2i}x_2) - \lambda_1 e - \lambda_2 \dot{e} + \ddot{y}_d$, the error dynamics becomes

$$\ddot{e} + \lambda_2 \dot{e} + \lambda_1 e = 0$$

which is stable for positive values of λ_1 and λ_2. The values of λ_1 and λ_2 are chosen as 30 and 20. Simulation results for all the control schemes are shown in Figure 8.5, which shows the better performance of the variable gain controllers as compared to the fixed gain one. A comparative performance based on the RMS tracking error is given in Table 8.8.

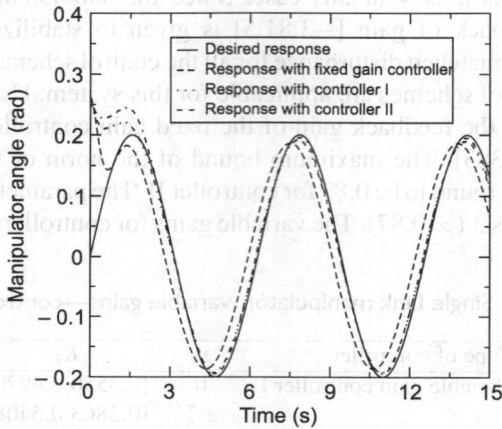

Figure 8.5 Single link manipulator: trajectory tracking

Table 8.8 Single link manipulator: comparative performance

Type of controller	RMS error
Fixed gain controller	0.055
Variable gain controller I	0.042
Variable gain controller II	0.038

8.6 CONTROLLER DESIGN USING DISCRETE T–S FUZZY SYSTEM

Let us consider a class of discrete time non-linear system given as

$$x(k + 1) = f(x(k), u(k)) \tag{8.70}$$

where x is the n-dimensional state vector and u is the p-dimensional input vector.

Like the continuous time non-linear systems, the above discrete time system can be effectively modelled as a discrete T–S fuzzy system where the jth fuzzy rule has the form

IF $x_1(k)$ is F_1^j and $\cdots$ and $x_n(k)$ is F_n^j THEN

$$x(k + 1) = A_j x(k) + B_j u(k) \tag{8.71}$$

where $x = [x_1, x_2 \ldots, x_n]^T$, $j = 1, \ldots, M$.

Given an input–output pair $(x(k), u(k))$, the fuzzy model around this operating point is constructed as the weighted average of the linear subsystems and has the form

$$x(k + 1) = \frac{\sum_{j=1}^{M} \mu_j (A_j x(k) + B_j u(k))}{\sum_{j=1}^{M} \mu_j} \tag{8.72}$$

where

$$\mu_j = \prod_{i=1}^{n} \mu_j^i(x_i) \tag{8.73}$$

$\mu_j^i(x_i)$ is the membership function of the fuzzy term F_i^j, $j = 1, 2, \ldots, M$. The fuzzy system (8.72) can be rewritten as

$$x(k + 1) = \sum_{j=1}^{M} \sigma_j (A_j x(k) + B_j u(k)) \tag{8.74}$$

$$\text{where} \quad \sigma_j = \frac{\mu_j}{\sum_{j=1}^{M} \mu_j} \quad \sum_{j=1}^{M} \sigma_j = 1$$

8.6.1 Linear State Feedback Controller for Discrete T–S Fuzzy System

In this section, we will represent the T–S fuzzy model in terms of a single linear plant and the rest of the plants would be treated as a disturbance to this linear plant so that the robust control theory can be applied to design a stabilizing controller for the system. In [68], Zak has proposed a fixed gain linear controller for a continuous time T–S fuzzy system using the same concept. In the previous section, we have proposed two different variable gain controllers for continuous time T–S fuzzy systems using a similar concept. As per our knowledge, so far this design technique has not been extended to a discrete time T–S fuzzy system. The discrete time T–S fuzzy model (8.74) can be rewritten as

$$x(k + 1) = \sum_{j=1}^{M} \sigma_j A_j x(k) + \sum_{j=1}^{M} \sigma_j B_j u(k)$$

$$= Ax(k) + Bu(k) + \sum_{j=1}^{M} \sigma_j (A_j - A) x(k) + \sum_{j=1}^{M} \sigma_j (B_j - B) u(k)$$

$$= Ax(k) + Bu(k) + F(x(k), u(k)) \tag{8.75}$$

The above equation is a representation of non-linear system dynamics in terms of a linear nominal plant $x(k + 1) = Ax(k) + Bu(k)$ and a non-linear disturbance term $F(x(k), u(k))$. The matrices A and B can be arbitrarily selected from the set $\{A_j\}$ and $\{B_j\}$, respectively. Further, the disturbance $F(x(k), u(k))$ can be expressed as

$$F(x(k), u(k)) = f(x(k)) + Bh_1(x(k)) + Bh_2(u(k)) \tag{8.76}$$

where $Bh_1(x(k)) = \sum_{j=1}^{M} \sigma_j (A_j - A) x(k)$ and $Bh_2(u(k)) = \sum_{j=1}^{M} \sigma_j (B_j - B) u(k)$.

The state disturbance term in (8.76) can be computed from Eqn (8.75) as

$$f(x(k)) + Bh_1(x(k)) = \sum_{j=1}^{M} \sigma_j(A_j - A)x(k) = \sum_{j=1}^{M} \sigma_j(A_{1_j} + BA_{2_j})x(k)$$

where $A_{1_j} \in R^{n \times n}$ and $A_{2_j} \in R^{m \times n}$ satisfy the following identity:

$$A_{1_j} + BA_{2_j} = (A_j - A) \tag{8.77}$$

Thus, $f(x(k)) = \sum_{j=1}^{M} \sigma_j A_{1_j} x(k)$ and $h_1(x(k)) = \sum_{j=1}^{M} \sigma_j A_{2_j} x(k)$. Similarly, the matched input disturbance is expressed as

$$Bh_2(u(k)) = \sum_{j=1}^{M} \sigma_j(B_j - B)u(k) = B \sum_{j=1}^{M} \sigma_j \overline{B}_j u(k)$$

where $\overline{B}_j \in R^{m \times m}$ satisfies the following identity:

$$B \overline{B}_j = (B_j - B) \tag{8.78}$$

The norm bound of the matched state disturbance is given by

$$\| h_1(x(k)) \| = \| \sum_{j=1}^{M} \sigma_j \overline{A}_j x(k) \| \leq \| \sum_{j=1}^{M} \sigma_j \overline{A}_j \| \| x(k) \|$$

$$\leq \sum_{j=1}^{M} \sigma_j \| \overline{A}_j \| \| x(k) \| \leq \alpha_{hx} \| x(k) \| \tag{8.79}$$

where $\alpha_{hx} = \max_j(\|\overline{A}_j\|)$. The norm bound of the unmatched state disturbance is given by

$$\| f(x(k)) \| = \| \sum_{j=1}^{r} \sigma_j A_{1_j} x(k) \| \leq \sum_{j=1}^{r} \sigma_j \| A_{1_j} \| \| x(k) \|$$

$$\leq \max_j \| A_{1_j} \| \| x(k) \| = \alpha_f \| x(k) \| \tag{8.80}$$

where $\alpha_f = \max_j \| A_{1_j} \|$. Similarly, the norm bound of the input disturbance is

$$\| h_2(u(k)) \| \leq \left(\sum_{j=1}^{M} \sigma_j \| \overline{B}_j \| \right) \| u(k) \| \leq \alpha_{hu} \| u(k) \| \tag{8.81}$$

where $\alpha_{hu} = \max_j(\|\overline{B}_j\|)$.

It is assumed that the matrix A in Eqn (8.75) is stable and satisfies the following matrix Lyapunov equation:

$$A^T PA - P = -Q$$

where P and Q are two positive definite matrices. Let us define the control action as follows:

$$u(k) = -\gamma B^T PAx(k)$$

To show the closed loop stability, let us consider a Lyapunov function candidate

$$V = x^T(k)Px(k)$$

The time derivative of the Lyapunov candidate $V = \boldsymbol{x}^T(k)P\boldsymbol{x}(k)$ evaluated on any trajectory of the closed-loop system is

$$
\begin{aligned}
\Delta V &= \boldsymbol{x}^T(k+1)P\boldsymbol{x}(k+1) - \boldsymbol{x}^T(k)P\boldsymbol{x}(k) \\
&= [A\boldsymbol{x} + B\boldsymbol{u} + f + Bh_1 + Bh_2]^T P[A\boldsymbol{x} + B\boldsymbol{u} + f + Bh_1 + Bh_2] \\
&\quad - \boldsymbol{x}^T P\boldsymbol{x} \\
&= \boldsymbol{x}^T A^T P A\boldsymbol{x} - \boldsymbol{x}^T P\boldsymbol{x} + \boldsymbol{u}^T B^T P A\boldsymbol{x} + f^T P A\boldsymbol{x} + h_1^T B^T P A\boldsymbol{x} + \\
&\quad h_2^T B^T P A\boldsymbol{x} + \boldsymbol{x}^T A^T P B\boldsymbol{u} + \boldsymbol{u}^T B^T P B\boldsymbol{u} + f^T P B\boldsymbol{u} + h_1^T B^T P B\boldsymbol{u} \\
&\quad + h_2^T B^T P B\boldsymbol{u} + \boldsymbol{x}^T A^T P f + \boldsymbol{u}^T B^T P f + f^T P f + h_1^T B^T P f + \\
&\quad h_2^T B^T P f + \boldsymbol{x}^T A^T P Bh_1 + \boldsymbol{u}^T B^T P Bh_1 + f^T P Bh_1 + \\
&\quad h_1^T B^T P Bh_1 + h_2^T B^T P Bh_1 + \boldsymbol{x}^T A^T P Bh_2 + \boldsymbol{u}^T B^T P Bh_2 + \\
&\quad f^T P Bh_2 + h_1^T B^T P Bh_2 + h_2^T B^T P Bh_2 \tag{8.82}
\end{aligned}
$$

Simplifying Eqn (8.82),

$$
\begin{aligned}
\Delta V &= -\boldsymbol{x}^T 2Q\boldsymbol{x} + 2\boldsymbol{x}^T A^T P B\boldsymbol{u} + 2\boldsymbol{x}^T A^T P f + 2\boldsymbol{x}^T A^T P Bh_1 + \\
&\quad 2\boldsymbol{x}^T A^T P Bh_2 + \boldsymbol{u}^T B^T P B\boldsymbol{u} + 2\boldsymbol{u}^T B^T P Bh_1 + 2\boldsymbol{u}^T B^T P Bh_2 + \\
&\quad h_1^T B^T P Bh_1 + h_2^T B^T P Bh_2 + f^T P f + 2h_1^T B^T P Bh_2 + \\
&\quad 2\boldsymbol{u}^T B^T P f + 2h_1^T B^T P f + 2h_2^T B^T P f \tag{8.83}
\end{aligned}
$$

To proceed further, we will use the following properties of matrices:

$$
\lambda_{\min}(Q)\|\boldsymbol{x}\|^2 \le \boldsymbol{x}^T Q\boldsymbol{x} \le \lambda_{\max}(Q)\|\boldsymbol{x}\|^2
$$
$$
\text{and therefore} \quad -\boldsymbol{x}^T Q\boldsymbol{x} \le -\lambda_{\min}(Q)\|\boldsymbol{x}\|^2
$$

Since we are dealing with only real matrices, the induced 2-norm of a matrix P is given as $\|P\|_{i2} = [\lambda_{max}(P^T P)]^{1/2}$. When P is symmetric, $P^T = P$. Thus, for a symmetric and positive definite P, its induced 2-norm is

$$
\|P\| = [\lambda_{\max}(P^2)]^{1/2} = \lambda_{\max}(P) \tag{8.84}
$$

The terms $\lambda_{\min}(.)$ and $\lambda_{\max}(.)$ are used to denote the minimum and maximum eigenvalues of a matrix.

Furthermore, $\boldsymbol{x}^T A^T P BB^T P A\boldsymbol{x} = \boldsymbol{x}^T A^T P B(\boldsymbol{x}^T A^T P B)^T = \|\boldsymbol{x}^T A^T P B\|^2$

$$
\tag{8.85}
$$

Taking all the above relations into account, putting the control law u in Eqn (8.83), and using norm bounds on uncertain elements, we get

$$
\begin{aligned}
\Delta V &\le -2\lambda_{\min}(Q)\|\boldsymbol{x}\|^2 + \gamma^2 \lambda_{\max}(\overline{P})\|\boldsymbol{x}^T A^T P B\|^2 - 2\gamma\|\boldsymbol{x}^T A^T P B\|^2 \\
&\quad + 2\alpha_{hx}\|\boldsymbol{x}^T A^T P B\| \|\boldsymbol{x}\| + 2\gamma\alpha_{hu}\|\boldsymbol{x}^T A^T P B\|^2 \\
&\quad + 2\gamma\alpha_{hx}\lambda_{\max}(\overline{P})\|\boldsymbol{x}^T A^T P B\| \|\boldsymbol{x}\| + 2\gamma^2\alpha_u\lambda_{\max}(\overline{P})\|\boldsymbol{x}^T A^T P B\|^2
\end{aligned}
$$

$$+ \alpha_{hx}^2 \lambda_{\max}(\overline{P})\|x\|^2 + \gamma^2 \alpha_{hu}^2 \lambda_{\max}(\overline{P})\|x^T A^T P B\|^2$$
$$+ 2\gamma \alpha_{hx}\alpha_{hu}\lambda_{\max}(\overline{P})\|x^T A^T P B\| \ \|x\| + 2\alpha_f \lambda_1 \|x\|^2 +$$
$$2\alpha_f \gamma \lambda_2 \|x^T A^T P B\| \ \|x\| + \alpha_f^2 \lambda_{\max}(P)\|x\|^2 +$$
$$2\alpha_{hx}\alpha_f \lambda_2 \|x\|^2 + 2\alpha_{hu}\alpha_f \gamma \lambda_2 \|x^T A^T P B\| \ \|x\|$$
$$= -2\overline{x}^T \overline{Q}\overline{x}$$

where $\overline{x} = \begin{bmatrix} \|x\| & \|x^T A^T P B\| \end{bmatrix}^T$, $\lambda_1 = \lambda_{\max}(A^T P)$, $\lambda_2 = \lambda_{\max}(B^T P)$, and $\overline{Q}$ is as follows:

$$\overline{Q} = \begin{bmatrix} \overline{q}_{11} & \overline{q}_{12} \\ \overline{q}_{21} & \overline{q}_{22} \end{bmatrix} \tag{8.86}$$

where

$$\overline{q}_{11} = \lambda_{\min}(Q) - \frac{1}{2}\alpha_{hx}^2 \lambda_{\max}(\overline{P}) - \lambda_1 \alpha_f - \frac{1}{2}\alpha_f^2 \lambda_{\max}(P) - \alpha_{hx}\alpha_f \lambda_2$$

$$\overline{q}_{12} = -\frac{1}{2}\alpha_{hx}(1 + \gamma \lambda_{\max}(\overline{P}) + \gamma \alpha_{hu}\lambda_{\max}(\overline{P})) - \alpha_f \gamma \lambda_2 - \alpha_{hu}\gamma \alpha_f \lambda_2$$

$$\overline{q}_{22} = \gamma(1 - \frac{1}{2}\gamma \lambda_{\max}(\overline{P}) - \alpha_{hu} - \gamma \alpha_{hu}\lambda_{\max}(\overline{P}) - \frac{1}{2}\gamma \alpha_{hu}^2 \lambda_{\max}(\overline{P}))$$

$$\overline{q}_{12} = \overline{q}_{21}$$

The system will be asymptotically stable if $\overline{Q}$ is positive definite. $\overline{Q}$ will be positive definite if the following conditions are satisfied.

$$\text{Condition I:} \quad q_{11} = \lambda_{\min}(Q) - \frac{1}{2}\alpha_{hx}^2 \lambda_{\max}(\overline{P}) - \lambda_1 \alpha_f -$$
$$\frac{1}{2}\alpha_f^2 \lambda_{\max}(P) - \alpha_{hx}\alpha_f \lambda_2 > 0 \tag{8.87}$$

$$\text{Condition II:} \quad \overline{q}_{11}\,\overline{q}_{22} - \overline{q}_{12}\,\overline{q}_{21} > 0 \tag{8.88}$$

Rearranging Condition II, we get

$$a\gamma^2 + b\gamma + c < 0 \tag{8.89}$$

where $a = (\dfrac{\alpha_{hx}^2}{4}\lambda_{\max}^2(\overline{P}) + \alpha_f^2 \lambda_2^2 + \alpha_{hx}\alpha_f \lambda_2 \lambda_{\max}(\overline{P}) +$

$$\frac{1}{2}q_{11}\lambda_{\max}(\overline{P}))(1 + \alpha_{hu})^2$$

$$b = (\frac{\alpha_{hx}^2}{2}\lambda_{\max}(\overline{P}) + \alpha_{hx}\alpha_f \lambda_2)(1 + \alpha_{hu}) + \overline{q}_{11}(\alpha_u - 1)$$

$$c = \frac{\alpha_{hx}^2}{4}$$

Two roots of equation $a\gamma^2 + b\gamma + c = 0$ are

$$\gamma_1 = \frac{-b - \sqrt{b^2 - 4ac}}{2a}, \qquad \gamma_2 = \frac{-b + \sqrt{b^2 - 4ac}}{2a} \tag{8.90}$$

To satisfy condition (8.89), γ should be chosen so that $\gamma_1 < \gamma < \gamma_2$. Again γ_1 and γ_2 must be real and one of these two roots must be positive so that we can choose a positive γ in between these two values.

The nominal plant (A, B) is so selected that A is asymptotically stable. If A is not asymptotically stable, a preliminary state feedback of the form $u = -Kx$ can be used to stabilize it before designing the controller. In such a case, the disturbance term will include the preliminary feedback as well. The overall control law in such case is given by

$$u = -(K + \gamma B^T P A)x \tag{8.91}$$

Discussion: The above described controller can be implemented only when the conditions (8.87) and (8.88) are satisfied which is not always the case. Moreover, these constraints are more stringent compared to the continuous time case.

Example 8.5 (Surge tank) Consider the dynamics of a surge tank as represented by the following differential equation:

$$\frac{dh(t)}{dt} = \frac{-c\sqrt{2gh(t)}}{A_r(h(t))} + \frac{1}{A_r(h(t))} u(t) \tag{8.92}$$

where $u(t)$ is the input flow (control input), which can be positive or negative, $h(t)$ is the liquid level of the tank (output of the system), $A_r(h(t))$ is the cross-sectional area of the tank, g is the gravitational acceleration, and $c = 1$ is the known cross-sectional area of the output pipe. Take $A_r(h(t)) = \sqrt{ah(t) + b}$, where $a = 3$ and $b = 1$. Discretize the system using the Euler approximation. Design a state feedback controller using Eqn (8.91) such that $h(t)$ follows a desired trajectory.

Solution
Using the Euler approximation to discretize the system, we have

$$h(k+1) = h(k) + T\left(\frac{-\sqrt{2gh(k)}}{\sqrt{3h(k)+1}} + \frac{u(k)}{\sqrt{3h(k)+1}}\right) \tag{8.93}$$

Using Eqn (8.93) (sampling time T is taken as 0.01 s) we have generated 3000 data points to identify the system. The output level range is taken from 0 and 10. Five fuzzy clusters have been used with centres evenly distributed between 0 to 10. The local subsystem parameters are obtained by updating the linear model weights using the gradient descent technique as described in the previous section with a learning rate $\eta = 0.001$. After identification, 3000 new data points are generated to validate the model. The result of model identification and prediction is shown in Figure 8.6 where a represents the response of the actual system and b represents response of the T–S fuzzy model. The RMS error for prediction is found to be 0.0302. Thus, the T–S fuzzy model of the system (8.93) is described by the following five rules:

Rule 1: If $x(k)$ is around 2
Then $x(k+1) = 0.989816\, x(k) + 0.003236\, u(k)$
Rule 2: If $x(k)$ is around 4
Then $x(k+1) = 0.991423\, x(k) + 0.003928\, u(k)$

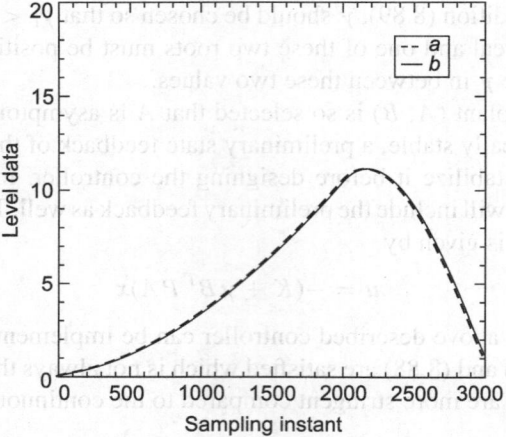

Figure 8.6 Model prediction for surge tank: 'a' represents the desired response and 'b' represents the actual one

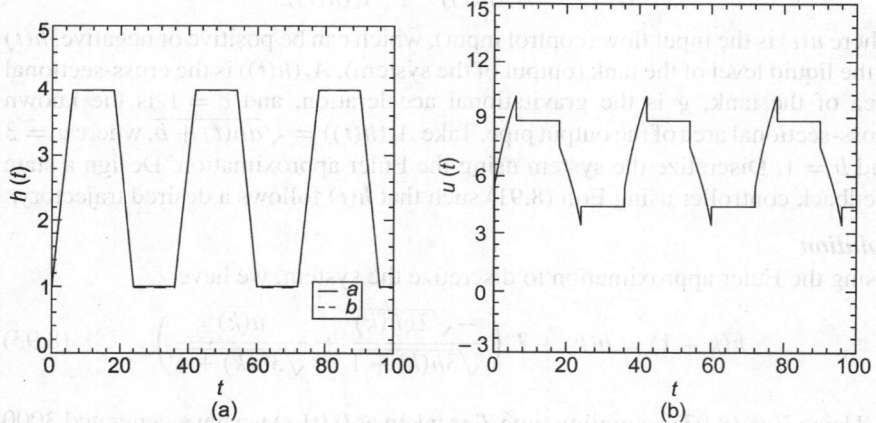

Figure 8.7 (a) Tracking: output level is plotted with respect to time, a is desired response, b is actual response and (b) the corresponding control input

Rule 3: If $x(k)$ is around 6
Then $x(k + 1) = 0.994009\, x(k) + 0.003383\, u(k)$
Rule 4: If $x(k)$ is around 8
Then $x(k + 1) = 0.995386\, x(k) + 0.002912\, u(k)$
Rule 5: If $x(k)$ is around 10
Then $x(k + 1) = 0.995711\, x(k) + 0.003163\, u(k)$

$x(k + 1) = 0.994009\, x(k) + 0.003383\, u(k)$ (*Rule* 3) has been taken as the nominal linear plant. As A matrix of this system is stable, preliminary feedback is not necessary. Solving the Lyapunov equation P is found to be 167.42 for $Q = 1$. The norm bounds α_f, α_{hx}, and α_{hu} are found to be 0, 1.2394, and 0.161, respectively. Condition (8.87) is satisfied for these values of α_X, P, and Q. Also

a, b, c in (8.89) are found to be 0.0013, -0.8359, and 0.384, respectively and the roots γ_1 and γ_2 are calculated as 0.4231 and 642.58. As both γ_1 and γ_2 are positive real, we can choose a γ in between these values so that controller (8.35) can be applied. γ has been chosen as 200. With this value of γ, the controller gain is computed to be -112.6. For tracking a reference signal, the feed-forward gain is found to be 114.4. Thus, the controller takes the following form:

$$u(k) = -112.4\,x(k) + 114.4\,r \qquad (8.94)$$

where r is the reference signal.

Simulation results are shown in Figure 8.7. Figure (a) shows tracking of the level output and (b) shows the required flow input. The RMS tracking error and mean square control input are computed as 0.022 and 54.7, respectively.

SUMMARY

Stabilization and controller design issues of non-linear systems approximated in terms of a T–S fuzzy model are discussed in this chapter. The design is accomplished by representing a non-linear system as a T–S fuzzy system. A parallel distributed compensator design technique is discussed where the compensator gains are computed by solving a set of LMI. When the T–S fuzzy system is viewed as a linear plant with non-linear disturbance terms, a fixed gain state feedback controller and two variable gain controllers have been presented which are shown to be the Lyapunov stable. Controllers have been implemented on different non-linear systems. The fixed gain controller has been extended for discrete time T–S fuzzy model. However the implementation constraints are more stringent compared to the continuous time counterparts. Readers are advised to refer to the following works [129, 130, 131, 132] for exploring other control design techniques using the T–S fuzzy model.

APPENDIX

Stabilization of Subsystems Dynamics with Preliminary Feedback

In this chapter, while designing variable gain state feedback controllers, we have mentioned that if the individual subsystem matrices A_is are not stable, a preliminary feedback of the following form can be given

$$u_p(t) = -\sum_{i=1}^{r} \sigma_i K_i x(t) \qquad (8.95)$$

which results in a cluster of stable linear systems. The state feedback $K_i x(t)$ are chosen so that it stabilizes the subsystem A_i such that $\overline{A}_i = A_i - BK_i$ is stable with desired dynamics. The proof of the above statement comes from Theorem 8.3 . The proof is derived again here along with the disturbance modelling for second method for completeness. In this case, the overall control action will be $u(t) = u_p(t) + u_c(t)$ where $u_c(t)$ is described in Chapter 8.5.2. The dynamics of the T–S model (8.4) with preliminary feedback can be written as

$$\dot{x}(t) = \sum_{i=1}^{r} \sigma_i A_i x(t) + Bu(t) + \sum_{j=1}^{r} \sigma_j (B_j - B)u(t)$$

$$= \sum_{i=1}^{r} \sigma_i A_i x(t) - B \sum_{i=1}^{r} \sigma_i K_i x(t) + Bu_c(t) +$$

$$\sum_{j=1}^{r} \sigma_j (B_j - B)(u_p(t) + u_c(t))$$

or

$$\dot{x}(t) = \sum_{i=1}^{r} \sigma_i (A_i - BK_i)x(t) + Bu_c(t) + \sum_{j=1}^{r} (B_j - B)(u_p(t) + u_c(t))$$

$$= \sum_{i=1}^{r} \sigma_i \overline{A}_i x(t) + Bu_c(t) + \sum_{j=1}^{r} \sigma_j (B_j - B)(u_p(t) + u_c(t))$$

$$= \overline{A}_i x(t) + \sum_{j=1}^{r} \sigma_j (\overline{A}_j - \overline{A}_i)x(t) + Bu_c(t)$$

$$+ \sum_{j=1}^{r} (B_j - B)(u_p(t) + u_c(t))$$

where $\overline{A}_i$s are stabilized individual subsystems.

EXERCISES

1. Consider the example of a ball–beam system whose dynamics are given by the following equations:

$$\ddot{x} = -\frac{5}{7}g\sin(\frac{r}{L}\theta)$$

$$\ddot{\theta} = \frac{K_g K_m}{J_{eq} R_m}\left(V_i - K_b K_g \dot{\theta}\right)$$

where x is the ball position and θ is the motor angular displacement. The system parameters are tabulated below.

Motor inertia J_m	3.87×10^{-7} kg m^2
Load inertia J_l	2.42×10^{-5} kg m^2
Equivalent inertia J_{eq}	0.0029 kg m^2

Gear ratio K_g	70 : 1
Motor resistance R_m	2.6 Ω
Back EMF constant K_b	0.00767 V/ (rad/s)
Motor torque constant K_m	0.00767 Nm/A
Beam length L	43.18 cm
Lever radius r	2.54 cm

The beam angle α is related to θ by the following relation:

$$\alpha = \frac{r}{L}\theta$$

(a) Considering $x, \dot{x}, \theta, \dot{\theta}$ as the system states and V_i the input, write down the state–space model of the system.

(b) Design tracking controllers based on the T–S fuzzy model of the system using the design schemes as described in Sections 8.4 and 8.5 such that the ball tracks following trajectory.

$$x_d = 0.2\sin(t)$$

Simulate the system response.

2. Consider the following dynamical equations of a VanderPol oscillator.

$$\dot{x}_1 = x_2$$
$$\dot{x}_2 = -x_1 + \mu(1 - x_1^2)x_2 + u$$

(a) Find out the T–S fuzzy model of the above system.

(b) Design tracking controllers based on the T–S fuzzy model of the system using the design schemes as described in Sections 8.4 and 8.5 so that x_1 oscillates between 0 and 1 with a frequency of 1 Hz. Simulate the system response.

3. Following are the dynamical equations of a magnetically levitated system

$$\dot{x} = v$$
$$\dot{v} = g - \frac{k_m k_a^2 V^2}{mx^2}$$

where x is the position and v is the velocity of the system. $g = 9.81 \ m/s^2$ is the acceleration due to gravity, $k_m = 0.0008$ N $(m/A)^2$ is the magnetic constant, $k_a = 0.397 \ f$ is the conductance $m = 0.02$ kg is the mass and V is the input voltage.

(a) Considering $u = V^2$, find out the T–S fuzzy model of the above system.

(b) Design a variable gain state feedback controller u based on the T–S fuzzy model of the system such that x converges to 0.

4. Consider the dynamics of a vertical plane two-link robotic manipulator which relate the joint torques $[\tau_1, \tau_2]$ to the joint angles $[\theta_1, \theta_2]$ of the links,

as

$$\tau_1 = [a_1 + a_2 \cos\theta_2]\, \ddot\theta_1 + [a_3 + \frac{a_2}{2}\cos\theta_2]\, \ddot\theta_2 + a_4 \cos\theta_1$$

$$-(a_2 \sin\theta_2)(\dot\theta_1\dot\theta_2 + \frac{\dot\theta_2{}^2}{2}) + a_5 \cos(\theta_1 + \theta_2)$$

$$\tau_2 = [a_3 + \frac{a_2}{2}\cos\theta_2]\, \ddot\theta_1 + a_3\, \ddot\theta_2 + (a_2 \sin\theta_2)\frac{\dot\theta_1{}^2}{2} + a_5 \cos(\theta_1 + \theta_2)$$

where $a_1 = 3.82$, $a_2 = 2.12$, $a_3 = 0.71$, $a_4 = 81.82$, and $a_5 = 24.06$. Design tracking controllers based on the T–S fuzzy model of the system using the design schemes as described in Sections 8.3, 8.4, and 8.5 such that the link angles θ_1 and θ_2 track the following trajectories.

$$\theta_{1d} = \frac{\pi}{6}\,\sin(2t) \qquad \theta_{2d} = \frac{\pi}{6}\,\cos(2t)$$

Simulate the system response.

[Hint: Use the transformation $\phi_1 = \frac{\pi}{2} - \theta_1$ to find out the state–space model of the system.]

9

Intelligent Control of a Pendulum on a Cart

This chapter presents two easily implementable control schemes for balancing a cart–pole system using an intelligent control framework. The proposed schemes use the Takagi–Sugeno (T–S) fuzzy model of a non-linear system. The concept of network inversion is used to design the controller for such a system. In one of the control schemes, the control input necessary to achieve a desired output is computed directly through iterative inversion of the fuzzy model. In the other scheme, the controller is parametrized as the T–S fuzzy average of local gains. The controller weight update laws are derived using both continuous and discrete time Lyapunov function synthesis. The controllers have been implemented on the cart–pole system both in simulation and real time.

9.1 T–S FUZZY MODEL REPRESENTATION

As described in Chapter 8, a discrete time non-linear system, $x(k + 1) = f(x(k), u(k))$, where $x(k) \in R^n$, $u(k) \in R^p$, can be effectively modelled as a T–S fuzzy system as

$$x(k + 1) = \sum_{j=1}^{r} \sigma_j (A_j x(k) + B_j u(k)) \tag{9.1}$$

where σ_j is the normalized fuzzy weight of the rule j. The methods to find A_js and B_js for each rule are also described in Chapter 8.

9.2 CONTROL USING THE T–S FUZZY MODEL

Given the T–S fuzzy model representation of a non-linear system, controller design techniques using LMI or a robust control approach are described in

Chapter 8. LMI technique depends on the existence of a common Lyapunov matrix for all local linear systems, and for some systems there may not exist any such matrix. In a robust control approach, implementation of the control law is constrained by the norm bound of a disturbance term. Thus, these techniques would not be applicable for all types of non-linear systems. To remove these applicability constraints, two easily implementable control schemes are provided in this chapter to design the controller for a discrete time T–S fuzzy model. These control techniques have been developed using the similar indirect adaptive control concepts given in [133]. The use of network inversion in designing neural control systems is illustrated in Chapter 4.

In the network inversion control scheme, the control input is iteratively updated between two sampling instants to match the system states with the desired states. The updated input is fed to the actual system at the next sampling instant. Since the control input is directly computed from the desired output, we term this design procedure 'direct inversion based control design'.

In the second control scheme, the control law is paramterized using a parallel distributed compensator [128] form, and the parameters are updated through the Lyapunov function synthesis.

The update laws for both the control schemes are derived using two approaches. In the first approach, the iterative update law between two sampling instants is derived using a continuous mode. A continuous time Lyapunov function is taken and the update law is derived such that the derivative of the Lyapunov function becomes negative definite during the update. In the second approach, a discrete mode is used to derive the iterative update law which makes the time difference of a discrete time Lyapunov function negative definite during the update as well as maintain the closed loop stability.

9.3 NETWORK INVERSION BASED CONTROL

The network inversion based control using the T–S fuzzy model is explained schematically in Figure 9.1. In the first step, a T–S fuzzy model of the cart–pole system is derived using the data generated from the actual non-linear model of the physical cart–pole. The control input to the T–S fuzzy model for the desired set point, i.e., both the cart and the pendulum are situated in the origin, is iteratively updated between two successive sampling instants until the T–S fuzzy model response reaches the desired states. This control input is actuated to the actual system.

9.3.1 Continuous-time Iterative Update

The iterative update law for control input u is derived using the Lyapunov function synthesis. The Lyapunov function is taken as

$$V = \frac{1}{2}\tilde{x}^T \tilde{x}$$

where $\tilde{x} = x_d - \hat{x}$. Here, x_d is the desired state vector, $\hat{x}$ is the predicted state vector obtained from the T–S model of the system. Control input u is predicted using iterative inversion of the T–S fuzzy model and fed to the actual system as shown in Figure 9.1. Since inversion is done between two sampling instants, the current states of the system as well as the desired states remain constant during the inversion process. This enables us to consider the T–S fuzzy model output $\hat{x}(k+1)$, during update of u, as a function of only u. The time derivative of the Lyapunov function can then be written as

$$\dot{V} = -\tilde{x}^T \frac{\partial \hat{x}}{\partial u} \dot{u}$$

$$= -\tilde{x}^T J \dot{u} \tag{9.2}$$

where $J = \frac{\partial \hat{x}}{\partial u} \in R^n$ which in this case is the same as the input matrix B of the T–S fuzzy model where $B = \sum_{j=1}^{M} \sigma_j B_j$.

Theorem 9.1   If an arbitrary initial input $u(0)$ is updated by

$$u(T) = u(0) + \int_0^T \dot{u}\, dt \tag{9.3}$$

where $\dot{u}$ is given by

$$\dot{u} = \frac{\|\tilde{x}\|^2}{\|J^T \tilde{x}\|^2} J^T \tilde{x} \tag{9.4}$$

then $\tilde{x}$ converges to zero provided $\dot{u}$ exists along the convergence trajectory.

Proof: Substituting $\dot{u}$ from Eqn (9.3) into (9.2), we get

$$\dot{V} = -\|\tilde{x}\|^2$$

which is negative definite. Thus, $\tilde{x}$ will converge to zero. The iterative inversion based input update law is given by the following equation:

$$u(i' + 1) = u(i') + \eta\, \dot{u} \tag{9.5}$$

where i' is the iterative index and η is a small positive constant which represents the update rate. The possible numerical instability associated with the weight update law can be avoided by adding a small positive constant ϵ in the denominator. In that case $\dot{u}$ modifies to

$$\dot{u} = -\frac{\|\tilde{x}\|^2}{\|J^T \tilde{x}\|^2 + \epsilon} J^T \tilde{x} \tag{9.6}$$

Using the above equation, $\dot{V}$ becomes

$$\dot{V} = -\|\tilde{x}\|^2 \frac{\|J^T \tilde{x}\|^2}{\|J^T \tilde{x}\|^2 + \epsilon} = -\alpha \|\tilde{x}\|^2$$

where $0 < \alpha < 1$. Since α is positive, $\dot{V}$ is negative semi-definite. Thus, V will decrease with the update of u, so as the tracking error $\tilde{x}$.

As mentioned earlier, once the update of u is over, input $u(k)$ at the kth instant is assigned to the updated value $u(i')$ and applied to the actual system.

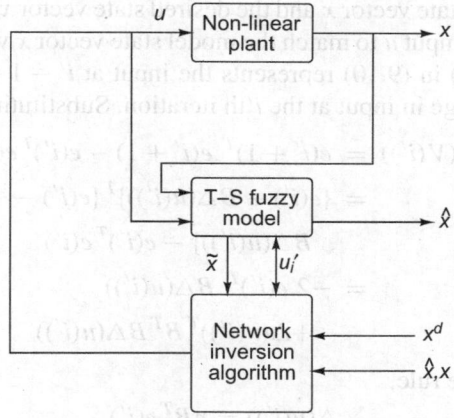

Figure 9.1 Direct inversion based control scheme

9.3.2 Discrete-time Update

In this section, we will use the discrete time Lyapunov approach to derive the update law as well as to maintain the closed loop stability. The update law in this case is chosen based on the popular gradient descent approach which will make the closed system Lyapunov stable for a range of learning rate. We will find the range of stable learning rate while analysing the closed loop stability of the system. The T–S fuzzy model of the system can be written as

$$\hat{x}(k+1) = Ax(k) + Bu(k) \tag{9.7}$$

where $A = \sum_{i=1}^{M} \sigma_i A_i$ and $B = \sum_{i=1}^{M} \sigma_i B_i$. To differentiate the actual system state $x(k)$ from the T–S fuzzy model output, we have used the notation $\hat{x}$ for the later. The objective is to find the iterative update law at each instant such that the learning is Lyapunov stable. The Lyapunov function is taken as

$$V(i') = e(i')^T e(i') \tag{9.8}$$

where, the error at the i'th iteration (during update of u) is described as $e(i') = x^d(k+1) - \hat{x}(i')$. $x^d(k+1)$ represents the desired state vector of the system at next time instant $k + 1$ and $\hat{x}(i')$ represents the T–S model state at the i'th iteration. The input update rule will be Lyapunov stable if

$$\Delta(V(i')) = V(i'+1) - V(i') \tag{9.9}$$
$$\leq 0$$

The error dynamics at each update is defined as

$$e(i'+1) = x^d(k+1) - \hat{x}(i'+1)$$
$$= x^d(k+1) - \{Ax(k) + Bu(i')\}$$
$$= x^d(k+1) - Ax(k) - B\{u(i'-1)$$
$$\quad +\Delta(u(i'))\}$$
$$= e(i') - B\Delta(u(i')) \tag{9.10}$$

The actual system state vector x and the desired state vector x_d remain unchanged while updating the input u to match the model state vector $\hat{x}$ with the desired state vector x^d. $u(i'-1)$ in (9.10) represents the input at $i'-1$ update and $\triangle(u(i'))$ represents the change in input at the i'th iteration. Substituting (9.10) in (9.9),

$$\triangle(V(i')) = e(i'+1)^T.e(i'+1) - e(i')^T e(i')$$
$$= \{e(i') - B\triangle(u(i'))\}^T\{e(i') -$$
$$B\triangle(u(i'))\} - e(i')^T e(i')$$
$$= -2.e(i')^T.B\triangle(u(i'))$$
$$+\triangle(u(i'))^T B^T B\triangle(u(i')) \tag{9.11}$$

Consider the update rule,

$$\triangle(u(i')) = \eta B^T e(i') \tag{9.12}$$

where η is the learning rate. It should be noted that the proposed update law is in the steepest descent direction of $V(i'+1)$. The objective then is to determine the limit of η such that the update law is Lyapunov stable.

(9.11) can be rearranged by substituting (9.12),

$$\triangle(V(i')) = -\frac{2}{\eta} \parallel \triangle(u(i')) \parallel^2 +\triangle(u(i'))^T B^T.B$$
$$\triangle(u(i'))$$
$$\leq -\frac{2}{\eta} \parallel \triangle(u(i')) \parallel^2 +$$
$$\lambda_{\max}(B^T B) \parallel \triangle(u(i')) \parallel^2 \tag{9.13}$$

where $\lambda_{\max}(.)$ represents the maximum eigenvalue. $\triangle(V(i'))$ (refer to (9.13)) is negative semidefinite if

$$\eta < \frac{2}{\lambda_{\max}(B^T B)} \tag{9.14}$$

After the update is over, $u(k)$ at a particular time instant k is assigned to the updated input $u(i')$ and the system states evolve accordingly.

Analogous to the continuous time, this scheme can also be used for stabilization. The Lyapunov function would be then considered as

$$V = \hat{x}(i'+1)^T \hat{x}(i'+1) \tag{9.15}$$

where $\hat{x}(i'+1) = Ax(k) + Bu(i')$ is the state vector representing the T–S fuzzy model output at the i'th iteration. The update rule in this case becomes

$$\triangle(u(i')) = -\eta.B^T \hat{x}(i') \tag{9.16}$$

9.4   T–S FUZZY CONTROLLER

Since the regulation problem associated with a cart–pole system can be solved using a state feedback control, the control input $u(k)$ can be parametrized using the parallel distributed compensator form as given by following equation.

$$u(k) = -\sum_{j=1}^{M} \sigma_j K_j x(k) \tag{9.17}$$

where K_j is the state feedback gain associated with the ith rule corresponding to the T–S fuzzy controller as shown in Figure 9.2. This is the traditional approach to indirect adaptive control as discussed in Chapter 4. The objective here is to update the feedback gains K_j such that the system states converges to zero.

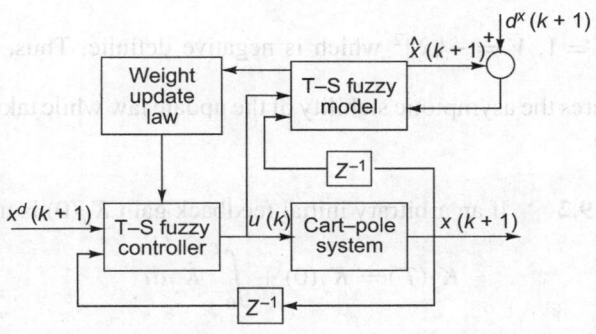

Figure 9.2 Indirect adaptive control using T–S fuzzy model

9.4.1 Continuous Time Weight Update Law

For a regulation problem, the Lyapunov function candidate is modified as

$$V = \frac{1}{2}\hat{x}^T \hat{x}$$

where $\hat{x}$ is the state output of T–S fuzzy model. We can get the required control input $u(k)$ by recursively updating K_j with the actual system state $x(k)$ fixed. Thus, the control action can be viewed as a function of feedback gains only. The time derivative of the Lyapunov function candidate can be derived as

$$\dot{V} = \hat{x}^T \dot{\hat{x}} = \hat{x}^T \frac{\partial \hat{x}}{\partial u} \left(\sum_{j=1}^{M} \sigma_j \frac{\partial u}{\partial K_j} \dot{K}_j\right)$$

$$= \sum_{j=1}^{M} \sigma_j (\hat{x}^T D_j \dot{K}_j)$$

where $D_j = \frac{\partial \hat{x}}{\partial u} \frac{\partial u}{\partial K_j}$. Since, in this case, $x(k)$ remains unchanged during the inversion process, D_j equals the following expression:

$$D_j = \left(\sum_{i=1}^{M} \sigma_i B_i\right) (-\sigma_j x(k)) \tag{9.18}$$

Let us select the incremental feedback gain as

$$\dot{K}_j = -\frac{\|\hat{x}\|^2}{\|D_j^T \hat{x}\|^2} D_j^T \hat{x} \tag{9.19}$$

Using the above update law for K_j, $\dot{V}$ becomes

$$\dot{V} = -\sum_{j=1}^{M} \sigma_j \|\hat{x}\|^2$$

Since $\sum_{j=1}^{M} \sigma_j = 1$, $\dot{V} = -\|\hat{x}\|^2$ which is negative definite. Thus, the following theorem ensures the asymptotic stability of the update law while taking the system states to zero.

> **Theorem 9.2**   If an arbitrary initial feedback gain $K_j(0)$ is updated by
>
> $$K_j(T) = K_j(0) + \int_0^T \dot{K}_j dt \tag{9.20}$$
>
> where $\dot{K}_j$ is given in (9.19), then $\hat{x}$ converges to zero provided $\dot{K}_j$ exists along the convergence trajectory.

The possible numerical instability associated with the weight update law can be avoided by adding a small positive constant in the denominator. This modification makes $\dot{V}$ negative semi-definite as illustrated in the previous section. Thus, the states decrease with time. At a particular time instant k, the parametrized input to the T–S fuzzy model is updated using the update law (9.20) for a number of iterations.

9.4.2   Discrete Time Weight Update Law

The direct inversion based control scheme can be extended to compute the gain values of a PDC (refer to Eqn (9.17)). The objective here is to find the update law for the controller gains. The parametrized input at the i'th iteration takes the following form:

$$u(i') = K(i')x(k) \tag{9.21}$$

where $K(i')$ represents the state feedback gain at the i'th iteration and $x(k)$ represents the system state at the kth time instant. For parallel distributed compensator $K(i') = \sum_{j=1}^{M} \sigma_j K_j(i')$ where K_j represents the feedback gain for the jth linear model. The parametrized update rule can be written as

$$u(i') = u(i' - 1) + \Delta(u(i'))$$
$$= K(i' - 1)x + \Delta(K(i'))x \tag{9.22}$$

where $\triangle(u(i')) = \triangle(K(i'))x$. x represents $x(k)$ which is constant during the iteration. The update law is described as

$$
\begin{aligned}
\triangle(u(i')) &= -\eta B^T \hat{x}(i') \\
&= -\eta B^T \{Ax + Bu(i' - 1)\} \\
&= -\eta B^T \{A + BK(i' - 1)\}x
\end{aligned}
\tag{9.23}
$$

The parametrized update law can be obtained by combining (9.22) and (9.23) as

$$
\triangle(K(i')) = -\eta B^T \{A + BK(i' - 1)\}
\tag{9.24}
$$

The parameter update law for the jth linear model can be written as

$$
\triangle(K_j(i')) = -\eta \sigma_j B^T \{A + BK(i' - 1)\}
\tag{9.25}
$$

u is updated using the above parameter update law. Once the update is over $u(k)$ is assigned to the updated value $u(i')$ and applied to the system at the next time instant.

Discussion: In this section, we have provided two iterative inversion based control algorithms which find the required control input for the system concerned using the T–S fuzzy model of the system. The iterative update laws have been derived using Lyapunov stability analysis. A popular approach to design controller for T–S fuzzy model is the LMI technique where the gains are designed by finding out a common Lyapunov matrix P. The existence of the common P is not always guaranteed. In those cases, LMI based techniques will fail to find out any feasible set of gains. But in the presented design technique no such sufficient condition is needed to be satisfied. The control action can be found either by direct inversion technique or in a parametrized form. The gains in the parametrized form can be found out online while maintaining the closed loop stability through the Lyapunov approach. We will show in the simulation results that the feedback gains designed using the proposed scheme actually satisfy the relaxed stability condition as given in [128]. Another advantage of the proposed inversion control scheme is that it can be used both for regulation and tracking, whereas the earlier design techniques are basically meant for regulation purpose only.

9.5 CART–POLE SYSTEM: SIMULATION AND EXPERIMENT

In this section, we will apply all the inversion based control design schemes for a cart–pole system and provide the simulation and experimental results. The performance of the inversion based controllers has been compared with LQR in real time. An inverted pendulum mounted on a motor-driven cart, as shown in Figure 9.5, is a model of the attitude control of a space booster on take-off. The objective of the attitude control problem is to keep the space booster in a vertical position. The inverted pendulum is unstable in the sense that it may fall over any time in any direction unless a suitable control force is applied. Here we consider only a two-dimensional problem in which the pendulum moves only in the plane of page. The control force u is applied to the cart. The dynamics of the system, as derived in Chapter 1, can be written as

$$\dot{x}_1 = x_2$$

$$\dot{x}_2 = \frac{g \, \sin(x_1) - amlx_2^2 \, \sin(2x_1)/2 - a \, F \cos(x_1)}{4l/3 - aml \, \cos^2(x_1)}$$

$$\dot{x}_3 = x_4$$

$$\dot{x}_4 = \frac{-\frac{mag}{2} \, \sin(2x_1) + \frac{4}{3}mla \, x_2^2 \sin(x_1) + \frac{4}{3}a \, F}{4/3 - am \, \cos^2(x_1)} \tag{9.26}$$

where x_1 is the angle of the pendulum from vertical, x_2 is the angular velocity of the pendulum, x_3 is cart position, x_4 is the cart velocity, and F is the input force applied to the system. The system parameters are m, the mass of the pendulum, M, the mass of the cart, $2l$, the length of the pendulum. g is the acceleration due to gravity. The above dynamics can be written in terms of voltage input u as

$$\dot{x}_1 = x_2$$

$$\dot{x}_2 = \frac{g \sin(x_1) - amlx_2^2 \sin(2x_1)/2 - c_1 a x_4 \cos(x_1)}{4l/3 - aml \cos^2(x_1)}$$

$$\qquad \frac{c_2 a \cos(x_1)}{4l/3 - aml \cos^2(x_1)} u$$

$$\dot{x}_3 = x_4$$

$$\dot{x}_4 = \frac{-\frac{mag}{2} \sin(2x_1) + \frac{4}{3}mlax_2^2 \sin(x_1) + \frac{4}{3}ac_1 x_4}{4/3 - am \cos^2(x_1)}$$

$$\qquad + \frac{\frac{4}{3}ac_2 u}{4/3 - am \cos^2(x_1)} \tag{9.27}$$

where the relation between F and u is as follows:

$$F = \frac{K_m K_g}{Rr} u - \frac{K_m^2 K_g^2}{Rr^2} x_4$$

In Eqn (9.27), a, c_1, and c_2 are three constants, given by $a = \dfrac{1}{M+m}$,

$c_1 = -\dfrac{K_m^2 K_g^2}{R r^2}$, $c_2 = \dfrac{K_m K_g}{R r}$, where K_m is the motor torque constant, K_g is the motor gear ratio, R is the motor armature resistance, and r is the radius of the gear connected to the shaft. To match the simulation and experimental results, the system parameters are taken as $m = 0.23$ kg, $M = 0.5$ kg, $l = 0.321$ m, $K_m = 0.00767$ Nm/A, $K_g = 3.71$, $R = 2.3585$ Ω and $r = 0.00635$ m which are the parameters of the real time experimental set up (refer Figure 9.5) provided by Quanser.

9.5.1 T–S Fuzzy Model of the Cart–Pole

To find the T–S fuzzy model of the cart–pole system, the pendulum angle x_1 is fuzzified into nine equally spaced regions within the operating zone $[-15°, \ 15°]$. The Gaussian function is chosen as the fuzzy membership function. Around the equilibrium point (0,0), we have linearized the continuous time system model using the standard Taylor series expansion. At other operating points, linear models are obtained by the technique, described in Chapter 8 of [7]. Then the subsystem models are discretized for the sampling interval of 0.01 s to get the discrete time T–S model of the system. Two typical subsystems are presented below, for clear understanding.

Rule 1: If $x_1(t)$ is around $0°$

$$\text{Then } \dot{x}(t) = \begin{bmatrix} 0 & 1 & 0 & 0 \\ 30.0147 & 0 & 0 & 35.7224 \\ 0 & 0 & 0 & 1 \\ -3.0323 & 0 & 0 & -15.2725 \end{bmatrix} x(t) + \begin{bmatrix} 0 \\ -7.9716 \\ 0 \\ 3.4081 \end{bmatrix} u(t)$$

Rule 2: If $x_1(t)$ is around $15°$

$$\text{Then } \dot{x}(t) = \begin{bmatrix} 0 & 1 & 0 & 0 \\ 29.0705 & 0 & 0 & 33.8045 \\ 0 & 0 & 0 & 1 \\ -2.8368 & 0 & 0 & -14.9624 \end{bmatrix} x(t) + \begin{bmatrix} 0 \\ -7.5436 \\ 0 \\ 3.3389 \end{bmatrix} u(t)$$

The corresponding discrete time models are obtained as

Rule 1: If $x_1(k)$ is around $0°$

$$\text{Then } x(k+1) = \begin{bmatrix} 1.0015 & 0.01 & 0 & 0.002 \\ 0.2951 & 1.00145 & 0 & 0.331 \\ -0.0001 & 0 & 1 & 0.009 \\ -0.0281 & -0.0001 & 0 & 0.858 \end{bmatrix} x(k)$$

$$+ \begin{bmatrix} -0.0004 \\ -0.0740 \\ 0.0002 \\ 0.0316 \end{bmatrix} u(k)$$

Rule 2: If $x_1(k)$ is around $15°$

$$\text{Then } x(k+1) = \begin{bmatrix} 1.0014 & 0.01 & 0 & 0.002 \\ 0.2862 & 1.00144 & 0 & 0.314 \\ -0.0001 & 0 & 1 & 0.009 \\ -0.0264 & -0.0001 & 0 & 0.861 \end{bmatrix} x(k)$$

$$+ \begin{bmatrix} -0.0004 \\ -0.0701 \\ 0.0001 \\ 0.0310 \end{bmatrix} u(k)$$

9.5.2 Control Systems Design

The control objective here is to stabilize the system states at the equilibrium point $x = [0\ 0\ 0\ 0]^T$. Once the system is expressed as a fuzzy cluster of nine linear models, the control input is computed using the continuous time direct iterative inversion approach as given in (9.4) and (9.5). The desired output is taken as the output of a reference linear model. The reference model is regulated using the

LQR control strategy. The maximum number of iterations to predict the control input is taken as 10. The simulation result is shown in Figure 9.3 which shows how the states converge to zero with time. Next, we have applied the iterative gradient search algorithm based on the discrete time Lyapunov approach (9.12). According to (9.14) the limit on η is found to be $\eta < 340.37$. In the simulation, we have taken η as 250 and the number of iterations as 5.

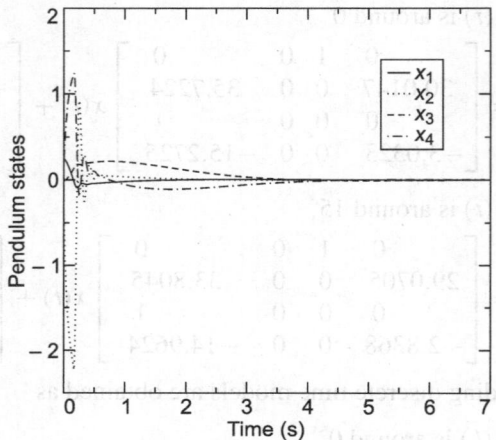

Figure 9.3 Simulation results for cart–pole system: direct based inversion algorithm is used to compute the control input

The parametrized control laws using both continuous and discrete time approach have also been applied to the system. The final updated gains for all the nine fuzzy regions are given in Table 9.1 for continuous time parameter update and in Table 9.2 for discrete time parameter update. It is found that for both cases, the final feedback gains are almost the same which is apparent from the tables. We thus provide results for parametrized control law using discrete time approach (9.22) – (9.25). The comparative performance of all the algorithms is shown in Figure 9.4, where (a) shows the pendulum angle and (b) shows the applied voltage to the system.

Table 9.1 Variation of feedback gain K for different operating regions for continuous time parametrized control

Fuzzy region	Feedback gain K
1	[217.147 18.133 24.601 38.793]
2	[166.028 13.360 18.746 29.162]
3	[165.573 13.373 18.710 29.112]
4	[164.938 13.324 18.661 29.132]
5	[164.493 13.296 18.637 29.099]
6	[164.897 13.325 18.661 29.121]
7	[165.283 13.363 18.680 29.157]
8	[166.053 13.386 18.742 29.163]
9	[207.031 17.410 23.123 36.949]

Table 9.2 Variation of feedback gain K for different operating regions for discrete time parametrized control

Fuzzy region	Feedback gain K
1	[218.294 18.282 24.814 39.083]
2	[166.654 13.956 18.942 29.862]
3	[165.685 13.873 18.830 29.712]
4	[165.093 13.823 18.761 29.620]
5	[164.893 13.806 18.738 29.589]
6	[165.093 13.823 18.761 29.620]
7	[165.685 13.873 18.830 29.712]
8	[166.657 13.956 18.942 29.862]
9	[206.931 17.330 23.523 37.047]

In the case of parametrized control law, it is observed that the feedback gains designed using the proposed control scheme (9.22)–(9.25) also satisfy the relaxed stability condition as described in [128]. Using the procedure given in [128], we have found that for the feedback gains given in Table 9.1, there exists a common Lyapunov matrix P. Using MATLAB LMI toolbox P is found to be

$$P = \begin{bmatrix} 92.7466 & 6.8973 & 11.0437 & 14.8817 \\ 6.8973 & 0.5843 & 0.8238 & 1.1394 \\ 11.0437 & 0.8238 & 1.4358 & 1.8058 \\ 14.8817 & 1.1394 & 1.8058 & 2.4823 \end{bmatrix}.$$

9.5.3 Experiment on a Cart–Pole System

The control algorithms have been tested on a real time set-up of the cart–pole system shown in Figure 9.5. The system has two potentiometers to read the cart position and pendulum angle and a 6 V dc motor to drive the cart. The module number is IP01. The cart and pendulum velocity are computed using the numerical differentiation technique. The mechanical system is interfaced with the host PC through NI-DAQ card 6025E. The entire set-up is controlled through C++ program at a sampling interval of 10 milli-sec. The system specifications are as follows:

Cart mass M: 0.5 kg

Pendulum mass m: 0.23 kg

Motor torque constant K_m: 0.00767 Nm/A

Motor gear ratio K_g: 3.71

Pendulum length from pivot to the centre of gravity l: 0.32065 m

Motor armature resistance R: 2.3585 Ω

radius of gear connected to motor shaft r: 0.00635 m

DC servo-motor rated voltage : 6V

DC servo-motor maximum continuous current: 1 A

Cart potentiometer sensitivity: 0.0931 m/V

Pole potentiometer sensitivity: 0.2482 rad/V

Potentiometer bias voltage: ±12 V

Cart potentiometer measurement range: ±5 V

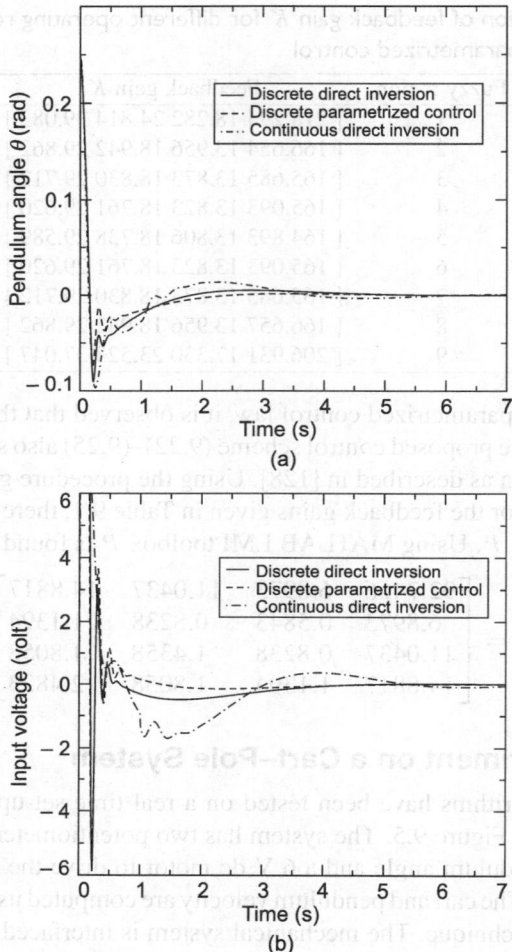

Figure 9.4 Comparative performance of inversion algorithms based on (i) discrete direct inversion based control (ii) discrete parametrized control (iii) continuous time direct inversion based control: (a) the pendulum angle and (b) applied voltage

   The control input obtained using the continuous time direct inversion method (9.4) and (9.5), discrete time direct inversion method (9.12), and discrete time parametrized control law (9.22)–(9.25) have been applied to the system. In the parameterized control scheme, the gains have not been updated in real time. Instead, the feedback gains (refer to Table 9.2) obtained from the simulation results are used. The performance is compared with the well established LQR control where the control input is obtained from the linear model at the equilibrium point. The LQR gain is found to be $[-198.9316 \quad -20.0555 \quad -22.3607 \quad -34.6575]$

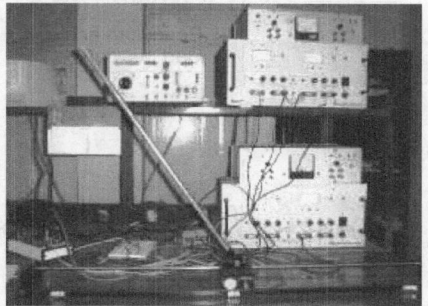

Figure 9.5 Experimental set-up

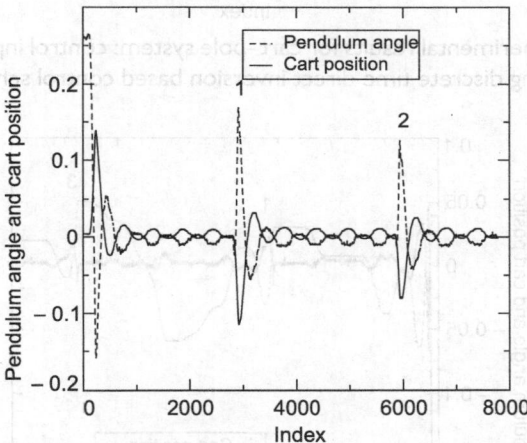

Figure 9.6 Experimental results for a cart–pole system: control input is computed using continuous time direct inversion based control scheme

for $Q = \begin{bmatrix} 15 & 0 & 0 & 0 \\ 0 & 0 & 0 & 0 \\ 0 & 0 & 0.25 & 0 \\ 0 & 0 & 0 & 0 \end{bmatrix}$ and $R = 0.0005$. The results of direct inversion,

parametrized inversion, and LQR are shown in Figures 9.6–9.9.

Table 9.3 Comparative performance on disturbance tolerance

Control algorithms	Max. disturbance in x_1
Direct inversion based control	0.22 rad
Parametrized control	0.07 rad
Linear quadratic regulator	0.07 rad

It is expected that the proposed controllers would work for a large initial angle, since the T–S fuzzy model represents the non-linear system for a wider range ($\pm 15°$, in the present case). This is observed in the experiment also. Since LQR is supposed to work only for a small region where the linear model is valid, the initial

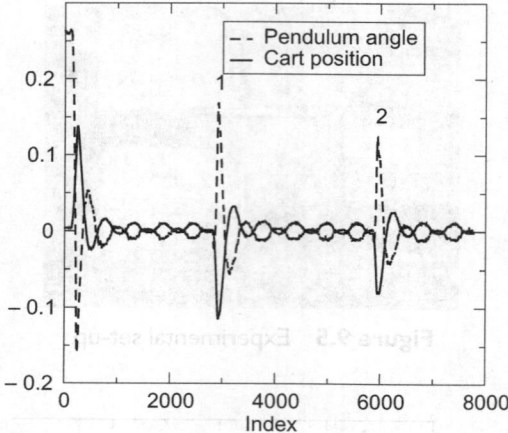

Figure 9.7   Experimental results for cart–pole system: control input is computed using discrete time direct inversion based control scheme

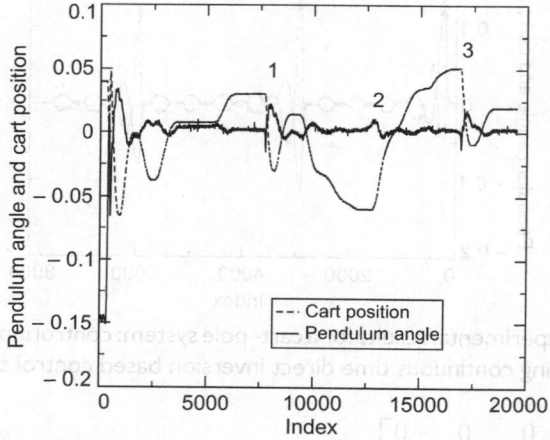

Figure 9.8   Experimental results for a cart–pole system: control input is computed using discrete time parameterized control

angle for LQR control was taken near to the origin. Figure 9.10 demonstrates how the proposed control algorithms balance the system for a large initial angle in real time. The disturbances of different magnitudes are given at regular intervals in real time to check the robustness of the proposed scheme. It can be easily noted from Figures 9.6–9.9 that the proposed direct inversion based control scheme is more robust in a sense that it could tolerate larger disturbance compared to LQR control. The instants where the system is disturbed are marked in the corresponding figures by the numbers. As the figures suggest, LQR takes longer time to stabilize after the disturbance. The disturbance rejection capability of all the algorithms are quantified in Table 9.3. The direct inversion based algorithms can tolerate a disturbance of magnitude 0.22 rad, whereas LQR and inversion based parameterized control can tolerate a disturbance of magnitude 0.07 rad.

Since we are not updating the gains online in the parameterized controller, this may be one of the reasons of getting poor disturbance rejection in this case. It can also be seen from Figures 9.8 and 9.9 that the graphs obtained from the parametrized control and LQR are similar in nature. Though the cart motion is smooth in both parametrized controller and LQR, an offset is present in the final cart position and the magnitude of the offset is greater in the case of LQR. The video of the experiment is available at

http://www.youtube.com/watch?v=L4uagc4bkzI

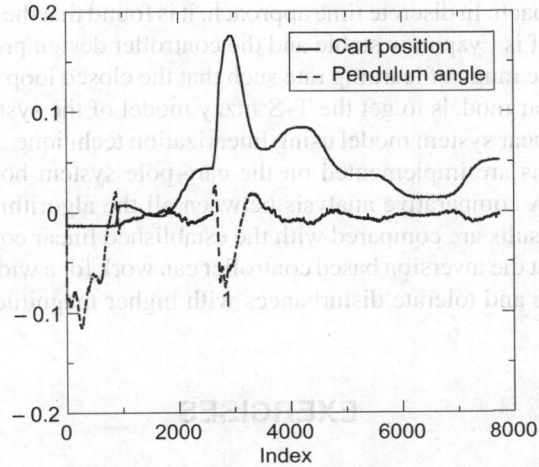

Figure 9.9 Experimental results for a cart–pole system: control input is computed using LQR control

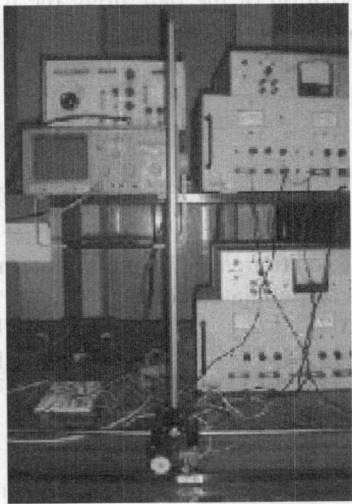

Figure 9.10 The inversion based controller balances the cart–pole system at its equilibrium point

SUMMARY

This chapter shows the application of an inversion based control scheme using the T–S fuzzy model of a system for which a cart–pole system is taken as a benchmark problem. In the first control scheme, the control input is computed directly through the inversion of the systems model. The other scheme parametrizes the input as a variable gain state feedback control where the feedback gains are updated by the inversion process. The update laws are derived using both continuous and discrete Lyapunov approach. In discrete time approach, it is found that the gradient descent update law itself is Lyapunov stable and the controller design problem simplifies to computing the range of learning rate such that the closed loop system becomes stable. The linear models to get the T–S fuzzy model of the system are obtained from the non-linear system model using linearization technique. All the inversion based algorithms are implemented on the cart–pole system both in simulation and real time. A comparative analysis between all the algorithms is given. The experimental results are compared with the established linear control law (LQR) and it is seen that the inversion based controller can work for a wide range of initial pendulum angle and tolerate disturbances with higher magnitudes compared to the LQR.

EXERCISES

1. Consider the example of a ball–beam system, where a servo motor is coupled to the ball and beam module. The motor drives a lever arm which is coupled to a track upon which a rolling ball rests. The purpose of design is to control the position of the ball along the track by manipulating the angular position of the servo. The simplified ball and motor dynamics are given as

$$\ddot{x} = -\frac{5}{7}g \sin(\frac{r}{L}\theta)$$

$$\ddot{\theta} = \frac{K_g K_m}{J_{eq} R_m}\left(V_i - K_b K_g \dot{\theta}\right)$$

where x is the ball position and θ is the motor angular displacement. The system parameters are tabulated below.

Motor inertia J_m	3.87×10^{-7} kg-m^2
Load inertia J_l	2.42×10^{-5} kg-m^2
Equivalent inertia J_{eq}	0.0029 kg-m^2
Gear ratio K_g	70 : 1
Motor resistance R_m	2.6 Ω
Back EMF constant K_b	0.00767 V/ (rad/s)
Motor torque constant K_m	0.00767 Nm/A
Beam length L	43.18 cm
Lever radius r	2.54 cm

The beam angle α is related to θ by the following relation:

$$\alpha = \frac{r}{L}\theta$$

(a) Considering $x, \dot{x}, \theta, \dot{\theta}$ as the system states and V_i the input, write down the state–space model of the system. Find out the discrete T–S fuzzy model representation of the system.

(b) Design a suitable inversion based control scheme for the above system such that the ball tracks a reference trajectory of $x_d = 0.2 \sin t$.

2. Consider the example of a horizontal plane two-degrees-of-freedom robot manipulator whose dynamics can be written explicitly as

$$a_1\ddot{q}_1 + (a_3C_{21} + a_4S_{21})\ddot{q}_2 - a_3S_{21}\dot{q}_2^2 + a_4C_{21}\dot{q}_2^2 = \tau_1$$

$$(a_3C_{21} + a_4S_{21})\ddot{q}_1 + a_2\ddot{q}_2 + a_3S_{21}\dot{q}_1^2 - a_4C_{21}\dot{q}_1^2 = \tau_2$$

where q_1 and q_2 are the link angles and τ_1 and τ_2 are the input torques. $C_{21} = \cos(q_2 - q_1)$, $S_{21} = \sin(q_2 - q_1)$. Take the four parameters $a_1, a_2, a_3,$ and a_4 as 0.15, 0.04, 0.03, and 0.025 kg m^2 respectively.

(a) Considering $q_1, \dot{q}_1, q_2, \dot{q}_2$ as the system states and τ_1 and τ_2 as the system inputs, express the system in the state–space form and find out the discrete T–S fuzzy model representation of the system.

(b) Design an inversion based controller for the above system to track the following reference trajectories:

$$q_{d1}(t) = 1.5 + 1.5\left(1. + 6e^{-t/3} - 8e^{-t/4}\right)$$

$$q_{d2}(t) = 1.0 + \left(1. + 6e^{-t/3} - 8e^{-t/4}\right)$$

CHAPTER

10

Visual Motor Control of a Redundant Manipulator

Guiding the end-effector position of a robot manipulator in its work space through visual feedback is commonly known as visual motor coordination (VMC). VMC is a special case of visual-servoing where manipulator dynamics is not taken into account. Readers can refer to [134], [135] for a survey on various visual-servoing techniques. Human way of hand–eye coordination is far superior to current practices of designing effective visual motor control tasks. Being inspired by superior visual motor control capability of biological species, various learning schemes have been proposed [136], [137], [138], [139] for inverse kinematic control of robot manipulators. VMC primarily involves two tasks, namely, extracting the coordinate information of robot end-effector and target point from camera images and then finding out the necessary joint angle movement to reach the target point. The first task necessitates camera calibration to find out an approximate model so that a mapping from real world coordinates to pixel coordinates is established. The Tsai algorithm [140] is one of the most widely used methods for this purpose. The second task is more commonly known as solving inverse kinematic problem.

This chapter presents two different learning frameworks where a redundant manipulator learns inverse kinematic solutions while interacting with the environment. First learning architecture has been developed using the Kohonen self-organizing map (KSOM) network, while the second learning model has been derived using the T–S fuzzy model.

10.1   SYSTEM MODEL
10.1.1   Experimental Set-up

The schematic diagram of a visual motor control set-up is shown in Figure 10.1. The set-up consists of a 7DOF robot manipulator, a pair of Ca-Zoom PTZ cameras,

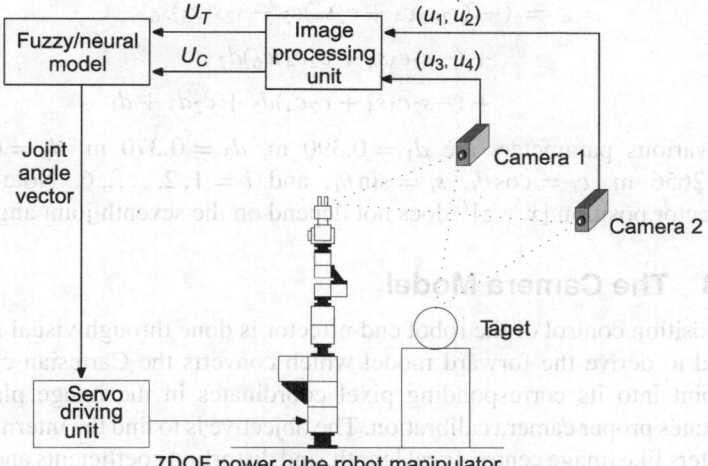

Figure 10.1 Schematic diagram of visual motor control

and a PC. The image processing part in Figure 10.1 is used to extract four dimensional image plane coordinate vectors for current end-effector position U_c and target point U_t. The fuzzy or neural model learns the inverse mapping of the manipulator to compute the necessary joint angle vector θ to reach the target U_t.

10.1.2 The Manipulator Model

The manipulator used for the experiment is a 7DOF PowerCube^{TM} robot manipulator from Amtec Robotics [141]. The manipulator model is derived from its link geometry. The D–H parameters [17] computed for this manipulator is provided in Table 10.1. Forward kinematics of the manipulator (the mapping from the joint angle space to the Cartesian space) is derived using the D–H parameters. The end-effector position is given by the following equations:

$$x = (-(((c_1 c_2 c_3 - s_1 s_3)c_4 - c_1 s_2 s_4)c_5$$

$$+ (-c_1 c_2 s_3 - s_1 c_3)s_5)s_6 + (-(c_1 c_2 c_3 - s_1 s_3)s_4$$

$$- c_1 s_2 c_4)c_6)d_7 + (-(c_1 c_2 c_3 - s_1 s_3)s_4$$

$$- c_1 s_2 c_4)d_5 - c_1 s_2 d_3$$

$$y = (-(((s_1 c_2 c_3 + c_1 s_3)c_4 - s_1 s_2 s_4)c_5$$

$$+ (-s_1 c_2 s_3 + c_1 c_3)s_5)s_6 + (-(s_1 c_2 c_3 + c_1 s_3)s_4$$

$$- s_1 s_2 c_4)c_6)d_7 + (-(s_1 c_2 c_3 + c_1 s_3)s_4$$

$$- s_1 s_2 c_4)d_5 - s_1 s_2 d_3$$

$$z = (-((s_2c_3c_4 + c_2s_4)c_5 - s_2s_3s_5)s_6$$
$$+ (-s_2c_3s_4 + c_2c_4)c_6)d_7$$
$$+ (-s_2c_3s_4 + c_2c_4)d_5 + c_2d_3 + d_1 \tag{10.1}$$

where various parameters are $d_1 = 0.390$ m, $d_3 = 0.370$ m, $d_5 = 0.310$ m, $d_7 = 0.2656$ m, $c_i = \cos\theta_i$, $s_i = \sin\theta_i$, and $i = 1, 2, \ldots, 6$. Note that the end-effector position $[x\ y\ z]^T$ does not depend on the seventh joint angle.

10.1.3 The Camera Model

Since position control of the robot end-effector is done through visual feedback, we need to derive the forward model which converts the Cartesian coordinate of a point into its corresponding pixel coordinates in the image plane. This necessitates proper camera calibration. The objective is to find the internal camera parameters like image centre, focal length, and distortion coefficients and external parameters corresponding to the position and orientation of a Cartesian coordinate system. The Tsai algorithm [140] for non-coplanar camera calibration is used to estimate 11 parameters—extrinsic (R_x, R_y, R_z, T_x, T_y, T_z) and intrinsic (f, κ, C_x, C_y, s_x).

Table 10.1   D–H parameters of PowerCube

Link (i)	α_{i-1}	a_{i-1}	d_i	θ_i
1	90	0	d_1	θ_1
2	−90	0	0	θ_2
3	90	0	d_3	θ_3
4	−90	0	0	θ_4
5	90	0	d_5	θ_5
6	−90	0	0	θ_6
7	0	0	d_7	$\theta 7$

The relationship between the position of a point P in world coordinates (x_w, y_w, z_w) and the point's image in the camera's frame buffer (x_f, y_f) can be established by a sequence of coordinate transformations. The image coordinate of point P is given by

$$x_f = \frac{s_x x_d}{d_x} + C_x$$

$$y_f = \frac{y_d}{d_y} + C_y \tag{10.2}$$

where s_x is the scaling factor, d_x and d_y are the x and y dimensions of camera's sensor element, respectively, C_x, C_y are pixels coordinates of image centre, and x_d and y_d are the true position on image plane considering radial distortions. The detailed derivation of these equations are available in [140] and [139].

10.2 VISUAL MOTOR CONTROL USING NEURAL NETWORKS

Several kinds of learning architectures have been suggested for learning the inverse kinematics of a robot manipulator. This includes multiple layer perceptrons (MLP) [142], radial basis functions (RBFs) [143, 144], locally weighted projection regression (LWPR) [145], reinforcement learning (RL), and self-organizing maps (SOM). Among these learning architectures, Kohonen's self-organizing map (KSOM) [146] has been used extensively for visuo-motor control of manipulators during the last two decades owing to its topology preserving property. A good survey on SOM based control methods for robot manipulator is available in [147]. Martinetz, et al. used three-dimensional KSOM-based neural architecture for visuo-motor control of non-redundant [137] and redundant manipulators [148]. Their method was later extended by Walter et al. [138] who introduced neural-gas algorithm along with an error correction scheme based on a Widrow–Hoff type learning rule. They also introduced neighbourhood concept in KSOM-based architecture to improve accuracy. Behera et al. [139] proposed a hybrid method to reduce the burden on the online training requirement. In this approach, a forward model comprising robot forward kinematics and camera model is used to generate data and train the network offline. The trained network is then used online to further fine-tune the parameters. KSOM-based VMC schemes have also been used for redundant manipulator where the redundancy is resolved by avoiding obstacles [149, 150], optimizing certain task oriented criteria [148]. Some more applications are reported in [151, 152].

10.2.1 Visual Motor Control with KSOM

The inverse kinematic relationship (10.21) is a one-to-many mapping and hence difficult to learn. In KSOM-based approach, the input space is mapped onto a 3-D lattice, where each lattice neuron represents a discrete cell. A *linear* map from the input space to the output space is learnt locally for each lattice neuron. The structure of the linear map for the lattice neuron γ is obtained by using the first-order Taylor series expansion of Eqn (10.21) given by

$$\theta - \theta_\gamma = A_\gamma(\mathbf{u} - \mathbf{w}_\gamma) \quad A_\gamma = \left.\frac{\partial \mathbf{g}}{\partial \mathbf{u}}\right|_{(\mathbf{w}_\gamma, \theta_\gamma)} \quad (10.3)$$

where $\mathbf{w}_\gamma$ is the centre of the lattice neuron, A_γ is the inverse Jacobian matrix, and θ_γ is the joint angle vector associated with γ lattice neuron. The discretization process using KSOM is illustrated pictorially in Figure 10.2. Each lattice neuron γ is represented by a triplet $(\mathbf{w}_\gamma, \theta_\gamma, A_\gamma)$, where $\mathbf{w}_\gamma$ discretizes the four-dimensional input space, θ_γ discretizes the six-dimensional output space and the matrix A_γ of dimension 6×4 represents a linear map from the input to the output space. Since each weight vector $\mathbf{w}_\gamma$ is associated with only one joint angle vector θ_γ, this method gives a unique inverse kinematic solution even for a redundant manipulator.

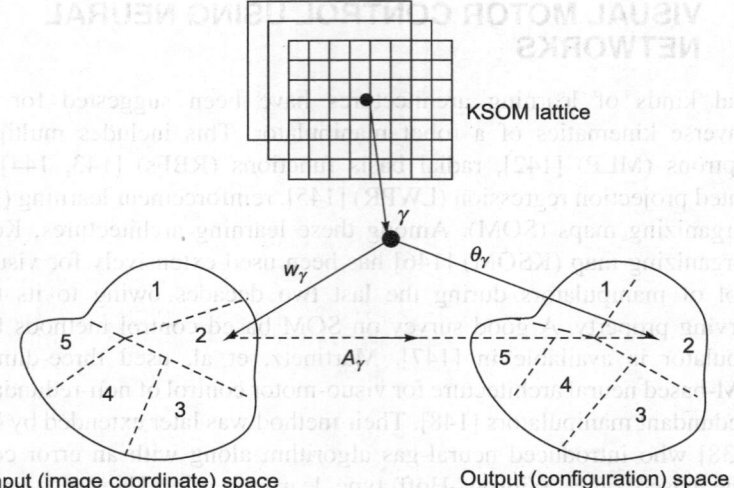

Figure 10.2   Discretization of input and output space using KSOM. $\mathbf{w}_\gamma$ discretizes the input space and θ_γ discretizes the output space. A_γ represents the linearized map between the input and output space

Given a target position $\mathbf{u}_t$, a winner neuron μ is selected based on its Euclidean distance metric in the image coordinate (input) space. The neuron whose weight vector is closest to the target is declared winner as shown below:

$$\mu = \arg\min_{\gamma} \|\mathbf{u}_t - \mathbf{w}_\gamma\| \qquad (10.4)$$

The arm is given a coarse movement θ_0^{out} obtained as the weighted average of the individual neuron outputs as shown below:

$$\theta_0^{\text{out}} = s^{-1} \sum_{\gamma} h_\gamma (\theta_\gamma + A_\gamma(\mathbf{u}_t - \mathbf{w}_\gamma)) \qquad (10.5)$$

where $s = \sum_{\gamma} h_\gamma$ and $h_\gamma = e^{-(\frac{\|\mu - \gamma\|}{2\sigma^2})}$ is the neighbourhood function that is used as a variable weighting coefficient for each individual neuron output. Because of this coarse movement, the end-effector reaches a position $\mathbf{v}_0$. A correcting fine movement θ_1^{out} is evaluated as follows:

$$\theta_1^{\text{out}} = \theta_0^{\text{out}} + s^{-1} \sum_{\gamma} h_\gamma A_\gamma(\mathbf{u}_t - \mathbf{v}_0) \qquad (10.6)$$

This corrective movement results in a final movement of the end-effector to a position $\mathbf{v}_1$. Although we can use several such corrective movements to increase the accuracy of tracking, usually one corrective movement is sufficient to obtain a good tracking accuracy. Note that θ_0^{out} and θ_1^{out} are the joint angle outputs obtained from the network which are applied directly to the robot manipulator. These are different from their internal representations θ_γ, which are the angle vectors associated with lattice neurons $\gamma = 1, 2, \ldots$.

Various network parameters are updated as follows [153, 138, 139]:

$$\Delta \mathbf{v} = \mathbf{v}_1 - \mathbf{v}_0 \tag{10.7}$$

$$\Delta \boldsymbol{\theta}^{\text{out}} = \boldsymbol{\theta}_1^{\text{out}} - \boldsymbol{\theta}_0^{\text{out}} \tag{10.8}$$

$$\Delta \boldsymbol{\theta}_\gamma = \frac{h_\gamma}{s} \left[\boldsymbol{\theta}_0^{\text{out}} - s^{-1} \sum_\gamma h_\gamma (\boldsymbol{\theta}_\gamma + A_\gamma (\mathbf{v}_0 - \mathbf{w}_\gamma)) \right] \tag{10.9}$$

$$\Delta A_\gamma = \frac{h_\gamma}{s \|\Delta \mathbf{v}\|^2} \left[\Delta \boldsymbol{\theta}^{\text{out}} - s^{-1} \sum_\gamma h_\gamma A_\gamma \Delta \mathbf{v} \right] \Delta \mathbf{v}^T \tag{10.10}$$

$$\mathbf{w}_\gamma \leftarrow \mathbf{w}_\gamma + \eta_w \, h_\gamma (\mathbf{u}_t - \mathbf{w}_\gamma) \tag{10.11}$$

$$\boldsymbol{\theta}_\gamma \leftarrow \boldsymbol{\theta}_\gamma + \eta_t \, \Delta \boldsymbol{\theta}_\gamma \tag{10.12}$$

$$A_\gamma \leftarrow A_\gamma + \eta_a \, \Delta A_\gamma \tag{10.13}$$

The parameters $\eta_w, \eta_t, \eta_a, \sigma$ are varied during the training as follows:

$$\eta = \eta_{\text{init}} \left(\frac{\eta_{\text{fin}}}{\eta_{\text{init}}} \right)^{(t/t_{\text{max}})} \tag{10.14}$$

In the above equation, $\eta \in \{\eta_w, \eta_t, \eta_a, \sigma\}$ is a user-defined parameter that varies between initial value η_{init} and η_{fin} as the iteration step t proceeds from 0 to t_{max} during the training phase.

The parameter update rules (10.11)–(10.13) may be explained as follows: The weight update law (10.11) is a simple KSOM clustering algorithm where the winner neuron and the neurons in its neighbourhood are moved towards the input vector $\mathbf{u}_t$. The update is aimed at reducing the error term $\|\mathbf{u}_t - \mathbf{w}_\gamma\|$. The parameter A_γ is updated so as to learn the linear mapping (10.3) which is rewritten as

$$\boldsymbol{\theta}^* = \boldsymbol{\theta}_\gamma + A_\gamma (\mathbf{u}_t - \mathbf{w}_\gamma) \tag{10.15}$$

where $\boldsymbol{\theta}^*$ is the desired joint angle necessary for reaching the target point $\mathbf{u}_t$. In order to learn this mapping, A_γ is updated so that the following cost function is minimized:

$$E_1 = \frac{1}{2} (\Delta \boldsymbol{\theta}^{\text{out}} - A_\gamma \Delta \mathbf{v})^2 \tag{10.16}$$

Applying the gradient–descent rule to the above equation, we obtain the Jacobian update law (10.13). The pre-factor $\|\Delta \mathbf{v}\|^{-2}$ determines the size of adaptation step. $\boldsymbol{\theta}_\gamma$ is the internal representation of the joint angle output $\boldsymbol{\theta}^{\text{out}}$ corresponding to each neuron γ. During the coarse movement, the robot joints are moved by $\boldsymbol{\theta}^{\text{out}}$ and its new position is recorded from the camera as $\mathbf{v}_0$. Now, $\boldsymbol{\theta}_\gamma$ is updated so as to reflect this new joint angle position at the nodal level. In other words, $\boldsymbol{\theta}_\gamma$ is updated so as to minimize the following error function given by

$$E_2 = \frac{1}{2} \left(\theta_0^{\text{out}} - \frac{\sum_\gamma h_\gamma [\theta_\gamma + A_\gamma (\mathbf{v}_0 - \mathbf{w}_\gamma)]}{\sum_\gamma h_\gamma} \right)^2 \qquad (10.17)$$

The update law for θ_γ given by Eqns (10.8) and (10.12) are obtained by applying the gradient–descent rule to the above error function.

The standard KSOM-based VMC scheme has the following limitations, which restrict its applicability to redundant manipulators.

- It is found that for a redundant manipulator with six or higher degrees of freedom, although the KSOM lattice neurons preserve topology of the input space as shown in Figure 10.3, the lattice fails to preserve the topology of the output (joint angle) space as shown in Figure 10.4. In Figure 10.3, it can be seen that the weight vectors ($\mathbf{w}_\gamma$) represented by square 'boxes' are spread out uniformly over the input space. On the other hand, in Figure 10.4, we find that the clusters in joint angle space (θ_γ) represented by square 'boxes' are concentrated at one location. It would be shown in the simulation section that because of the fact that the network fails to capture the output topology, the positioning accuracy attained using the standard KSOM algorithm is sensitive to initial conditions.

- The standard KSOM algorithm returns a unique inverse kinematic solution for any target in the manipulator workspace. This might not be desirable in the case of redundant manipulators where we would like to choose a different configuration to satisfy some additional requirement. Even though the training data sets are replete with redundant solutions, there is no provision to *preserve* this redundancy during the evolution of parameters.

To summarize, the standard KSOM-based visual motor control schemes are not efficient to learn the inverse-kinematic relation for redundant manipulators. In the following section, we propose a new architecture in order to overcome the above mentioned drawbacks of conventional VMC algorithms.

10.2.2 Simulation and Experimental Results

Network Architecture and Workspace Dimensions

A three-dimensional neural lattice with $7 \times 7 \times 7$ neurons is selected for the task. Training data are generated using the forward kinematic model (10.1) and the camera model (10.2). A Cartesian work-space of dimension of 600 mm $\times$ 500 mm $\times$ 500 mm is considered for both simulation and experiment. All points within this work space are visible through both the cameras of the stereo-vision system. Joint angle values are generated randomly within the physical limits of the manipulator and only those input–output pairs are retained where the end-effector positions are visible by both the cameras simultaneously. The ranges of input and output spaces are given in Table 10.2. Since end-effector positions in the camera plane and joint angles have different range of values, data points are normalized within ± 1.

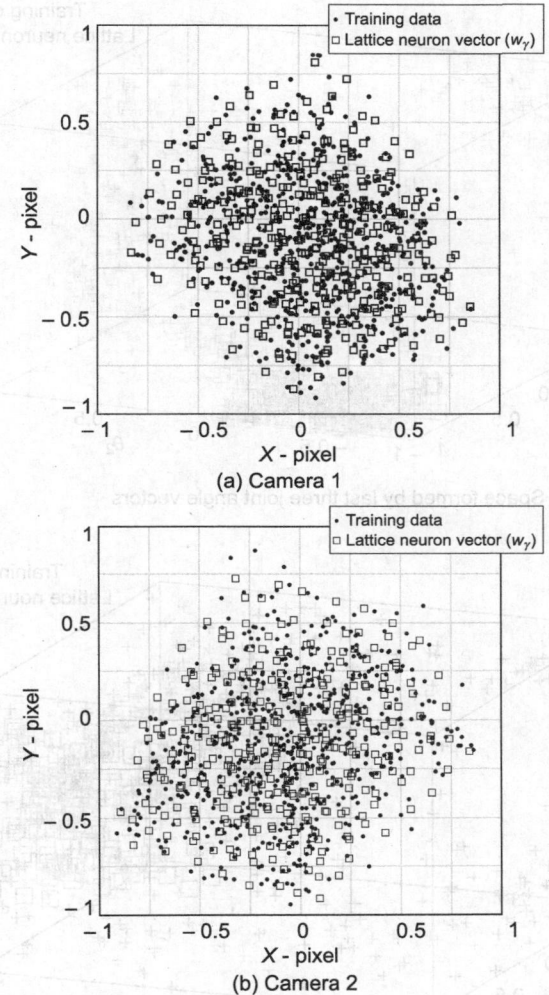

Figure 10.3 Clustering in image-coordinate (input) space. Lattice neurons capture the topology of the input space during training. The circular dots denote the actual input data generated during the training and the squares represent the cluster centres $\mathbf{w}_\gamma$.

10.2.3 Training

The network is trained offline using 50000 data generated using forward kinematic model (10.1) and (10.2). The training can be carried out 'online' which would necessitate generating data by moving the robot continuously. Generating such a large number of data on a real system might not be convenient. Hence, we follow the hybrid approach proposed by Behera et al. [139] where the network is trained off-line using approximate models and then it is fine-tuned during the online operation.

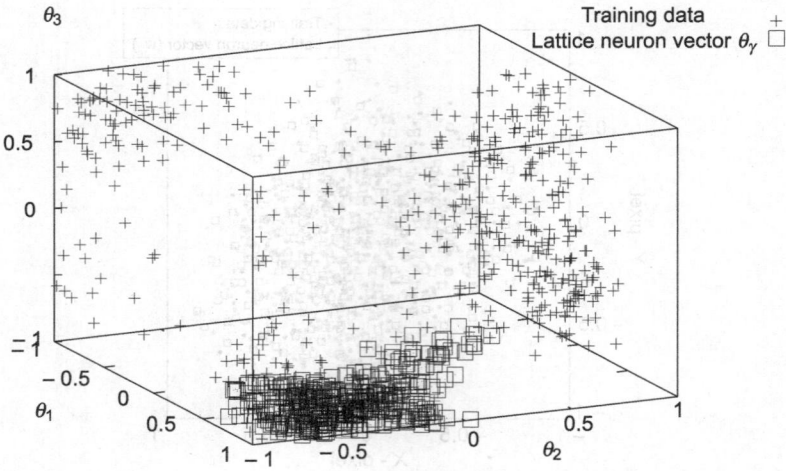

(a) Space formed by last three joint angle vectors

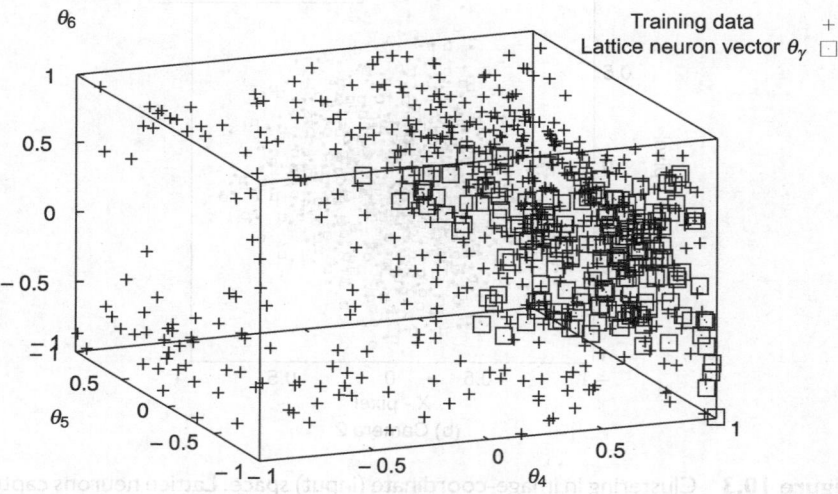

(b) Space formed by the last three joint angle vectors

Figure 10.4   The topology of the output space is not captured by the original VMC
algorithm during the evolution of parameters (training phase). While
the training data shown by '+' signs are distributed across the entire
volume, the cluster centres θ_γ are collected at one location

10.2.4 Testing

The following tasks were performed to demonstrate the efficacy and usefulness
of the proposed schemes.

Table 10.2 Ranges of input and output spaces

Ranges of joint angle	Range of Cartesian work space
$-160° \leq \theta_1 \leq 160°$	$-0.3 \text{ m} \leq x \leq 0.3 \text{ m}$
$-95° \leq \theta_2 \leq 95°$	$0.3 \text{ m} \leq y \leq 0.8 \text{ m}$
$-160° \leq \theta_3 \leq 160°$	$0.0 \text{ m} \leq z \leq 0.5 \text{ m}$
$-50° \leq \theta_4 \leq 120°$	
$-90° \leq \theta_5 \leq 90°$	
$-120° \leq \theta_6 \leq 120°$	
$-360° \leq \theta_7 \leq 360°$	

Tracking a Straight Line Trajectory

The desired straight line trajectory in the Cartesian space is given by

$$y = \frac{5}{6}x + \frac{11}{20}$$

$$z = \frac{5}{6}(x + 0.3) \tag{10.18}$$

where $-0.3 \text{ m} \leq x \leq 0.3 \text{ m}$. A total of 600 points are generated sequentially on this line and the joint angles for each point is computed using only one fine movement. Since learning algorithm has been designed so that it can be used online, the parameters are updated even during the testing phase. This helps improve the tracking accuracy. A typical tracking result obtained using the sub-clustering scheme is shown in Figure 10.5. In this case, the redundancy is resolved using the lazy-arm criterion. The corresponding joint angle trajectory is shown in Figure 10.5. It is seen that for a continuous trajectory in the Cartesian space, the joint angle movement is continuous and hence the inverse kinematic solution is conservative in nature.

Tracking an Elliptical Trajectory

The desired trajectory to be traversed is given by

$$x = 0.2 \sin t$$
$$y = 0.5 + 0.2 \cos t$$
$$z = \frac{5}{6}(x + 0.3) \tag{10.19}$$

where t varies from 0 to 2π. A total of 628 points are generated sequentially on this trajectory; the joint angles are computed in one step for each point. A typical trajectory obtained using the lazy-arm criterion and the sub-clustering technique is shown in Figure 10.6(a). The corresponding trajectories for joint angles are shown in Figure 10.6(b). This reaffirms our previous assertion that the inverse kinematic solution obtained is conservative.

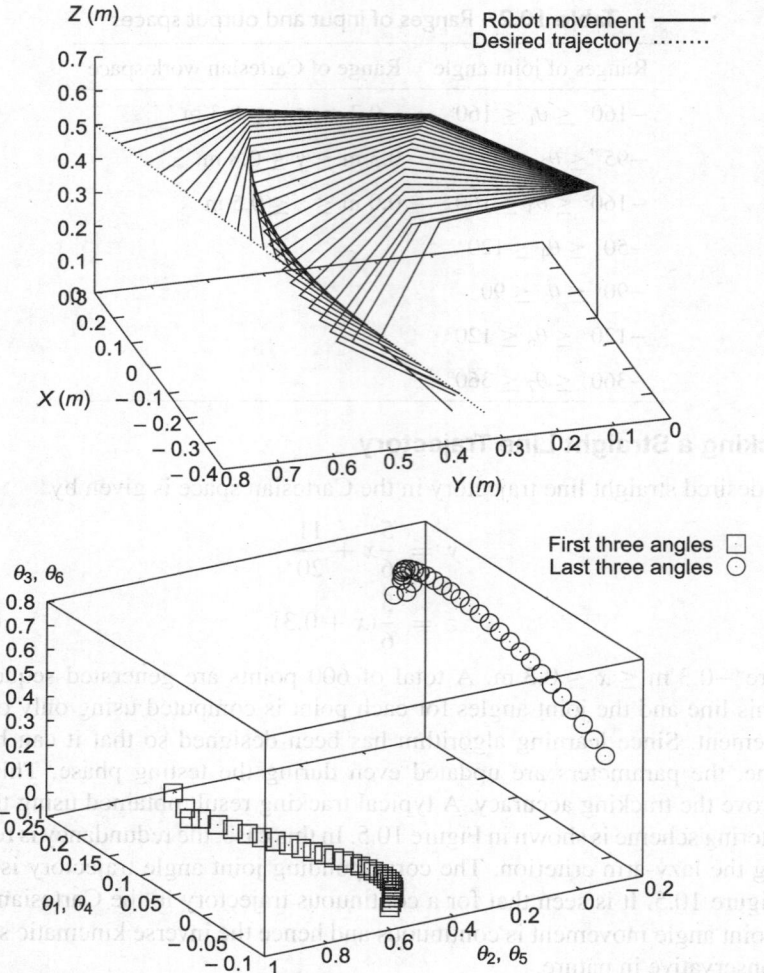

Figure 10.5 Tracking a straight line using the lazy-arm criteria. A continuous trajectory in the task space gives rise to a continuous trajectory in the joint angle space

10.2.5 Real-time Experiment

The actual set-up used for experiment is shown in Figure 10.7. The main hardware components of the set-up consist of a computer, PowerCube robot manipulator [141], and a pair of Ca-Zoom PTZ cameras [154] along with its image acquisition facility. PowerCube has separate servo controllers, one for each revolute joint. The robot is programmed using C APIs provided by the company. It is possible to provide the joint angles directly to each joint. The velocity and acceleration of joints are fixed beforehand. The Ca-Zoom PTZ cameras are interfaced with the PC through USB ports. The Cartesian workspace visible by both cameras has a dimension of 600 mm × 600 mm × 500 mm. The image frame has a dimension

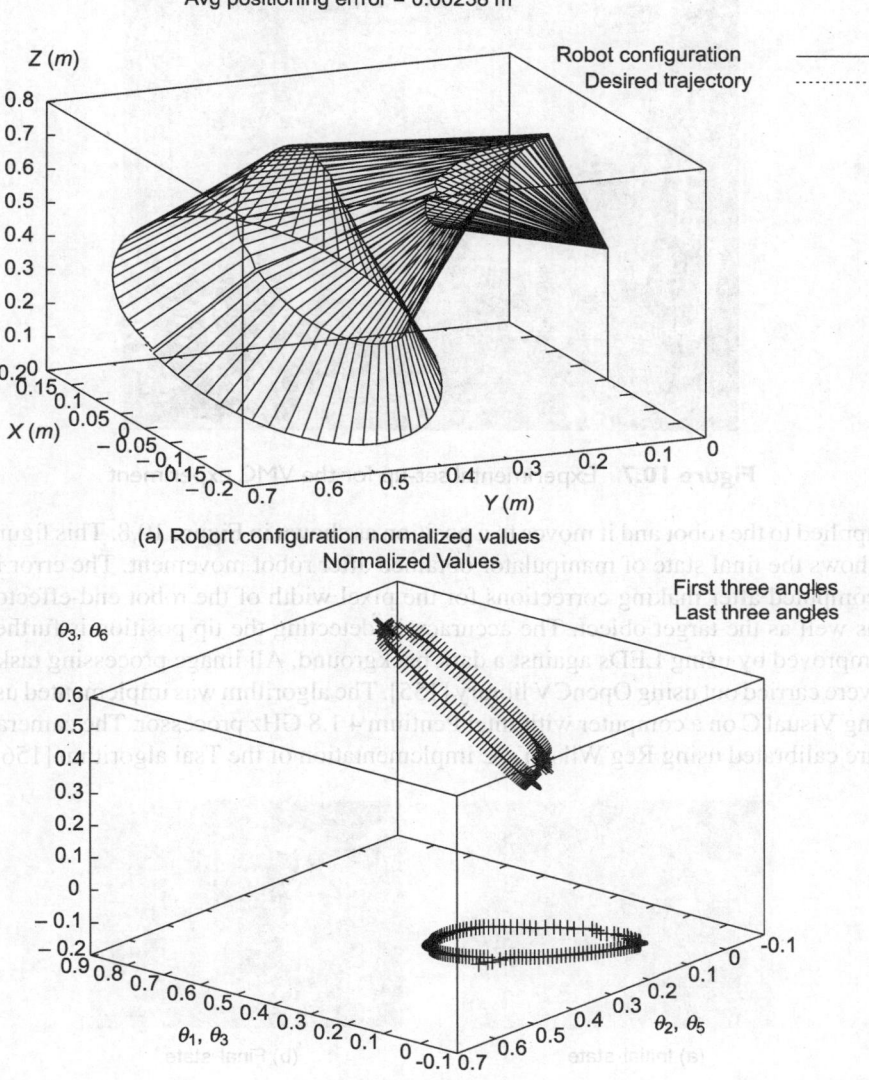

Avg positioning errror = 0.00238 m

(a) Robort configuration normalized values

(b) Joint angle trajectories

Figure 10.6   Tracking an elliptical trajectory using the lazy arm movement. The inverse kinematic solution is conservative in the sense that a closed loop trajectory in the task space gives rise to a closed trajectory in the configuration space

of 320×240 pixels. The target and the robot tip are identified with yellow and pink colours, respectively. The initial location of the robot end-effector and target in the image plane is shown in Figure 10.8. The regions of interest are extracted using thresholding and filtering operations. The centroid of the region is used by the VMC algorithm to compute necessary joint angles. These joint angles are

Figure 10.7 Experimental set-up for the VMC experiment

applied to the robot and it moves to a position as shown in Figure 10.8. This figure shows the final state of manipulator obtained after robot movement. The error is computed after making corrections for the pixel width of the robot end-effector as well as the target object. The accuracy in detecting the tip position is further improved by using LEDs against a dark background. All image processing tasks were carried out using OpenCV library [155]. The algorithm was implemented using Visual C on a computer with Intel Pentium 4 1.8 GHz processor. The cameras are calibrated using Reg Wilson's C implementation of the Tsai algorithm [156].

(a) Initial state   (b) Final state

Figure 10.8 The manipulator is visually guided to reach a target. Initial state is the state before robot movement. The final state refers to the state obtained after robot movement

The trained network was used online to compute joint angle vectors for 20 random locations in the manipulator workspace. Since it was not possible to accurately measure the manipulator tip positions in world coordinate, the distance error was measured directly in pixel coordinates. Only one fine step was used to compute the necessary joint angle vector for each point. The average distance error in the image plane is computed to be 12 pixels. This error can be reduced by taking multiple fine steps. It takes 15 steps, on an average, for reaching an accuracy of about 1 pixel for a given target point in image plane.

10.3 VISUAL MOTOR CONTROL USING A FUZZY NETWORK

In the context of visual-servoing of manipulators, fuzzy logic has been used for identifying targets and grasping tasks [157], [158], [159]. Kumaresan et al. [160] suggest a fuzzy logic based scheme where the need for re-calibrating cameras is eliminated in the case of any inadvertent camera displacements. While there are several other indirect applications of fuzzy logic based controllers for solving visual motor coordination problem, there is hardly any work where fuzzy logic has been used to learn the inverse kinematics directly from its input and output data sets. Prochazka [161], [162] argued that ensembles of sensory input 'vote' for motor outcomes according to the behaviour-dependent weighting factors, analogous to fuzzy control. This forms the basis of this work where a fuzzy clustering architecture is proposed to learn the inverse kinematic between the end-effector position viewed through a pair of cameras and the joint angle vector.

In this chapter, the inverse mapping is derived in terms of a fuzzy cluster of local linear maps. The input space is discretized into a number of clusters whose centres are formed using the fuzzy c-mean clustering algorithm [163], [164]. The centre locations are determined by the distribution of the input data patterns themselves and hence these clusters preserve the topology of the input space. For each centre, a linear inverse kinematic relationship is learnt between the end-effector position and the joint angle vectors. The fuzzy clustering allows each data vector to belong to every cluster with a fuzzy membership value between 0 and 1. Thus, the required joint angle vector for a desired target position can be calculated as a weighted average of the joint angles corresponding to all cluster centres.

Another issue that is dealt with in this chapter is the *conservative property of inverse kinematic solution*. Pseudo-inverse based methods are not conservative [165] in the sense that a closed continuous trajectory in the Cartesian space (or the image space) does not yield a closed trajectory in the joint angle (configuration) space. While it is not necessary when we are only worried about reaching an isolated point, it might be necessary while following a trajectory in the task space. In order to preserve this conservative property, an incremental learning based scheme is adopted for the architecture described above. Moreover, the advantage of this method over single step learning is that one can always improve the positioning accuracy based on the current error.

Manipulator forward kinematic model (10.1) and camera model (10.2) relate the seven-dimensional joint angle space to the four-dimensional image coordinate space. However, the manipulator end-effector position does not depend on the last joint angle, which is only a rotational motion for the wrist. Thus, the end-effector position can be seen as a function of only six joint angles. The forward map between six-dimensional joint angle space to the four-dimensional image coordinate space may be represented as

$$\mathbf{u} = F(\boldsymbol{\theta}) \tag{10.20}$$

where $\mathbf{u} = [u_1 \, u_2 \, u_3 \, u_4]$ is the four-dimensional image coordinate vector obtained from two cameras using Eqn (10.2) and $\theta = [\theta_1 \; \theta_2 \; \theta_3 \; \theta_4 \; \theta_5 \; \theta_6]^T$ is the six-dimensional joint angle vector. In the visual motor control, we seek a feasible joint angle vector θ for a given end-effector position. Thus, the inverse kinematics problem may be formulated as

$$\theta = G(\mathbf{u}), \quad G = F^{-1} \quad (10.21)$$

where $\mathbf{u}$ is the 4×1 target vector in the image coordinate plane to be reached by the robot end-effector. θ is a 6×1 joint angle vector to be computed. Since the dimension of the task space is less than the dimension of the joint-angle space, we will have more than one solution for θ for a particular $\mathbf{u}_t$. Thus, we need to resolve redundancy while solving the inverse kinematic problem. The objective of this work is to learn this mapping $G = F^{-1}$ from the manipulator task space to the joint space using a fuzzy cluster of local linear models, which solves the redundancy in the joint space as well.

Eqn (10.21) can be linearized at an operating point $(\theta_0, \mathbf{u}_0)$ by using the first-order Taylor's series approximation:

$$(\theta - \theta_0) = \left.\frac{\partial \overline{G}}{\partial \mathbf{u}}\right|_{(\theta_0 \, u_0)} (\mathbf{u} - \mathbf{u}_0) \quad (10.22)$$

If we denote $\frac{\partial G}{\partial \mathbf{u}}$ by $\overline{A}$, then Eqn (10.21) can be expressed as a fuzzy cluster of M local linear models where jth fuzzy rule has the following form:

IF x_1 is $x_{1_{0_j}}$ and y_1 is $y_{1_{0_j}}$ and x_2 is $x_{2_{0_j}}$ and y_2 is $y_{2_{0_j}}$ THEN

$$(\theta - \theta_{0_j}) = \overline{A}_j \, (\mathbf{u} - \mathbf{u}_{0_j}) \quad (10.23)$$

where $\mathbf{u}_{0_j} = [x_{1_{0_j}} \; y_{1_{0_j}} \; x_{2_{0_j}} \; y_{2_{0_j}}]^T$ and $\overline{A}_j \in R^{4 \times 6}$.

Given an input–output pair $(\theta, \mathbf{u})$, the fuzzy model around this operating point is constructed as the weighted average of all the linear systems and has the form

$$\theta = \frac{\sum_{j=1}^{M} \mu_j \left(\theta_{0_j} + \overline{A}_j(\mathbf{u} - \mathbf{u}_{0_j})\right)}{\sum_{j=1}^{M} \mu_j} \quad (10.24)$$

where μ_j is the membership value of the jth fuzzy region. Eqn (10.24) can be rewritten as

$$\theta = \sum_{j=1}^{M} \sigma_j \left(\theta_{0_j} + \overline{A}_j(\mathbf{u} - \mathbf{u}_{0_j})\right) \quad (10.25)$$

where

$$\sigma_j = \frac{\mu_j}{\sum_{j=1}^{M} \mu_j} \quad \sum_{j=1}^{M} \sigma_j = 1 \quad (10.26)$$

The unknown parameters in the inverse kinematic model given in Eqn (10.25) are θ_{0_j}, the output joint angle vector, $\overline{A}_j$, the inverse Jacobian matrix; σ_j, the normalized fuzzy membership value, and u_{0_j}, the cluster centre. All these parameters are associated with the jth fuzzy cluster. Similarly, there are N such fuzzy clusters and these parameters are learned online using the training data. While training, the robot manipulator is guided to a random position $\mathbf{u}$ as observed by two cameras by actuating a joint angle vector θ randomly selected from the feasible joint angle space.

10.3.1 Fuzzy C-Mean Clustering

A fuzzy C-mean clustering based algorithm has been proposed where both the fuzzy centres and output joint angle vectors are learned online.

The main difference between a normal clustering and a fuzzy clustering is that in normal clustering, a particular data point belongs to a specific cluster whereas in fuzzy clustering a data point can belong to more than one clusters. The belongingness of each cluster will depend on the corresponding membership value. In real applications, it is difficult to find sharp boundaries between the clusters and hence it is natural to expect the fuzzy clusters to represent such data distributions in a better way. The fuzzy C-mean (FCM) clustering algorithm was introduced by Dunn in 1973 [163] and was improved by Bezdek in 1981 [164]. FCM finds out the cluster centres by minimizing the total weighted mean-square error of the data points and cluster centres. Let $\{x_{(q)} : q = 1, \ldots, Q\}$ be a set of Q data points. Each data point $x_{(q)} = (x_{(q)}^1, \ldots, x_{(q)}^n)$ has n components. The process of clustering is to assign the Q feature vectors into K clusters $\{c_{(k)} : k = 1, \ldots, K\}$ usually based on minimization of the following objective function:

$$J = \sum_{q=1}^{Q} \sum_{k=1}^{K} \mu_{qk}^m \|x_q - c_k\|^2 \qquad 1 \leq m < \infty \qquad (10.27)$$

where m is any real number greater than 1, x_q is the qth data of the n-dimensional data set, μ_{qk} is the degree of membership of x_q in the cluster k, c_k is the n-dimension centre of the cluster. It should be noted that the summation over the membership values of all clusters should be 1 for each data, i.e., $\sum_{k=1}^{K} \mu_{qk} = 1$ for each q.

Fuzzy partitioning is carried out through an iterative optimization of the objective function shown in (10.27), where the update laws for membership value μ_{qk} and the cluster center c_k are as follows:

$$\mu_{qk} = \frac{1}{\sum_{j=1}^{K} \left(\frac{\|x_q - c_k\|}{\|x_q - c_j\|} \right)^{\frac{2}{m-1}}} \qquad (10.28)$$

$$c_k = \frac{\sum_{q=1}^{Q} \mu_{qk}^m x_q}{\sum_{q=1}^{Q} \mu_{qk}^m} \qquad (10.29)$$

The iteration is stopped when a specific termination criterion is met. This procedure converges to a local minimum of J. As mentioned earlier, FCM allows each data vector to belong to every cluster with a fuzzy truth value between 0 and 1, which is computed using (10.28).

The work space in the image plane coordinate is divided into a number of clusters and the centres of the clusters ($\mathbf{u}_{0_j}$ and θ_{0_j} in Eqn (10.25)) are computed using the C-mean clustering algorithm.

The FCM algorithm has also been extended for a three-dimensional (3D) lattice. In this case the cost function to be minimized is modified as follows:

$$J = \sum_{q=1}^{Q} \sum_{i=1}^{I} \sum_{j=1}^{J} \sum_{k=1}^{K} \mu_{qijk}^{m} \|x_q - c_{ijk}\|^2 \quad 1 \le m < \infty \quad (10.30)$$

where i, j, and k are the number of nodes in the respective directions. The distance is now calculated in a 3D lattice and the centres and the membership values are modified based on the newly measured distance parameter. It will be shown in the simulation section that the fuzzy 3D cluster provides better results compared to a 1D cluster.

The fuzzy C-mean clustering algorithm as proposed has been implemented for the 7DOF robot manipulator considered in this chapter. The corresponding cluster locations are shown in Figure 10.10. It is seen that there are more clusters in denser regions than in thinner regions.

10.3.2 Multi-step Incremental Learning

While tracking a trajectory, to maintain the conservativeness of the inverse relationship [166], an incremental learning scheme is proposed which update the Jacobian matrices and centres of each fuzzy cluster online. Martinetz et al. [137] first proposed an incremental learning algorithm for a Kohonen–SOM based neural architecture which retains the conservativeness in the inverse mapping. The single-step learning is purely supervised, whereas the multi-step learning takes into account how closely the end-effector is following the target. Using this concept, an error correcting mechanism for a fuzzy cluster based inverse mapping is implemented in this section. As shown in Figure 10.9, the error between the current position and the target position is used to provide further incremental steps in the joint angle so that the position tracking accuracy is improved. Thus in this case, along with the target position, the error between the target position and end-effector position is also fedback through the stereo-vision system. The algorithm modifies the parameters of the fuzzy cluster at every incremental step.

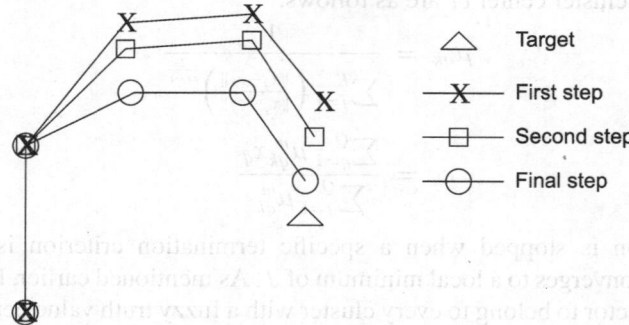

Figure 10.9 Multi-step learning: the error between the current position and the target position decides the corrective action

The algorithmic steps are as follows.

- For a target vector u_t a number of fuzzy zones will be fired with a membership value of σ_j for jth centre. The robot arm is given a coarse movement θ_0^{out}, which is determined by Eqn (10.25). Suppose that θ_0^{out} takes the end-effector to a position v_0. A correcting fine movement θ_1^{out} is computed as

$$\theta_1^{out} = \theta_0^{out} + \sum_j \left[\sigma_j \bar{A}_j (u_t - v_0) \right] \qquad (10.31)$$

With this correcting movement the final end-effector position becomes v_1. Several such correcting movements can be used to increase the position tracking accuracy. The parameters involved with each centre can be updated at each error correcting movement.

- The learning scheme used in this work is as follows:

$$\Delta v = v_1 - v_0 \qquad (10.32)$$

$$\Delta \theta^{out} = \theta_1^{out} - \theta_0^{out} \qquad (10.33)$$

$$\Delta \theta_{0_j} = \theta_0^{out} - \theta_{0_j} - \bar{A}_j (v_0 - u_{0_j}) \qquad (10.34)$$

$$\Delta \bar{A}_j = \|\Delta v\|^{-2} (\Delta \theta^{out} - \bar{A}_j \Delta v) \Delta v^T \qquad (10.35)$$

$$u_{0_j} = u_{0_j} + \eta \, \sigma_j (u_t - u_{0_j}) \qquad (10.36)$$

$$\theta_{0_j} = \theta_{0_j} + \eta \, \sigma_j \, \Delta \theta_{0_j} \qquad (10.37)$$

$$\bar{A}_j = \bar{A}_j + \eta \, \sigma_j \, \Delta \bar{A}_j \qquad (10.38)$$

where σ_j is defined in (10.26). The learning parameter η can be varied during the training time using the following equation:

$$\eta = \eta_{init} \left(\frac{\eta_{fin}}{\eta_{init}} \right)^{t/t_{max}}$$

where η_{init} is the initial learning rate, η_{fin} is the final learning rate, t is the current iteration, and t_{max} is the maximum iteration. The first weight update law (10.36) is aimed at reducing the error term $\|u_t - u_{0_j}\|$. The parameter $\bar{A}_j$ is updated such that the linear mapping $\theta^* = \sum_j \sigma_j [\theta_{0_j} + \bar{A}_j (u_t - u_{0_j})]$ is learned properly where θ^* is the desired joint angle vector. In order to learn this mapping, the following cost function is minimized:

$$E = \frac{1}{2} (\Delta \theta^{out} - \bar{A}_j \Delta v)^2$$

This leads to the update law (10.38) where $\|\Delta v\|^{-2}$ is a weighing factor. During the coarse movement, joint angles are moved by θ_0^{out} and its new end-effector position is recorded by the cameras as v_0. Hence θ_{0_j} is updated in such a way that this new joint angle position is reflected in the update law. The corresponding update law is given by (10.37).

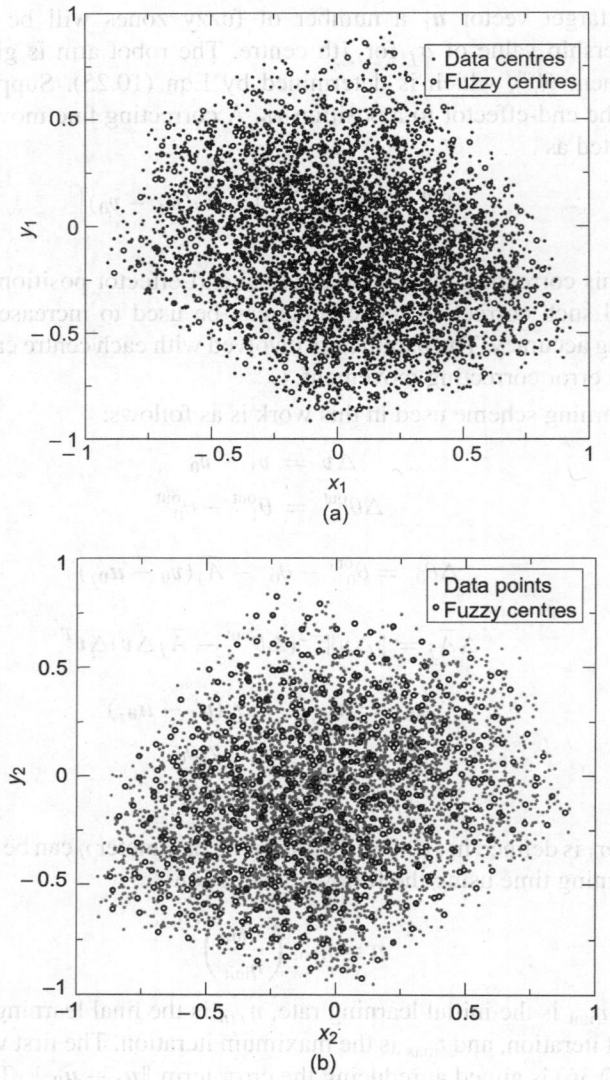

Figure 10.10 Fuzzy centres are obtained in the image coordinate space during training using the proposed fuzzy cluster: (a) camera 1, (b) camera 2

10.3.3   Simulation and Experimental Results

In this section, we provide simulation as well as experimental results for visual motor coordination using the fuzzy learning paradigms as described in Section 10.3.1.

During the training phase, joint angles are randomly generated within its physical limits in such a way that the end-effector positions lie in a region visible by both the cameras simultaneously. This corresponds to a cuboidal region in the Cartesian space. The physical boundaries of the input and output space is given in Table 10.3. The training data are generated using the forward kinematic Eqns (10.1) and the camera model Eqn (10.2).

Table 10.3 Boundary of input and output space

Physical limits on joint angles	Ranges of Cartesian coordinates
$-250° < \theta_1 < 70°$	$-0.3\ m < x < 0.3\ m$
$-95° < \theta_2 < 95°$	$0.3\ m < y < 0.8\ m$
$-160° < \theta_3 < 160°$	$0 < z < 0.5\ m$
$-50° < \theta_4 < 120°$	
$-95° < \theta_5 < 95°$	
$-120° < \theta_6 < 120°$	

10.3.4   VMC Using Incremental Learning

As discussed earlier, the positioning accuracy attained in a single step might not be sufficient for many tasks. An incremental learning scheme is desirable so that the accuracy can be improved by taking multiple steps. It can be considered as an online error correcting mechanism where the parameters are also updated between two consecutive movements. Unlike the single-step learning, the error between the target position and the current end-effector position is also fedback through the camera system to provide the next incremental step. The incremental learning scheme is described in Section 10.3.2. The learning parameters for incremental learning are given below:

$$\eta_{\text{init}} = 0.5, \quad \eta_{\text{fin}} = 0.01$$

For a 1D lattice, the number of fuzzy clusters is chosen as 729. For a 3D lattice, the number of nodes taken in each direction is 9, thus yielding a total number of 729 clusters. The lattice nodes capture the topology of the input space as shown in Figure 10.10. The dots represent the actual data points generated during training and the circles represent the cluster centres.

It is possible to attain arbitrary accuracy using incremental learning. The number of steps necessary for attaining an accuracy of 1 mm for 50 points is shown in Figure 10.11. The average number of steps necessary for attaining 1 mm accuracy is found to be 13 in the case of a 3D fuzzy cluster, whereas for standard KSOM [138] it is 17. Another important finding is that in the case of the standard KSOM the joint angles often crosses the physical boundaries, whereas in the case of a fuzzy cluster joint angles stay within the physical limit. A hard limiter is used for the former case to limit the joint angles within their bounds.

Trajectory Tracking:
The proposed fuzzy based learning scheme is also used to achieve a desired trajectory in the Cartesian space. The desired trajectories for the $X - Y - Z$

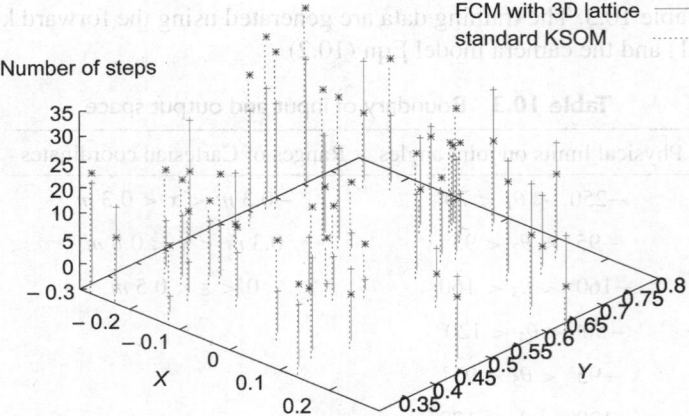

Figure 10.11 Incremental learning: number of movements necessary for attaining an accuracy of 1 cm. For the same accuracy, the average number of steps for 50 random points is less in the case of FCM with 3D lattice.

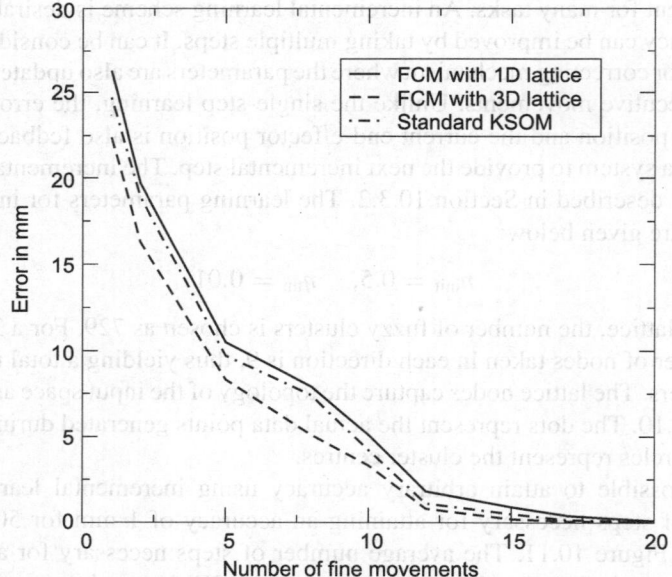

Figure 10.12 Incremental learning performance comparison. The figure shows how the error varies as the number of steps increases. FCM with 3D lattice provides a faster improvement in accuracy.

coordinates are taken as follows:

$$X = 0.1 + 0.2 \ \sin(\alpha)$$
$$Y = 0.3 + 0.2 \ \cos(\alpha)$$
$$Z = 0.2$$

where α varies from 0 to 2π with an incremental step of 0.07. The result of trajectory tracking is shown in Figure 10.13. The average error attained in a single-step is 3 mm. The different parameters used for trajectory tracking are as follows:

$$\eta_{init} = 0.3 \quad \eta_{fin} = 0.005$$

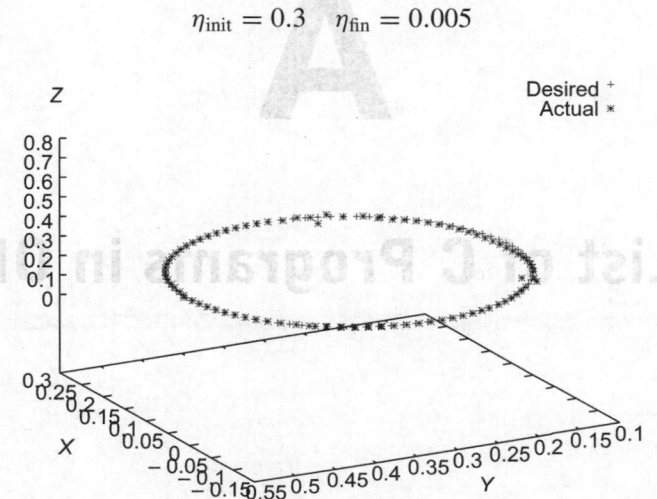

Figure 10.13 Trajectory tracking in the Cartesian space using a 1D fuzzy cluster. All quantities in the figure are in meter. The average tracking error is 3 mm.

SUMMARY

This chapter has presented two different learning schemes for visual motor coordination scheme for a 7 DOF robot manipulator. Since the orientation of the manipulator end-effector is not specified at the target position, the manipulator is redundant for the task. In other words, the task is to find a seven-dimensional joint angle vector for a given four-dimensional target position viewed through a stereo-vision system.

The basic idea is to discretize the input space into a number of discrete cells and in each cell, the non-linear inverse kinematic relationship is expressed as a linear map between the input and output space using a parametrized inverse Jacobian matrix. For a given target point, the final solution is expressed as a weighted average of individual contributions from each discrete cells.

In the first approach, the KSOM network has been used, while in the second a T–S fuzzy model has been used. Although the learning models have been derived using the data generated by the explicit arm-camera model, the learned model works in real time with very little fine tuning in the real time. Thus, the manipulator is being visually guided using learning architecture that has effectively modelled local inverse Jacobian matrices.

APPENDIX

A

List of C Programs in ORC

1. C program corresponding to Example 2.4.
2. C program corresponding to Example 3.10.
3. C program corresponding to Example 4.2.
4. C program corresponding to Example 4.3.
5. C program corresponding to Example 4.4.
6. C program corresponding to Example 4.5.
7. C program corresponding to Example 5.2.
8. C program corresponding to Exercises 1 and 2 of Chapter 5.
9. C program corresponding to Example 5.3.
10. C program corresponding to Example 5.4.
11. C program corresponding to Example 5.5.
12. C program corresponding to Example 6.2.
13. C programs corresponding to the example given in Section 7.2.
14. C programs corresponding to the example given in Section 7.3.
15. C program corresponding to the example given in Section 7.5.1.
16. C program corresponding to Example 8.2.
17. C program corresponding to Example 8.3.
18. C program corresponding to Example 8.4.
19. C program corresponding to Example 8.5.
20. C program corresponding to Exercises 1 and 4 of Chapter 8.
21. C programs corresponding to the simulation results of the example given in Section 9.5.

22. C programs corresponding to the simulation results of the example given in Section 10.2.2.

23. C programs corresponding to the simulation results of the example given in Section 10.3.3.

24. Video of control experiments on the Cart-pole system as discussed in Section 9.5.3 of Chapter 9.

25. Video of visual control experiments on the redundant manipulator as discussed in Section 10.2.5 of Chapter 10.

References

[1] A.T. Fuller. Maxwell´s treatment of siemens´s hydraulic governor. *Int. J of Control*, 60:861–884, 1994.

[2] Michael Athans. *Optimal Control: An Introduction to the Theory and its Applications*. McGraw Hill, New York, 1966.

[3] P.R. Kumar and Pravin Varaiya. *Stochastic systems : estimation, identification and adaptive control*. Prentice Hall, London, 1986.

[4] Kemin Zhou and John C. Doyle. *Essentials of Robust Control*. Prentice Hall, New Jersey, 1998.

[5] Karl Johan Astrom and Bjorn Wittenmark. *Adaptive Control*. Addison-Wesley, 1989.

[6] A. Yesildirek and F.L. Lewis. Feedback linearization using neural networks. *Automatica*, 31(11):1659–1664, 1995.

[7] M. C. M. Teixeira and S. H. Zak. Stabilizing controller design for uncertain nonlinear systems using fuzzy models. *IEEE Trans. on Fuzzy Syst.*, 7(2):133–142, 1999.

[8] G.F. Franklin, J.D. Powell, and M.L. Workman. *Digital Control of Dynamic Systems*. Prentice Hall, 1997.

[9] D. Gross and C.M. Harris. *Fundamentals of Queueing Theory*. John Wiley and Sons, 1998.

[10] Ali Zilouchian and Mo Jamshidi. *Intelligent Control Systems Using Soft Computing Methodologies*. CRC Press, 2001.

[11] Intelligent Control of Robotic Systems. *D. Katic and M. Vukobratovic*. Springer, 2003.

[12] M. Vidyasagar. *Nonlinear Systems Analysis*. Prentice-Hall, Inc., 1978.

[13] J.J. E. Slotine and W. Li. *Applied Nonlinear Control*. Prentice Hall, New Jersey, 1991.

[14] H.K. Khalil. *Nonlinear Systems*. Prentice Hall, New Jersey, 2002.

[15] M. Krstic, I. Kanellakopoulos, and P. Kokotovic. *Nonlinear and Adaptive Control Design*. John Wiley and Sons Inc., 1995.

[16] E. Kreyszig. *Advanced Engineering Mathematics*. John Wiley & Sons Ltd, 1999.

[17] J.J. Craig. *Introduction to Robotics*. Pearson Education, Inc., 1989.

[18] M.W. Spong and M. Vidyasagar. *Robot Dynamics and Control*. John Wiley and Sons, 1989.

[19] Chiman Kwan and F.L. Lewis. Robust backstepping control of nonlinear systems using neural networks. *IEEE Transactions on Systems, Man and Cybernetics - Part A*, 30(6):753–766, November 2000.

[20] F. Rosenblatt. The perceptron: A probabilistic model for information storage and organization in the brain. *Psychological Review*, 65:386–408, 1958.

[21] Marvin Minsky and Seymour Papert. *Perceptrons*. MIT Press, Cambridge, Massachusetts, 1969.

[22] D.E. Rumelhart, G.E. Hinton, and R.J. Williams. *Parallel Distributed Processing: Explorations in the Microstructure of Cognition*, volume 1, chapter Learning internal representations by error propagation. MIT Press, 1986.

[23] P.J. Werbos. *Beyond regression: New tools for prediction and analysis in the behavioral sciences*. PhD thesis, Harvard University, Cambridge, MA, 1974.

[24] D.B. Parker. Learning-logic. Technical Report TR-47, MIT, Cambridge, MA, 1985.

[25] D.S. Broomhead and D. Lowe. Multivariate functional interpolation and adaptive networks. *Complex Systems*, 2:321–355, 198.

[26] R.P. Lippmann. An introduction to computing with neural networks. *IEEE ASSP Magazine*, 4(2), April 1987.

[27] K.S. Narendra and K. Parthasarathy. Gradient methods for optimisation of dynamical systems containing neural networks. *IEEE Transactions on Neural Networks*, 5(2):255–266, 1991.

[28] S. Haykin. *Neural Networks. A Comprehensive Foundation*. Macmillan College Publishing, New York, 1994.

[29] G.C. Goodwin and K. S. Sin. *Adaptive filtering, prediction and control*. Englewood Cliffs, NJ: Prentice Hall, 1991.

[30] Simon Haykin. *Adaptive Filter Theory*. Pearson Education Asia, fourth edition edition, 2002.

[31] Simon Haykin. *Neural Networks*. Pearson Education Inc., 1999.

[32] Dilip Sarkar. Methods to speed up error back propagation learning algorithm. *ACM Computing Surveys*, 27(4), 1995.

[33] C. Charalambous. Conjugate gradient algorithm for efficient training of artificial neural networks. *IEE Proceedings*, 139(3):301–310, 1992.

[34] Stanislaw Osowski, Piotr Bojarczak, and Maciej Stodolski. Fast second order learning algorithm for feedforward multilayer neural network and its applications. *Neural Networks*, 9(9):1583–1596, 1996.

[35] Youji Iiguni, Hideaki Sakai, and Hidekatsu Tokumaru. A real-time learning algorithm for a multilayered neural netwok based on extended kalman filter. *IEEE Transactions on Signal Processing*, 40(4):959–966, April 1992.

[36] Jaroslaw Bilski and Leszek Rutkowski. A fast training algorithm for neural networks. *IEEE Transactions on Circuits and Systems-II: Analog and Digital signal processing*, 45(6):749–753, June 1998.

[37] Martin T. Hagan and Mohammad B. Mehnaj. Training feedforward networks with marquardt algorithm. *IEEE Transactions on Neural Networks*, 5(6):989–993, November 1994.

[38] G. Lera and M. Pinzolas. Neighborhood based levenberg-marquardt algorithm for neural network training. *IEEE Transactions on Neural Networks*, 13(5):1200–1203, September 2002.

[39] Bogdan M. Wilamowski, Serder Iplikci, Okyay Kaynak, and M. Onder Efe. An algorithm for fast convergence in training neural networks. *IEEE*, 2001.

[40] Toledo A., Pinzolas M., Ibarrola J.J., and Lera G. Improvement of the neighborhood based levenberg-marquardt algorithm by local adaptation of the learning coefficient. *IEEE Transactions on Neural Networks*, 16(4):988–992, July 2005.

[41] Laxmidhar Behera, Swagat Kumar, and Awhan Patnaik. On adaptive learning rate that guarantees convergence in feed-forward neural networks. *IEEE Trans. on NN*, 17(5), September 2006.

[42] Xinghuo Yu, M. Onder Efe, and Okyay Kaynak. A general backpropagation algorithm for feedforward neural networks learning. *IEEE Transactions on Neural Networks*, 13(1):251–254, January 2002.

[43] Wen Yu, Alexander S. Poznyak, and Xiaoou Li. Multilayer dynamic neural networks for non-linear system on-line identification. *International Journal Of Control*, 74(18):1858 – 1864, 2001.

[44] Miroslav Krstic and Ioannis Kanellakapoulos. *Non Linear and Adaptive control design*. John Wiley & Sons, Inc., 1995.

[45] G. Qiu, M.R. Varley, and T.J. Terrel. Accelerated training of bp using adaptive momentum steps. *IEEE Electronic letters*, 28(4), February 1992.

[46] X.H. Yu, G.A. Chen, and S.X. Cheng. Acceleration of bp learning using optimized learning rate and momentum. *IEEE Electronic letters*, 29(14), July 1993.

[47] T. Tollenaere. Supersab: Fast adaptive back propagation with good scaling properties. *Neural Networks*, 3(5):561–573, 1990.

[48] R. Williams and D. Zipser. Experimental analysis of the real-time recurrent learning algorithm. *Connection Science*, 1:87–111, 1989.

[49] R. Williams and D. Zipser. A learning algorithm for continually running fully recurrent neural networks. *Neural Computation*, 1:270–280, 1989.

[50] B. Pearlmutter. Learning state space trajectories in recurrent neural networks. *Neural Computation*, 1:263–269, 1989.

[51] B. Sato. A real time learning algorithm for recurrent analog neural networks. *Biological Cybernatics*, 62:237–241, 1990.

[52] T. Kohonen. *Self-organizing Maps*. Springer, Heidelberg, 2001.

[53] K.S. Narendra and K. Parthasarathy. Identification and control of dynamical systems using neural networks. *IEEE Transactions on Neural Networks*, 1(1):4–27, 1990.

[54] Z. Uykan, C. Gunzelis, M.E. Celebi, and H. N. Koivo. Analysis of input-output clustering for determining center of rbfn. *IEEE Trans. on Neural Network*, 11(4), 2000.

[55] L. Behera and N. Kirubanandan. A hybrid neural control scheme for visual-motor coordination. *IEEE Control Systems*, 1999.

[56] T.J. Ross. *Fuzzy Logic with Enginering Applications*. Mc-GrawHill Inc., 1995.

[57] A. Konar. *Comutational Intelligence, Principles, Techniques and Applications*. Springer, 2005.

[58] L.A. Zadeh. A fuzzy-algorithmic approach to the definition of complex or imprecise concepts. *Int. Jour. Man-Machine Studies*, 8:249–291, 1976.

[59] B. Kosko. Fuzzy systems as universal approximators. In *IEEE International Conference on Fuzzy Systems*, pages 1153–1162, 1992.

[60] L.A. Zadeh. Fuzzy logic = computing with words. *IEEE Trans. Fuzzy Syst.*, 4(2):103–112, 1996.

[61] W.J.M. Kickert and E. H. Mamdani. Analysis of a fuzzy logic controller. *Fuzzy Sets and Systems*, 1:29–44, 1978.

[62] E.H. Mamdani. Application of fuzzy algorithms for the control of a dynamic plant. *Proc. IEEE*, 121:1585–1588, 1974.

[63] T. Takagi and M. Sugeno. Fuzzy identification of systems and its application to modelling and control. *IEEE Tran. SMC*, 15:116–132, 1985.

[64] M. Mizumoto. Realization of pid controls by fuzzy control methods. *Fuzzy Sets and Systems*, 70:171–182, 1995.

[65] J. Zhang and A.J. Morris. Fuzzy neural networks for nonlinear systems modeling. *IEE Proc. Control Theory Appl.*, 142:551–561, 1995.

[66] L. Behera and K.K. Anand. Guaranteed tracking and regulatory performance of nonlinear dynamic systems using fuzzy neural networks. *IEE Proc. Control Theory Appl.*, 146(5), 1999.

[67] H.O. Wang, K. Tanaka, and M.F. Griffin. An approach to fuzzy control of nonlinear systems: Stability and design issues. *IEEE Trans. Fuzzy Syst.*, 4(1):14–23, 1996.

[68] Stainslaw H. Zak. Stabilizing fuzzy system models using linear controllers. *IEEE Tran. Fuzzy Systems*, 7:236–240, 1999.

[69] S. Mehta and J. Chiasson. Nonlinear control of a series dc motor: Theory and experiment. *IEEE Trans. on Industrial Electronics*, 45(1):134–141, 1998.

[70] K.S. Narendra and K. Parthasarathy. Identification and control of dynamical systems using neural networks. *IEEE TNN*, March 1990.

[71] L. Behera, M. Gopal, and S. Chaudhury. Inversion of rbf networks and applications to adaptive control of nonlinear systems. *IEE Proceedings On Control Theory and Applications*, 142(6):617–624, 1995.

[72] L. Behera, M. Gopal, and S. Chaudhury. On adaptive control of a robot manipulator using inversion of its neural emulator. *IEEE Trans Neural Networks*, 7(6):1401–1414, 1996.

[73] A. S. Poznyak, E. N. Sanchez W. Wu, and J. P. Perez. Nonlinear adaptive trajectory tracking using dynamic neural networks. *IEEE Trans Neural Networks*, 10(6):1402–1411, 1999.

[74] M. S. Ahmed. Neural net based mrac for a class of nonlinear plants. *Neural Networks*, 13:111–124, 2000.

[75] D. Wang and P. Bao. Enhancing the estimation of plant jacobian for adaptive neural inverse control. *Neurocomputing*, 34:99–115, 2000.

[76] S. Chen, C.F.N. Cowans, S. A. Billings, and P.M. Grant. Parallel recursive prediction error algorithm for training layered neural networks. *Int. J. Control*, 51(6):1215–1228, 1990.

[77] Y. Iiguni, H. Sakai, and H. Tokumaru. A real-time learning algorithm for a multilayered network based on extended kalman filter. *IEEE Trans on Signal Processing*, 40, 1992.

[78] S. Haykin. *Adaptive Filter Theory*. Prentice Hall, Englewood Cliffs, NJ, 1991.

[79] S. Haykin. *Neural Networks, A Comprehensive Foundation*. Macmillon College Publishing Company, NY, 1994.

[80] G.C. Goodwin and K.S. Sin. *Adaptive Filtering, Prediction and Control*. Prentice Hall, Englewood Cliffs, NJ, 1991.

[81] C.A. Jensen, R. D. Reed, R. J. Marks, M. A. El-sharkawi, , J-B Jung, R. T. Miyamoto, G. M. Anderson, and C. J. Eggen. Inversion of feedforward neural networks: Algorithms and applications. *Proc. IEEE*, 87(9):1536–1549, 1999.

[82] D.A. Hoskins, J. N. Hwang, and J. Vagners. Ijerative inversion of neural networks and its applications to adaptive control. *IEEE Trans. Neural Net*, 3(2), 292-301, 1992.

[83] A. Linden and J. Kindermann. Inversion of multilayer nets. In *IJCNN, Washington, DC*, June 1989.

[84] Sukhan Lee and R. M. Kill. Inverse mapping of continuous functions using local and global information. *IEEE Trans Neural Networks*, 5(3):409–423, 1994.

[85] D. Stokes and G.S. May. Indirect adaptive control of reactive ion etching using neural networks. *IEEE Trans. on Robotics and Automation*, 17(5):650–657, 2001.

[86] Jung-Wook Park, R.G. Harley, and G.K. Venayagamoorthy. Indirect adaptive control for synchronous generator: comparison of mlp/rbf neural networks approach with lyapunov stability analysis. *IEEE Trans. on Neural Nets.*, 15(2):460–464, 2004.

[87] H.N. Nounou and K.M. Passino. Stable auto-tuning of adaptive fuzzy/neural controllers for nonlinear discrete-time systems. *IEEE Trans. on Fuzzy Systems*, 12(1):70–83, 2004.

[88] J.J.E. Slotine and W. Li. Composite adaptive control of robot manipulators. *Automatica*, 25(4):509–519, 1989.

[89] J.Y. Choi and J.A. Farrell. Nonlinear adaptive control using networks of piecewise linear approximators. *IEEE Trans. on Neural Net.*, 11(2):390–401, 2000.

[90] Indrani Kar and Laxmidhar Behera. Direct adaptive neural control for affine nonlinear systems. *Applied Soft Computing*, 9:756–764, 2009.

[91] S. Jagannathan and F.L. Lewis. Multilayer discrete-time neural-net controller with guaranteed performance. *IEEE Trans. NN*, 7(1):124–130, 1996.

[92] S. Jagannathan. Discrete-time cmac nn control of feedback linearizable nonlinear systems under a persistence of excitation. *IEEE Trans. NN*, 10(1):128–137, 1999.

[93] F.L. Lewis, S. Jagannathan, and A. Yesildirek. *Neural network control of robot manipulators and nonlinear systems*. Taylor & Francis, 1999.

[94] C. Kwan, F.L. Lewis, and D.M. Dawson. Robust neural-network control of rigid-link electrically driven robots. *IEEE Trans on Neural Networks*, 9(4):581–588, 1998.

[95] Hua Deng, Han-Xiong Li, and Yi-Hu Wu. Feedback-linearization-based neural adaptive control for unknown nonaffine nonlinear discrete-time systems. *IEEE Trans. on Neural Nets.*, 19(9):1615–1625, 2008.

[96] Jang-Hyun Park, Sung-Hoe Huh, Seong-Hwan Kim, Sam-Jun Seo, and Gwi-Tae Park. Direct adaptive controller for nonaffine nonlinear systems using self-structuring neural networks. *IEEE Trans. on Neural Nets.*, 16(2):414–422, 2005.

[97] S.S. Ge, G.Y. Li, and T.H. Lee. Adaptive nn control for a class of strict feedback discrete-time nonlinear systems. *Automatica*, 39:807–819, 2003.

[98] S.S. Ge, C. Yang, and T. H. Lee. Adaptive predictive control using neural network for a class of pure-feedback systems in discrete time. *IEEE Trans. on Neural Nets.*, 19(9), 2008.

[99] C. Kwan and F.L. Lewis. Robust backstepping control of nonlinear systems using neural networks. *IEEE Trans. on SMC A*, 30:753–766, 2000.

[100] D. Wang and J. Huang. Neural network-based adaptive dynamic surface control for a class of uncertain nonlinear systems in strict-feedback form. *IEEE Trans. on Neural Nets.*, 16(1):195–202, 2005.

[101] Chun-Fei Hsu, Chih-Min Lin, and Tsu-Tian Lee. Wavelet adaptive backstepping control for a class of nonlinear systems. *IEEE Trans. on Neural Nets.*, 17(5):1175–1183, 2006.

[102] Jin Young Choi and J.A. Farrell. Adaptive observer backstepping control using neural networks. *IEEE Trans. on Neural Nets.*, 12(5):1103–1112, 2001.

[103] S.D. Senturia. *Microsystem Design*. Kluwer Academic Publishers, 2001.

[104] R. Padhi, N. Unnikrishnan, X. Wang, and S.N. Balakrishnan. A single network adaptive critic (snac) architecture for optimal control synthesis for a class of nonlinear systems. *Neural Networks*, 19:1648–1660, 2006.

[105] A.E. Bryson and Y.C. Ho. *Applied Optimal Control.* Taylor and Francis, 1975.

[106] P.J. Werbos. *Approximate Dynamic Programming for real-time control and neural modeling.* In D. A. White and D. A. Sofge (Eds.), *Handbook of intelligent control,* Multiscience Press, 1992.

[107] S. Ferrari and R.F. Stengel. *Model based adaptive critic designs.* In Jennie Si, A. G. Barto, W. B. Powell, and D. Wunsch II (Eds.), *Handbook of learning and Approximate Dynamic Programming,* IEEE Press, 2004.

[108] D.V. Prokhorov and D.C. Wunsch II. Adaptive critic designs. *IEEE Transactions on Neural Networks,* 8(5):997–1007, September 1997.

[109] Chuan-Kai Lin. Adaptive critic autopilot design of bank-to-turn missiles using fuzzy basis function networks. *IEEE Trans. on SMC B,* 35(2):197–207, 2005.

[110] T. Hanselmann, L. Noakes, and A. Zaknich. Continuous-time adaptive critics. *IEEE Trans. on Neural Nets.,* 18(3):631–647, 2007.

[111] M.S. Iyer and D.C. Wunsch II. Dynamic re-optimization of a fed-batch fermentor using adaptive critic designs. *IEEE Trans. on Neural Nets.,* 12(6):1433–1444, 2001.

[112] S. Mohagheghi, G.K. Venayagamoorthy, and R.G. Harley. Adaptive critic design based neuro-fuzzy controller for a static compensator in a multimachine power system. *IEEE Trans. on Power Systems,* 21(4):1744–1754, 2006.

[113] A.K. Deb, Jayadeva, M. Gopal, and S. Chandra. Svm-based tree-type neural networks as a critic in adaptive critic designs for control. *IEEE Trans. on Neural Nets.,* 18(4):1016–1030, 2007.

[114] G.K. Venayagamoorthy, R.G. Harley, and D.C. Wunsch. Comparison of heuristic dynamic programming and dual heuristic programming adaptive critics for neurocontrol of a turbogenerator. *IEEE Trans. on Neural Nets.,* 13(3):764–773, 2002.

[115] D.V. Prokhorov. Training recurrent neurocontrollers for real-time applications. *IEEE Trans. on Neural Nets.,* 18(4):1003–1015, 2007.

[116] E.H. Mamdani. Application of fuzzy algorithms for the control of a dynamic plant. *Proc. IEEE,* 121(12):1585–1588, 1974.

[117] C.C. Lee. Fuzzy logic in control systems: Fuzzy logic controller,parts i and ii. *IEEE Trans. Syst. Man Cybern.,* 20(2):404–435, 1990.

[118] L.A. Zadeh. Outline of a new approach to the analysis of complex systems and decision processes. *IEEE Trans. Syst. Man Cybern.,* SMC3:28–44, 1973.

[119] M. Margaliot and G. Langholz. Fuzzy control of a benchmark problem: A computing with words approach. *IEEE Tran. Fuzzy Syst.,* 12:230–235, 2004.

[120] M.N. Uddin and M.A. Rahman. High-speed control of ipmsm drives using improved fuzzy logic algorithms. *IEEE Trans. on Industrial Electronics*, 54(1):190–199, 2007.

[121] T. Yalcinoz and H. Altun. Power economic dispatch using a hybrid genetic algorithm. *IEEE Power Engineering Review*, 21(3):59–60, 2001.

[122] T. Orlowska-Kowalska and K. Szabat. Optimization of fuzzy-logic speed controller for dc drive system with elastic joints. *IEEE Trans. on Industry Applications*, 40(4):1138–1144, 2004.

[123] Y.C. Choi and Chang-Hun Kim. Cdp servo system control using fuzzy logic control. *IEEE Trans. on Consumer Electronics*, 53(4):1314–1321, 2007.

[124] G. Zhiqiang, T.A. Trautzsch, and J.G. Dawson. A stable self-tuning fuzzy logic control system for industrial temperature regulation. *IEEE Trans. on Industry Applications*, 38(2):414–424, 2002.

[125] Heinz Muehlenbein and Thilo Mahnig. *Foundations of Real-World Intelligence*, chapter Evolutionary Computation and Beyond. CSLI Publications, 2001.

[126] M. Sugeno and G.T. Kang. Structure identification of fuzzy model. *Fuzzy Sets and Systems*, 28(1):15–33, 1988.

[127] P. Premkumar, Indrani Kar, and Laxmidhar Behera. Variable gain controllers for nonlinear systems using t-s fuzzy model. *IEEE Trans. SMC-B*, 36:1442–1449, 2006.

[128] K. Tanaka, T. Ikeda, and H.O. Wang. Fuzzy regulators and fuzzy observers: Relaxed stability conditions and lmi based designs. *IEEE Tran. on Fuzzy Systems*, 6(2), 1998.

[129] Kuang-Yow Lian, Hui-Wen Tu, and Jeih-Jang Liou. Stability conditions for lmi-based fuzzy control from viewpoint of membership functions. *IEEE Trans. on Fuzzy Systems*, 14(6):874–884, 2006.

[130] Ho Jae Lee, Jin Bae Park, and Guanrong Chen. Robust fuzzy control of nonlinear systems with parametric uncertainties. *IEEE Trans. on Fuzzy Systems*, 9(2):369–379, 2001.

[131] Chang-Woo Park and Young-Wan Cho. T-s model based indirect adaptive fuzzy control using online parameter estimation. *IEEE Trans. on SMC B*, 34(6):2293–2302, 2004.

[132] Chung-Shi Tseng, Bor-Sen Chen, and Huey-Jian Uang. Fuzzy tracking control design for nonlinear dynamic systems via t-s fuzzy model. *IEEE Trans. on Fuzzy Systems*, 9(3):381–392, 2001.

[133] L. Behera. Query based model learning and stable tracking of a robot arm using radial basis function network. *Computers and Electrical Engineering*, 29:553–573, 2003.

[134] S. Hutchinson, G.D. Hager, and P.I. Corke. A tutorial on visual servo control. *IEEE Transactions on Robotics and Automation*, 12(5):651–670, October 1996.

[135] Danica Kragic and et al. Survey on visual servoing for manipulation. http://citeseer.ist.psu.edu/484743.html.

[136] M. Kuperstein. Adaptive visual-motor coordination in multijoint robots using parallel architecture. *Proc. IEEE International conference on Robotics and Automation*, 4:1595–1602, March 1987.

[137] T.M. Martinetz, H.J. Ritter, and K.J. Schulten. Three-dimensional neural net for learning visual motor coordination of a robot arm. *IEEE Transactions on Neural Networks*, 1(1):131–136, March 1990.

[138] J.A. Walter and K.J. Schulten. Implementation of self-organizing neural networks for visual-motor control of an industrial robot. *IEEE Transactions on Neural Networks*, 4(1):86–95, January 1993.

[139] L. Behera and N. Kirubanandan. A hybrid neural control scheme for visual-motor coordination. *IEEE Control System Magazine*, 19(4):34–41, 1999.

[140] R.Y. Tsai. A versatile camera calibration technique for high-accuracy 3d machine vision metrology using off-the-shelf tv cameras and lenses. *IEEE Journal of Robotics and Automation*, RA-3(4):323–344, August 1987.

[141] Amtec robotics. http://www.amtec-robotics.com/.

[142] F. Pourboghrat. Neural networks for learning inverse kinematics of redundant manipulators. In *Proc. of 32nd Midwest symposium on Circuits and Systems*, pages 760–762. IEEE, August 1989.

[143] G. Sun and B. Scassellati. A fast and efficient model for learning to reach. *International Journal of Humanoid Robotics*, 2(4):391–413, December 2005.

[144] R.V. Mayorga and P. Sanongboon. A radial basis function network approach for inverse kinematics and singularities prevention of redundant manipulators. In *Proc. of Int. Conf. on Robotics and Automation (ICRA)*, pages 1955–1960. IEEE, May 2002.

[145] S. Vijayakumar, A. D'Souza, T. Shibata, J. Conradt, and S. Schaal. Statistical learning for humanoid robots. *Autonomous Robots*, 12:55–69, 2002.

[146] T. Kohonen. *Self-Organization and Associative Memory*. Springer-Verlag, New York, 1984.

[147] G.A. Barreto, A.F.R. Araujo, and H.J. Ritter. Self-organizing feature maps for modeling and control of robotic manipulators. *Journal of Intelligent and Robotic Systems*, 36:407–450, 2003.

[148] T. Martinetz, H. Ritter, and K. Schulten. Learning of visuomotor-coordination of a robot arm with redundant degrees of freedom. In *Proc. of the Int. Conf. on Parallel Processing in Neural Systems and Computers (ICNC)*, pages 431–434, Dusseldorf, Amsterdam, 1990. Elsevier.

[149] M. Han, N. Okada, and E. Kondo. Coordination of an uncalibrated 3-d visuo-motor system based on multiple self-organizing maps. *JSME International Journal Series C*, 49(1):230–239, 2006.

[150] H. Zha, T. Onitsuka, and T. Nagata. A self-organization learning algorithm for visuo-motor coordination in unstructured environment. *Artificial life and robotics*, 1(3):131–136, September 1997.

[151] G. Asuni, G. Teti, C. Laschi, E. Guglielmelli, and P. Dario. A bio-inspired sensory-motor neural model for a neuro-robotic manipulation platform. In *12th Int. Conf. on Advanced Robotics (ICAR)*, pages 607–612. IEEE, 2005.

[152] L. Sun and Ch. Doeschner. Visuo-motor coordination of a robot manipulator based on neural network. In *Int. Conf. on Robotics and Automation*, pages 1737–1742, Leuven, Belgium, 1998. IEEE.

[153] N. Kumar and L. Behera. Visual motor coordination using a quantum clustering based neural control scheme. *Neural Processing Letters*, 20:11–22, 2004.

[154] Everest vit. http://www.everestvit.com/.

[155] Open source computer vision library. http://www.intel.com/technology/computing/opencv/.

[156] Reg Wilson. Tsai camera calibration software. http://www.cs.cmu.edu/ rgw/TsaiCode.html.

[157] R.D. Giuseppe, F. Taurisano, C. Distante, and A. Anglani. Visual servoing of a robotic manipulator based on fuzzy logic controller. In *Proc. of Int. conf. on Robotics and Automation*, pages 1487–1494, Detroit, Michigan, 1999. IEEE.

[158] C.S. Kim, W.H. Seo, S.H. Han, and O. Khatib. Fuzzy logic control of a robot manipulator based on visual servoing. In *Proc. of Int. Symposium on Industrial Electronics*, pages 1597–1602, Pusan, South Korea, 2001. IEEE.

[159] D. Kragic and H.I. Christensen. Cue integration in visual servoing. *IEEE Transactions on Robotics and Automation*, 17(1):18–27, February 2001.

[160] S. Kumaresan, Hua. Li, and Xing min Li. *Fuzzy logic and intelligent systems*, chapter Robot hand-eye coordination based on fuzzy logic, pages 245–269. Springer, Netherlands, 1995.

[161] A. Prochazka. The fuzzy logic of visuomotor control. *Canadian Journal of Physiology and Pharmacology*, 74(4):456–462, 1996.

[162] A. Prochazka and D. Gillard. Sensory control of locomotion. In *Proc. of the American Control Conference*, pages 2846–2850, Albuquerque, New Mexico, 1997.

[163] J.C. Dunn. A fuzzy relative of the isodata process and its use in detecting compact well-separated clusters. *Journal of Cybernetics*, 3:32–57, 1973.

[164] J.C. Bezdek. *Pattern Recognition with Fuzzy Objective Function Algoritms*. Plenum Press, New York, 1981.

[165] A. D'Souza, S. Vijayakumar, and S. Schaal. Learning inverse kinematics. In *Int. Conf. on Intelligent Robots and Systems*, pages 298–303, Maui, Hawai, 2001. IEEE.

[166] G. Tevatia and S. Schaal. Inverse kinematics for humanoid robots. In *IEEE Int. Conf. on Robotics and Automation*, 2000.

Index